PRENTICE-HALL VOCATIONAL AGRICULTURE SERIES

BEEF PRODUCTION. Diggins, Bundy, and Christensen

CROP PRODUCTION. Delorit, Greub, and Ahlgren

DAIRY PRODUCTION. Diggins, Bundy, and Christensen

EXPLORING AGRICULTURE. Evans and Donahue

JUDGING LIVESTOCK, DAIRY CATTLE, POULTRY, AND CROPS. Youtz and Carlson

LEADERSHIP TRAINING AND PARLIAMENTARY PROCEDURE FOR FFA. Gray and Jackson

LIVESTOCK AND POULTRY PRODUCTION. Bundy, Diggins, and Christensen

MODERN FARM POWER. Promsberger, Bishop, and Priebe

PROFITABLE SOIL MANAGEMENT. Knuti, Korpi, and Hide

SWINE PRODUCTION. Bundy, Diggins, and Christensen

USING ELECTRICITY. Hamilton

FOURTH EDITION

SWINE PRODUCTION

CLARENCE E. BUNDY

Professor Emeritus
Department of Agricultural Education
Iowa State University
Ames, Iowa

RONALD V. DIGGINS

Former Vocational Agriculture Instructor
Eagle Grove, Iowa

VIRGIL W. CHRISTENSEN

Chairman, Agriculture and Natural Resources Department
Hawkeye Institute of Technology
Waterloo, Iowa

PRENTICE-HALL, INC., ENGLEWOOD CLIFFS, NEW JERSEY

SWINE PRODUCTION, Fourth Edition

Clarence E. Bundy, Ronald V. Diggins, and Virgil W. Christensen

ISBN 0-13-879783-8

Cover photograph by Grant Heilman

10 9 8 7 6 5 4 3 2

PRENTICE-HALL INTERNATIONAL, INC., London
PRENTICE-HALL OF AUSTRALIA, PTY. LTD., Sydney
PRENTICE-HALL OF CANADA, LTD., Toronto
PRENTICE-HALL OF INDIA PRIVATE LTD., New Delhi
PRENTICE-HALL OF JAPAN, INC., Tokyo

PREFACE

In the United States and Canada, hogs are an important source of farm income. In the Corn Belt and on specialized farms in other areas, moreover, swine production has become a big and complex business. Contributing to this size and complexity are new developments and new technology in breeding, housing, nutrition, disease and parasite control, waste management, and marketing.

Beef and poultry are strong competitors of pork for the consumer dollar. There is some competition, too, from veal, lamb, and fish. Swine producers, therefore, must be alert to the situation, or they may lose a part of the market for pork products.

We present in *Swine Production*, Fourth Edition, the latest developments in production and marketing. We have relied heavily on the experience gained from our 45-year connection with the livestock industry. We learned about the needs and problems of swine producers and farm youth through association with them in educational and practical situations. Our participation in the programs of the National Pork Producers Council, the state pork-producer associations, the swine testing programs, and the National Barrow Show has also been of much value in determining the problems and developments in the swine industry.

Every phase of swine production has been handled in a systematic manner. Each chapter deals with a specific, logically organized subject. Much new information has been provided. The illustrations and tables have been increased in number and their contents updated. Following each chapter are a summary, questions for study, and a bibliography to which the reader may refer for further information.

Swine producers, vocational and technical agriculture students, 4-H Club members, and interested lay persons will find this book helpful in understanding the pork industry and in solving the many problems of pork production. The information is complete, up-to-date, accurate, and simply stated.

The results of experimentation by the U.S. Department of Agriculture and by cooperating state agricultural experiment stations have been reviewed. The research specialists of the meat packing industry have provided valuable assistance. The secretaries of the various purebred swine breed associations, as well as representatives of the organizations that produce inbred and hybrid breeding stock, have been most cooperative. We are especially grateful to L. N. Hazel, E. A. Kline, L. L. Christian, and Emmett Stevermer of the Iowa State University; A. J. Muehling, A. D. Leman, and G. R. Carlyle of the University of Illinois; Charles Christians of the University of Minnesota; D. E. Ullrey and J. A. Hoefer of Michigan State University; Richard F. Wilson and Howard S. Teague of the Ohio State University and of the Ohio Agricultural Research and Development Center; Berle A. Kock of Kansas State University; I. T. Omtvedt and Wm. G. Luce of Oklahoma State University; Ernest R. Peo, Jr., and Wm. T. Ahlschwede of the University of Nebraska; R. F. Behlow of North Carolina State University; Donald B. Hudman of Texas A & M University; John C. Rea and Ralph Ricketts of the University of Missouri; R. H. Grummer of the University of Wisconsin; H. W. Jones of Purdue University; Richard Wahlstrom and James H. Bailey of South Dakota State University; M. D. Whiteker of the University of Kentucky; John Sink of Pennsylvania State University; T. J. Cunda of the University of Florida; Wm. A. Curry of the University of Maryland; Ellis A. Pierce and W. G. Pond of Cornell University; Hubert Heitman of the University of California; and Sam L. Hansard and Frank B. Masincupp of the University of Tennessee for their assistance in the preparation of the manuscript for *Swine Production,* Fourth Edition.

We appreciate greatly the assistance of a large number of business and commercial organizations, meat packing companies, and farm magazines for illustrations provided for the Fourth Edition of *Swine Production.* We are especially grateful to the *National Hog Farmer, Successful Farming,* and *Wallaces' Farmer.*

We are grateful to David C. Opheim, Francis Telshaw, and Linda and Jerry Geisler for the excellent drawings made especially for our use. We also wish to thank the Morrison Publishing Company, Doane Agricultural Services, Inc., Dr. Leonard W. Schruben of Kansas State University, and John Phillips of George Hormel and Company for special permission to reproduce copyrighted materials.

CLARENCE E. BUNDY

RONALD V. DIGGINS

VIRGIL W. CHRISTENSEN

CONTENTS

1 PORK PRODUCTION IS BIG BUSINESS

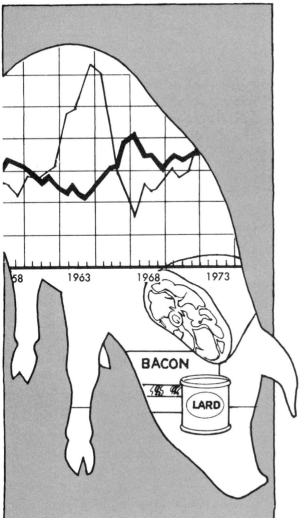

The brood sow has often been described as the "mortgage lifter" on the farms of the nation. This statement is probably true. The livestock and meat industry has a dominating position in our agricultural economy, and the hog enterprise is one of the most important livestock enterprises on the farms in many states and areas.

On December 1, 1973, there were 61 million hogs on the nation's farms. These hogs had a total value of 3.7 billion dollars. Iowa led the states in both number and value of hogs with 14.7 million hogs valued at 970 million dollars. Illinois ranked second with 7.4 million hogs valued at 437 million dollars. Other high states were Indiana, Missouri, Minnesota, and Nebraska. The ten Corn Belt states, Ohio, Indiana, Illinois, Wisconsin, Minnesota, Iowa, Missouri, South Dakota, Nebraska, and Kansas accounted for 76.5 percent of the nation's hog and pig inventory.

Hogs were produced on 62 percent of the farms in Iowa in 1973, but on less than 10 percent of the farms in 13 other states. Hog enterprises were found on 35 or more percent of the farms in South Dakota, Nebraska, Illinois, Missouri, and Georgia.

In 1949 hogs were raised on 56 percent of the farms in the nation and on 61 percent of the farms in the north central states. In 1973 only 27 percent of U. S. farmers and 37 percent of north central states farmers were raising hogs.

About one-fourth of the farm income of some of the Corn Belt states is derived from the sale of hogs. In Iowa where 75 cents of every farm dollar comes from livestock, nearly 35 cents comes from the sale of hogs. In 1972 the swine industry in Iowa was a 1,270-million-dollar business and in Illinois a 687-million-dollar business.

As shown in Figure 1-2, hog sales accounted for 15.1 percent of the nation's farm livestock cash receipts in 1972. The enterprise ranked third among the livestock enter-

prises in the percentage of total farm cash receipts produced.

While more than 60 percent of the hogs marketed in this country are produced in six states, hog enterprises on the farms in the other states are important to the welfare of the families involved. In some areas many of the hogs produced are grown for farm slaughter.

It is estimated that 20 to 30 percent of the corn grown in the Corn Belt is fed to hogs. Barley, oats, wheat, and grain sorghums are fed in large quantities in some states outside the Corn Belt, and garbage is fed to hogs on some farms that are adjacent to our metropolitan centers.

FIGURE 1-1 (*Above*). The 2.3 million persons in the U.S. consume large quantities of ham. It is a favorite meat in the daily diet and for special occasions. (*Courtesy Land O Lakes, Felco Division.*) FIGURE 1-2 (*Below*). Commodities as a percentage of total U.S. livestock income, 1972. Source: U.S.D.A. data.

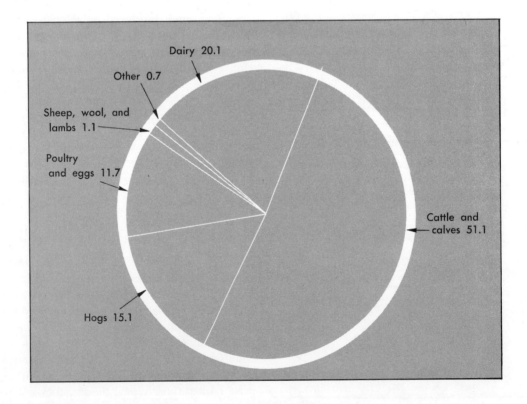

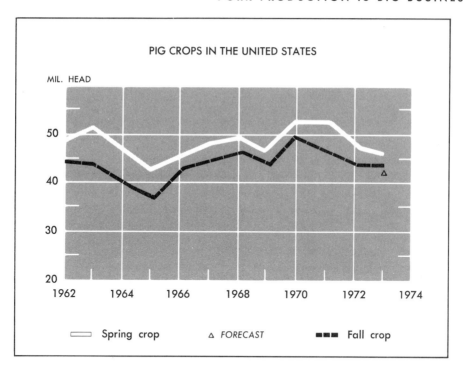

FIGURE 1-3. The spring crop is larger than the fall crop, but the difference has been decreasing. (*Courtesy U.S.D.A.*)

TRENDS IN SWINE PRODUCTION

There is considerable variation in the number of hogs produced annually in areas and states. Shown in Figure 1-3 is a summary of U. S. pig crops from 1962 to 1973. In 1934 and 1935 only about 60 million hogs were produced in this country, due largely to feed shortages resulting from widespread drought. The peak production of nearly 122 million, which about doubled the 1934 production, came in 1943 during World War II. Hog production in this country since 1945 has varied from 86.8 million produced in 1945 to about 95.7 million produced in 1955, with a high production of 101.8 million during 1951 and a low of 77.9 million hogs produced in 1953.

During the 1960's pig crops varied from a low of 76 million in 1965 to 88.3 million in 1969. Production since 1969 has varied from a high of 101.9 million in 1970 to 90.8 million in 1972 and 88.1 million in 1973. Much of the increase during the 1970's was due to the use of confinement feeding and farrowings throughout the year. Note in Figure 1-3 that the number of pigs farrowed between June and November (the fall crop) now about equals the number farrowed between December and May (the spring crop).

States Leading in Hog Production

Corn and hog production go hand in hand, so the leading hog-producing states are in the Corn Belt. Shown in Table 1-1 is a ranking of 50 states according to the numbers of pigs produced in 1973.

Iowa, Illinois, Missouri, Indiana, Minnesota, and Nebraska were the six high states in number of pigs produced. These six states

T A B L E 1-1   RANK OF STATES IN NUMBER OF PIGS PRODUCED, 1973

State	Number (1,000 Head)	State	Number (1,000 Head)
1. Iowa	19,062	27. Montana	357
2. Illinois	11,228	28. Maryland	309
3. Missouri	6,843	29. Louisiana	249
4. Indiana	6,731	30. California	218
5. Minnesota	6,074	31. Oregon	160
6. Nebraska	4,896	32. Idaho	135
7. Ohio	3,187	33. Arizona	133
8. South Dakota	3,132	34. New York	126
9. Kansas	3,033	35. Washington	115
10. North Carolina	2,946	36. New Mexico	99
11. Wisconsin	2,826	37. West Virginia	87
12. Georgia	2,437	38. Massachusetts	85
13. Kentucky	2,053	39. Delaware	85
14. Texas	1,532	40. Hawaii	69
15. Tennessee	1,527	41. Utah	68
16. Alabama	1,402	42. New Jersey	56
17. Michigan	1,118	43. Wyoming	49
18. South Carolina	866	44. Nevada	17
19. Mississippi	774	45. New Hampshire	16
20. Pennsylvania	765	46. Maine	11
21. Virginia	755	47. Rhode Island	10
22. North Dakota	560	48. Connecticut	10
23. Oklahoma	515	49. Vermont	6
24. Colorado	507	50. Alaska	1
25. Arkansas	504		
26. Florida	401	U.S. Total	88,145

Source: Economic Research Service, U.S. Department of Agriculture.

accounted for 62.2 percent of the nation's 1973 pig crop. Iowa alone produced more pigs than the 39 low producing states, and nearly twice as many pigs as Illinois, the second largest producing state. Iowa produced about 22 percent of the nation's pig crop in both 1972 and 1973. The 1972 production for this state was 20.5 million head.

Shifts in Hog Production

About 71 percent of the nation's hogs were produced in the 12 north central states during the ten-year period 1942–1951. Ten of these states produced in 1973 almost 76 percent of the nation's hogs.

Southern section. The southern states produced about 23 percent of the total hog crop during the 1942–1951 period, but this section produced only 11.4 percent in 1973.

North Atlantic section. This section produced 2 percent of the nation's pig crop during the 1942–1951 period, but was responsible for the production of only 1.3 percent of the 1973 crop.

Western section. For the ten-year period, this section produced 3.5 percent of the pigs produced in the nation. It produced only 2.1 percent in 1973.

Canada. There were 5.7 million hogs in Canada in 1969 and 7.4 million in 1972. These numbers are not large when compared with

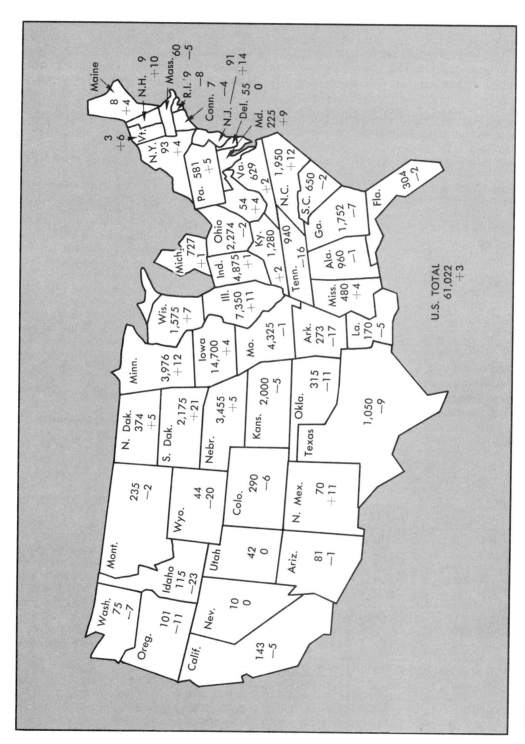

FIGURE 1-4. Hogs and pigs inventory by states as of December 1, 1973. The first figure in each state is the number of pigs in thousands; the second is the percent of difference from the previous year. (*Courtesy U.S.D.A.*)

the 62.5 million hogs in the U. S. in January 1972. Canada has greatly expanded her swine enterprises both in quantity and in quality. Even so Canada provides the United States with a good export market. We exported more than 11 million pounds of lard to Canada both in 1972 and 1973. Canada purchased 46.2 million pounds of pork from the U. S. in 1973.

Factors Affecting Hog Production

The changes which come about in the production of pigs from year to year in a state or area are caused by a number of economic factors. Most important of them are the following:

1. Supply and price of live hogs on the market.
2. Consumer demand for pork and lard.
3. Available feed supplies.
4. Supply and price of competing meat products on the consumer market.
5. Hog-corn price ratios.
6. Swine disease outbreaks.
7. World economic conditions.
8. Labor supply.

These factors individually or collectively determine to a large extent the seasonal production of hogs and the production from year to year. Hog producers must be students of economics in order to manipulate their swine production programs so that they will be in harmony with these economic factors.

CONSUMER DEMAND FOR PORK AND LARD

The future of pork production in this country will to a large extent be determined by the desire of American people for pork and lard, by their willingness to pay a reasonable price for these products, and by export demand. Pork and lard must compete with other animal and vegetable products in palatability, in nutritive value, in availability, in ease of merchandising and storage, and in price.

Meat and Lard Consumption

The average person in this nation consumed in 1972 a total of 240.9 pounds of

TABLE 1-2   MEAT AND LARD CONSUMPTION—UNITED STATES, 1920-1973
(Pounds Consumed per Person)

Year	Pork	Lard	Beef	Veal	Lamb and Mutton	Chicken	Turkey	All Meats (Excluding Lard)
1920	63.5	12.4	59.1	8.0	5.4	13.7	—	149.7
1930	67.0	12.9	48.9	6.4	6.7	15.7	1.5	146.2
1940	73.5	14.6	54.9	7.4	6.6	14.1	2.9	159.4
1950	69.2	14.0	63.4	8.0	4.0	20.6	4.1	169.3
1960	65.2	10.7	85.2	6.2	4.8	28.0	6.1	195.5
1970	66.4	4.7	113.7	2.9	3.3	41.5	8.2	236.0
1971	73.0	4.3	113.0	2.7	3.1	41.4	8.5	241.7
1972	67.4	3.8	116.0	2.2	3.3	42.9	9.1	240.9
1973	61.1	3.4	110.9	1.8	2.8	42.4	8.9	227.9

Source: Economic Research Service, U.S. Department of Agriculture.

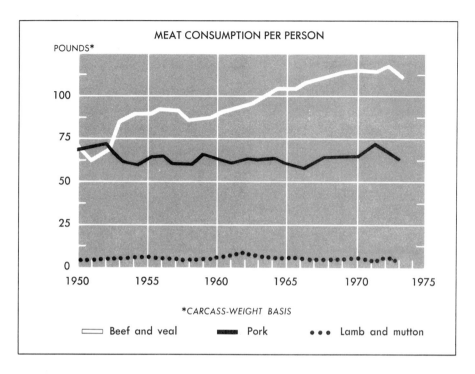

FIGURE 1-5. Per capita consumption of beef, pork, and lamb. (*Courtesy U.S.D.A.*)

meat and 3.8 pounds of lard. This was the highest per capita meat consumption on record. Lard consumption reached a high in 1923 when each person consumed an average of 14.5 pounds. As shown in Table 1-2, the average American in 1973 consumed approximately 61 pounds of pork, 110 pounds of beef, 1.8 pounds of veal, nearly 3 pounds of lamb and mutton, 42.4 pounds of chicken, nearly 9 pounds of turkey, and 3.4 pounds of lard.

As shown in Figure 1-5, per capita consumption of pork since 1952 has been considerably lower than that of beef. Between 1915 and 1952, consumption of pork was greater than that of beef with the exception of periods during World War I and during the drought in the Corn Belt in the mid-thirties.

The American family is eating more chicken and turkey than it ate in 1940. In

1940 the average person consumed 14.1 pounds of chicken and 2.9 pounds of turkey. Consumption of chicken has increased nearly 340 percent, and turkey consumption has increased 320 percent during the 33-year period.

Surplus Lard Problem

The substitution of vegetable oils and fats in cooking and the decrease in use of animal fats in soap-making have lowered the demand for lard. The price paid by the packer for live hogs is determined by the consumer demand for the pork cuts and for lard, and by the supply available. Today, consumers are eating less pork than beef, and they are eating the lean cuts, the hams, loins, picnics, and Boston butts. They buy less fat cuts and lard. They insist that the fat on the hams, picnics, and chops be

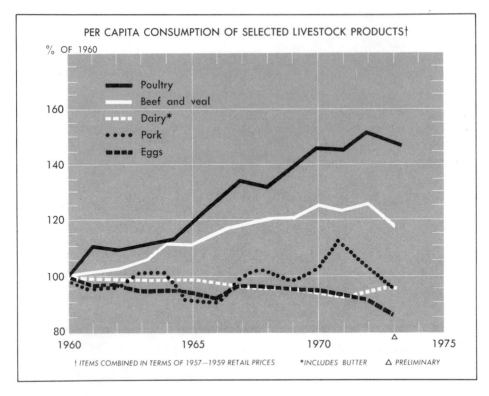

PER CAPITA CONSUMPTION OF SELECTED LIVESTOCK PRODUCTS†

% OF 1960

- ━━━ Poultry
- ━━━ Beef and veal
- ━ ━ Dairy*
- • • • Pork
- ■ ■ ■ Eggs

† ITEMS COMBINED IN TERMS OF 1957—1959 RETAIL PRICES   *INCLUDES BUTTER   △ PRELIMINARY

FIGURE 1-6. In 1973 the consumer bought more poultry and beef and less of the other livestock products than in 1960. (*Courtesy U.S.D.A.*)

trimmed. As a result we are producing more lard than we can sell profitably, even though we export lard to other countries.

A summary of this nation's production and use of lard from 1930 to 1972 is shown in Table 1-3. Note that we exported in 1960 about 26 percent of our lard. We exported only about 12 percent in 1972. Lard represents 10 to 12 percent of the weight of a live hog, but only about 7 cents of each dollar spent for pork products is spent for lard. The future of pork production rests, in part at least, upon our ability to produce hogs with more red meat and less lard, upon our ability to obtain a better market for lard, or upon both.

Per capita consumption of lard has dropped off considerably during the last 27 years. In 1946 we consumed an average of

11.8 pounds of lard, 10.2 pounds of shortening, and 6.4 pounds of vegetable oils. In 1973 we consumed an average of only 3.4 pounds of lard, but consumption of shortening and vegetable oils had increased to 16.7 and 19.6 pounds respectively.

We consumed in 1946 an average of 10.5 pounds of butter and 3.9 pounds of margarine. In 1972 the average person consumed only 5 pounds of butter and 11.3 pounds of margarine. We now consume more than five times as much margarine, shortening, and vegetable oils as we do butter and lard combined. Lard is in direct competition with vegetable oils.

Shown in Table 1-4 is a summary of fat and oil production in this nation in 1950, 1960, 1970, and 1971. Note that we produced in 1971 nearly four times as much soybean

T A B L E 1-3   U.S. RENDERED LARD—PRODUCTION, EXPORT, AND DOMESTIC
CONSUMPTION
(Million Pounds)

Year	Production	Export	Direct Use as Food	Use in Manufacture of Shortening and Margarine
1930	2,227	674	1,559	25
1940	2,288	232	1,901	23
1950	2,631	523	1,891	221
1960	2,562	681	1,358	556
1970	1,913	419	939	520
1971	1,960	345	880	679
1972	1,558	189	795	560
1973	1,254	113	713	413

Source: *Agricultural Statistics, 1973*, U.S. Department of Agriculture.

T A B L E 1-4   U.S. PRODUCTION OF FATS AND OILS—1950-1971

Kind	1950 (Million Pounds)	1960 (Million Pounds)	1970 (Million Pounds)	1971 (Million Pounds)
Butter	1,386	1,373	1,137	1,144
Lard	2,631	2,562	1,913	1,960
Tallow	2,272	3,507	4,905	5,210
Cottonseed oil	1,197	1,808	1,211	1,275
Peanut oil	186	104	253	259
Soybean oil	2,454	4,420	8,265	7,892

Source: *Agricultural Statistics, 1973*, U.S. Department of Agriculture.

oil as lard. We produced almost as much cottonseed oil and nearly three times as much tallow as lard. Lard actually represented only about 11 percent of the total of these fats and oils produced in this country in 1971.

Lard is no longer used in the manufacture of soaps but it is used extensively in the manufacture of shortening and margarine, as shown in Table 1-3. More than 413 million pounds of lard were used in the manufacture of shortening and margarine in 1973.

Soybean oil production. The soybean crop has played havoc with our lard market.

We harvested for seed less than five million acres of soybeans in this country in 1940. In 1972 we grew 47 million acres. Soybean production is likely to increase. Unless soybean oil is used more extensively in industry, hog producers can expect continued strong competition from this product.

Population Trends

The population of our nation is increasing at the rate of about 1.6 million persons each year. In 1950 our population was 152 million persons and we produced over 99 million hogs. We had in 1973 a population of 209 million and produced about 88 million hogs, an average of about one hog per 2.4 persons. If our population continues at the present rate, it will be necessary for us to raise about 110 million hogs to provide each person in the year 2000 with the same amount of pork and lard now supplied. A typical American by the time he reaches age 70 will have consumed the equivalent of 150 head of cattle, 310 hogs, 225 lambs, 2,400 chickens, and produce of 25 acres of grain crops and 50 acres of fruits and vegetables.

HOG-CORN PRICE RATIOS

Since corn is the basic feed in producing hogs, and feed costs represent from 60 to 70 percent of the production costs, it is usually possible to determine the extent to which hog production will be profitable by comparing the price of hogs with the price of corn. The term *hog-corn ratio* represents the

FIGURE 1-7. These meaty pigs are fed rations conducive to the production of low-fat carcasses. (*Larry Day photograph. Courtesy Fairall and Co.*)

TABLE 1-5   HOG-CORN PRICE RATIOS DURING FALL (U.S. AND NORTH CENTRAL REGION) AND FARROWINGS FOLLOWING SPRING, 1966 TO 1972

Year	HOG-CORN PRICE RATIO United States	HOG-CORN PRICE RATIO North Central Region	Number of Sows Farrowing Following Spring (1,000 head)	Increase or Decrease from Preceding Spring in Sows Farrowing (Percent)
1966	15.7	16.0	6,559	5.7
1967	17.0	17.3	6,659	1.5
1968	17.9	18.3	6,323	−5.0
1969	22.7	23.4	7,134	12.8
1970	12.7	12.9	7,303	2.4
1971	18.3	19.0	6,510	−10.9
1972	22.3	22.8	6,535	−0.4

Source: *Livestock and Meat Situation*, U.S. Department of Agriculture, Oct., 1973.

relationship between the two, based upon the number of bushels of corn that can be bought for the price of 100 pounds of pork.

Break-even Hog-corn Price Ratio

A break-even hog-corn price ratio is about 18 to 1. That is, 100 pounds of live hog should bring the price of 18 bushels of corn. With corn at $2.00, hogs should bring $36.00. With corn at $2.50 and hogs at $50.00, the hog-corn ratio stands at 20 to 1. This is considered a satisfactory conversion ratio for efficient producers but is not high enough to make possible a profitable enterprise for the less efficient producer.

Effect of Hog-corn Ratios Upon Production

It is a common practice for farmers to reduce the number of animals in their breed-ing herds during periods when narrow hog-corn ratios exist. Shown in Table 1-5 is a yearly summary of the hog-corn ratios during the September to December periods from 1966 to 1972 and the changes in number of sows farrowing spring pigs. In 1969 the ratio was 22.7 to 1, and there was an increase of 12.8 percent in the number of sows farrowing spring litters. In 1970 the hog-corn ratio narrowed to 12.7, and the number of sows farrowing spring litters increased only 2.4 percent. The influence of hog-corn ratios on spring farrowing is shown in Figure 1-8.

Effect of Hog-corn Ratios Upon Slaughter

The hog-corn price ratio usually foretells future changes in hog slaughter. The ratio was high in 1965 when the price of 100 pounds of hogs paid for 21.5 bushels of corn. This ratio was followed by an approximately

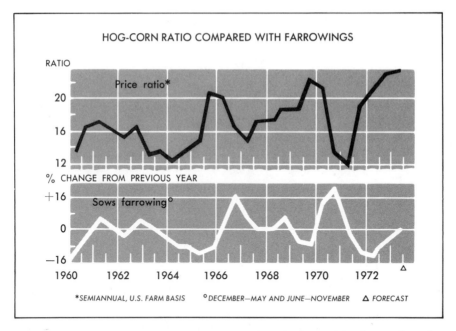

FIGURE 1-8 (*Above*). Hog-corn ratio compared with changes in the number of sows farrowing. (*Courtesy U.S.D.A.*) FIGURE 1-9 (*Below*). Small changes in production bring about bigger changes in price. (*Courtesy U.S.D.A.*)

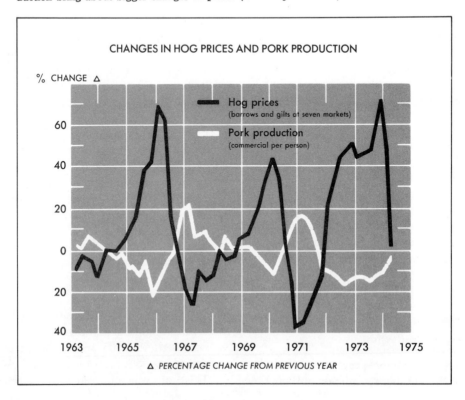

16 percent increase in the number of pigs saved in 1966. Hog slaughter turned upward in July, 1966, and continued high through early 1967.

The big problem in making effective use of hog-corn ratios is that of predicting the price that hogs will bring 9 months after the sows are bred or 5 months after the pigs have been farrowed. Changes in pork prices and pork production, 1963 to 1973, are shown in Figure 1-9.

Hog-feed Price Ratio

In areas where barley or other grains are fed to hogs, hog-feed price ratios are used. During one period the hog-barley price ratio in Oregon was seven to one. This type of ratio is used in the same manner as the hog-corn ratio, which was described in the preceding paragraphs.

ADVANTAGES AND DISADVANTAGES OF HOG PRODUCTION

It is necessary to consider many factors in selecting the livestock enterprises best suited to an individual farm and farm operator. Most important of these factors are the following:

1. Size of farm.
2. Type and productivity of the soils.
3. Kinds and quantities of crops grown.
4. Types, size, and condition of buildings.
5. Availability of water.
6. Markets available.
7. Transportation available.
8. Breeding and feeding stock available.
9. Investment required in breeding or feeding stock.
10. Investment required in housing and equipment.
11. Labor requirements.
12. Diseases and other hazards.
13. Rapidity of income.
14. Income per $100 invested.
15. Efficiency of animals to convert feeds into foods for human consumption.
16. Stability of demand for the products.
17. Personal preference.

Some farmers diversify their programs by engaging in two or more livestock programs. In this way they protect themselves against a heavy loss when an individual enterprise is unprofitable.

Advantages of Hog Production

Hog farming is well adapted to both specialized and diversified farming, and the returns come much more quickly than do those from many other enterprises. The investment in swine breeding stock and in equipment is relatively small, and it is possible to get in and out of the business in a comparatively short time. Feeding hogs on corn and other grains is usually a profitable method of marketing these grains. It is also an efficient way of producing meat. A pound of pork can be produced on as little as two and one-half to three pounds of feed. The young chicken is the only other farm animal which can produce a pound of meat on this amount of feed. No other farm animal excels the hog in ability to produce high-quality fat.

Hogs can be raised on small or large farms, and in small or large numbers. They make excellent use of pasture but can be produced profitably in confinement. They do not require expensive housing and equipment if raised on pasture.

Breeding stock and feeding stock are available in almost all areas where grain crops are grown. Hog slaughter plants are located in all production areas. The market for pork has been quite stable, more so than the market for either beef or poultry.

The labor requirements in producing hogs are lower than those in dairy and poultry production. One man can manage a large swine breeding herd.

FIGURE 1-10. Self-feeders help simplify the job of raising pigs in confinement. (*Courtesy* Hog Farm Management.)

According to Illinois University data, hogs returned $178 for each $100 worth of feed consumed on well-managed Illinois farms during the 1968–1972 period. Dairy cows returned $205, and poultry earned $165 for each $100 invested in feed. Beef cow herds returned about $171 for each $100 invested in feed.

During 1973, dairy herds in Illinois returned $177 for each $100 invested in feed; poultry returned $151; hogs, $192; feeder cattle, $120; and beef cow herds, $184.

Disadvantages of Hog Production

During recent years a number of diseases have caused heavy losses on some farms and in some areas. Rhinitis, brucellosis, erysipelas, gastroenteritis, anemia, and necrotic enteritis losses have been especially severe. Hog producers must follow a good sanitation program and feed adequate rations if they are to have a profitable enterprise.

The fact that farmers can get in and out of hog production in a comparatively short time may prove disadvantageous at times. When hog-corn price ratios are wide, 22 to 1 or above, farmers may flock in, and increased production may crowd the market, resulting in an unprofitable selling situation.

Another disadvantage of the hog enterprise in the past has been that farmers produced too many lardy hogs. As a result, there has been a surplus of lard, and housewives have decreased their purchases of fatty pork cuts. This disadvantage of pork production, like those mentioned previously, is being overcome by use of good breeding, feeding, and marketing practices.

EFFICIENCY IN PORK PRODUCTION

The profit in producing hogs is determined largely by the efficiency of the grower in production and marketing practices. It is estimated that the average farmer loses at least 25 percent of the pigs farrowed. The cost of maintaining a brood sow from breeding time until the pigs are weaned is about the same regardless of the number of pigs farrowed and weaned. More emphasis must be given to the saving of large litters of pigs.

In 1972, 1,029 participants in the Illinois Farm Business Record Program farrowed an average of 9 pigs per litter, and weaned 7.1. The number of litters farrowed did not affect either number farrowed or weaned. Central Iowa Farm Business Association members in 1972 weaned an average of 7.4 pigs. The most profitable farms in the Minnesota Farm Business Management Education Program farrowed an average of 9.3 pigs and saved 7.4. The low-profit producers farrowed an average of 8.5 and saved 6.5 pigs. Seven pigs per litter in the past has been the break-even point. Profitable production necessitates 8 or more pigs weaned per litter. The 25 Iowa

Master Swine Producers in 1973 marketed an average of 9.4 pigs per litter.

The loss of pigs due to disease is great. Hogs are subject to a large number of diseases. Some animals die; others are weakened. Considerably more feed is required to feed out diseased hogs, additional time is necessary to get them on the market, and they cannot be sold when the market is high. Often the carcass of the animal must be condemned entirely or in part, and sometimes the packer must offer a lower price for the diseased animals. The efficient producer employs measures which prevent disease outbreaks.

Care must also be taken to prevent losses due to swine parasites. Worms and mange probably cause some loss on most farms.

The efficient producer of hogs must be very careful in the selection of breeding stock and in the breeding of these animals. Pro-

lificness is inherited, as are growth rate and carcass quality. Good feeding and disease control cannot entirely offset losses due to poor breeding stock or to the use of poor breeding methods.

From 60 to 70 percent of the cost of producing a pound of pork goes for feed. Efficient producers select feeds carefully and feed them in proper balance. They make adequate use of home-grown feeds and forages, and uses protein supplements, minerals, antibiotics, and vitamins to supplement the home-grown feeds. They try to produce a pound of pork on 3 to 4 pounds of feed.

The farmers in the Illinois Farm Bureau Farm Management Service program produced 100 pounds of pork in 1972 on 426 pounds of feed.

It is possible to produce profitably and at the same time market inefficiently. Markets are better in some seasons than they

FIGURE 1-11. An increasing percentage of U.S. hogs are reared in confinement. This Indiana facility will accommodate 1,500 to 2,000 pigs each year. (*Agricultural Associates photograph. Courtesy Starcraft Swine Equipment.*)

FIGURE 1-12. A healthy pig crop fed by use of a mechanized feeding facility. (*Courtesy A. O. Smith Harvestore.*)

are in others. Packers pay higher prices for hogs of certain weights than they do for others. Some packers pay higher prices for hogs of high quality than do other buyers. The grading systems used by various buyers are not the same. Some dock more heavily for heavy hogs, or for lardy hogs, than do others. The high-income farmers in the Minnesota program in 1972 received $2.24 per hundredweight more for their hogs than did the low-income group of farmers. Through careful planning it is possible to have ready for market the kind and weight of hogs that will top the market. Efficiency in marketing is essential.

SUMMARY

Hog production is big business in this nation, especially in the Corn Belt states. Iowa, Illinois, Missouri, Indiana, Minnesota, and Nebraska produced 62.2 percent of the nation's hog crop in 1973. Iowa alone produces more hogs than the 39 low-producing states.

The numbers of hogs produced annually are affected by the hog-corn price ratio, the feed supply, the supply of hogs at the mar-

kets, consumer demand, world economic conditions, supplies of meats which compete with pork, price, and the swine disease situation.

The average person in the nation consumed in 1973 a total of 227.9 pounds of meat, of which 61.1 pounds was pork. The per capita lard consumption was 3.4 pounds.

Lard represents 10 to 12 percent of the weight of a live hog, but only 7 cents of each dollar spent for pork products is spent for lard.

A hog-corn price ratio represents the relationship between the two, based upon the number of bushels of corn that can be bought for the price of 100 pounds of pork. A break-even hog-corn price ratio is about 18 to 1.

Hog production is a profitable enterprise that requires a comparatively small investment and enjoys rather quick returns. Hogs are efficient converters of feed into food for human consumption.

The market for lard has not kept pace with the market for pork. Pork producers must find a way to produce hogs with more lean meat and less fat.

Profitable hog production is dependent upon how well farmers do the following:

1. Use good breeding stock.
2. Properly manage the breeding herd.
3. Feed and manage sows properly during gestation.
4. Properly care for sows and litters at farrowing time.
5. Feed and manage sows and litters properly during the suckling period.
6. Provide adequate legume pasture if pigs are not confined.
7. Provide satisfactory housing.
8. Maintain efficient feeding and watering equipment.
9. Feed adequate growing rations.
10. Control those diseases and parasites that attack swine.
11. Keep records of production and costs.
12. Follow sound practices in marketing their hogs.
13. Make efficient use of labor.

QUESTIONS

1 What percentage of the income on your farm is derived from the sale of hogs?

2 What percentage of the grains produced on your farm is fed to hogs?

3 What percentage of the farm income in your state is obtained from the sale of hogs?

4 How much meat is consumed in a year by the average American? How much of this amount is pork?

5 What are the leading states in hog production?

6 How many pounds of lard can you buy at your local meat market for the price of a pound of pork chops?

7 What fats are competing with lard at the consumer market?

8 What do you think can be done to decrease the lard produced on our farms without reducing the production of high-priced pork cuts?

9 Which livestock enterprise on your farm will produce the largest returns per $100 invested in feed?

10 What are the advantages and disadvantages of hog production on your farm?

11 What are the essentials of a profitable swine enterprise? Explain.

12 What is the influence of hog-corn ratio on pork production?

REFERENCES

Bundy, Clarence E., Ronald V. Diggins, and Virgil W. Christensen, *Livestock and Poultry Production,* 4th ed. Englewood Cliffs, N.J.: Prentice-Hall, Inc., 1974.

Ensminger, M. E., *Swine Science,* 4th ed. Danville, Ill.: The Interstate Printers and Publishers, 1970.

U. S. Department of Agriculture, *Agricultural Statistics 1973.* Washington, D.C.: Bureau of Agricultural Economics, 1974.

———, *The Livestock and Meat Situation.* Washington, D.C.: Agricultural Marketing Service, November 1973 and February 1974.

———, *Livestock and Meat Statistics.* Washington, D.C.: Agricultural Marketing Service, July 1973.

2 TYPES AND BREEDS OF SWINE

We have a number of breeds of hogs in this country, and it is sometimes difficult to determine the breed or breeds that will do best in a breeding program. Each breed has a loyal group of supporters, and there is rivalry among the producers of the various breeds. Each of the breeds has desirable characteristics, and all of them have some weaknesses. Usually there are as many differences among the individuals within a breed as there are between breeds.

FACTORS IN SELECTING A BREED

Certain factors are important in selecting a breed of hogs for a given program, just as other factors are important in buying a tractor or in purchasing seed corn. The goal in hog production is to produce large litters of pigs that can be grown out rapidly and economically, and, when sold, will command the top market price. Pork producers must decide which breed or breeds will fit best in their breeding programs.

The following factors must be given careful consideration in the selection of a breed of hogs:

1. **Availability of good breeding stock.** Breeding stock of good quality should be available in the community or nearby.

2. **Prolificness.** The ability of the sows to produce and nourish large litters of healthy pigs is very important in selecting breeding stock.

3. **Growth ability.** There are differences among the breeds in their ability to make rapid gains.

4. **Temperament.** The animals should be active but should have a good disposition and be easily handled.

5. **Carcass quality.** Some differences exist among breeds in their ability to produce carcasses that are high in the lean cuts and low in lard and fat cuts.

6. **Efficient use of feed.** Some breeds are more efficient in converting feed to pork.

7. Nicking ability. Some breeds nick better than other breeds when used in cross-breeding programs.

8. Market demand. The extent to which the breed is in demand in the community is important if breeding stock is to be sold. It is also important in selling market hogs. Some breeds produce better carcasses of certain weights than do some other breeds.

9. Disease resistance. While there is little difference in disease resistance among some of the older breeds, there are differences among other old breeds, and among the newer breeds.

10. Feeds available. Some breeds are thought to be better rustlers than others and, as a result, do better on poor pastures and on limited rations.

FIGURE 2-1. The Grand Champion Duroc Boar at the 1973 National Barrow Show. He was exhibited by Forkner Farms, Richards, Missouri. He sold for $38,000 and was purchased by Soga-No-Ya Swine Business of Japan. (*Courtesy Geo. A. Hormel & Co.*)

11. Personal likes and dislikes. We are more likely to do a good job of caring for animals of a breed we like, than of animals of a breed we do not like.

Some research data are available on the merits of the various breeds. Most comparisons previously have involved few lines of the various breeds. The popularity of a breed as shown by (1) the numbers of purebred animals recorded, (2) the demand for breeding stock by good swine producers, (3) the price that packers are willing to pay for market hogs of the breed, (4) the rating of the breed in market barrow and carcass shows, and (5) the results of boar testing and other feed and carcass quality testing programs, may provide our best indications of the value of the breed in a breeding program.

CLASSIFICATION OF BREEDS

Swine breeds were classified at one time as either *lard type* or as *bacon type*. Most

T A B L E 2-1   PHYSICAL CHARACTERISTICS OF BREEDS

Breed	Predominant Color of Hair	Type of Ears
OLD ESTABLISHED BREEDS		
Landrace	White	Large, slightly drooping
Berkshire	Black with white feet, face, switch	Erect
Chester White	White	Drooping
Duroc	Red	Drooping
Hampshire	Black with white belt	Erect
Hereford	Red with white head, feet, underline, and switch	Drooping
Mulefoot	Black or black with white feet	Erect to slightly drooping
OIC	White	Drooping
Poland China	Black with white on face, feet, legs, and switch	Drooping
Spot	Black-and-white spotted	Drooping
Tamworth	Red	Erect
Yorkshire	White	Erect
NEW BREEDS		
Beltsville No. 1	Black with white spots	Drooping
Beltsville No. 2	Light red	Erect
Lacombe	White	Drooping
Maryland No. 1	Black with white spots	Erect
Minnesota No. 1	Red	Slightly erect
Minnesota No. 2	Black with white spots	Slightly erect
Minnesota No. 3	Light red with black spots	Slightly erect
Montana No. 1	Black	Slightly drooping
Palouse	White	Slightly erect to drooping
San Pierre	Black and white	Erect

of the hogs produced in this country were of the lard type. We had a ready market for lard. Conditions have changed. Our lard market is partly gone, and we have vegetable fats competing with lard on the shortening market. As a result, swine breeders have focused their attention on the production of *meat-type* hogs.

While we can no longer classify hogs as lard type or as bacon type, we still find a considerable variation both within and among breeds in the extent that they produce carcasses high in the lean cuts and low in lard and fat cuts. Since the animals within a breed vary in this respect, it appears best to classify the breeds on physical characteristics and recency of origin.

Shown in Table 2-1 is a classification of the most popular old established breeds and of the new breeds according to the predominant color of hair and type of ears.

The Landrace is listed as an old breed since the Landrace bloodlines obtained originally from Denmark, Norway, and Sweden have now been recorded by the American Landrace Association, one of the newer swine record associations.

A summary of the numbers of purebred animals of the various breeds that were recorded by the respective breed associations in 1970, 1971, 1972, and 1973 is shown in Table 2-2.

More Durocs were recorded than any other breed in both 1972 and 1973. The

T A B L E 2-2   RECORDINGS OF SWINE BREED ASSOCIATIONS

Breed	1970	1971	1972	1973
Berkshire	8,012	7,210	5,529	5,144
Chester White	19,934	16,456	20,387	18,993
Duroc	76,394	62,830	66,647	71,435
Hampshire	74,101	55,180	56,110	51,135
Landrace	8,810	6,123	6,420	6,518
Poland China	16,102	11,191	9,079	7,958
Spot	13,974	11,117	14,824	17,471
Tamworth	1,424	1,000	1,415	923
Yorkshire	56,506	42,135	40,373	43,433

Hampshire breed ranked second and the Yorkshire breed ranked third in recordings in those years. In 1970 the Duroc breed led all others in number of animals recorded. The Hampshire, Yorkshire, and Chester White breeds ranked second, third, and fourth in recordings in 1970.

The recently developed breeds of inbred swine are recorded by the Inbred Livestock Registry Association of Noblesville, Indiana.

The new breeds of hogs are not raised in large numbers. Less than 0.5 percent of all hogs recorded in 1972 were of the new breeds. The origin of each of these breeds is presented later in this chapter.

BREED DIFFERENCES IN CARCASS QUALITY

The market hogs exhibited at the National Barrow Show held annually at Austin, Minnesota, come from many states and represent the bloodlines of the respective breeds. The data obtained from the carcasses of the slaughtered live show winners and from the winners of the carcass contest give some indication of the ability of the various breeds to produce desirable carcasses.

National Barrow Show Results

Shown in Table 2-3 is a summary of the data obtained from the carcasses of the 916 market hogs that were slaughtered at the 1971, 1972, and 1973 National Barrow Shows. The number includes the first five winners in each of three weight divisions in the live barrow shows for the eight breeds and the crossbreds for 1971 and 1972 and all the barrows exhibited in 1973.

In length of carcass the Yorkshire, and Hampshire breeds and the crossbreds ranked highest. The Poland China, Chester White, and Berkshire carcasses were the shortest. In backfat thickness the Hampshire, Duroc, and crossbred carcasses ranked best with less than 1.09 inches. The Landrace and Spot carcasses had most backfat.

The Landrace and Duroc carcasses had the best belly grades. The Chester White and Hampshire breeds had the lowest grades. In percentage of ham the Chester White, crossbred, and Spot excelled. The Berkshire and Landrace carcasses had the least ham. The crossbred, Hampshire, and Poland China carcasses had the largest loin eyes (5.12 square inches or higher). Smallest loin eyes were found on the carcasses of Landrace and Berkshire barrows.

In ham and loin index, the crossbred, Duroc, and Poland China carcasses ranked high. The Berkshire, Landrace, and Yorkshire carcasses rated lowest.

The crossbred, Duroc, and Poland China barrows reached a weight of 220 pounds in

T A B L E 2-3   COMBINED SUMMARY OF CARCASS RESULTS OF LIVE SHOW
BARROWS–1971, 1972, AND 1973 NATIONAL BARROW SHOWS*

Breed	Num-ber	Length (Inches)	Backfat (Inches)	Belly Grade	Ham (Percent)	Loin Eye (Square Inches)	Ham & Loin Index	Age at 220 Pounds (Days)
Crossbred	87	31.3	1.08	1.8	17.13	5.39	125.1	169
Duroc	163	30.9	1.06	1.5	17.11	5.07	121.8	175
Poland China	80	30.2	1.18	1.7	16.99	5.12	121.7	175
Spot	106	30.9	1.19	1.6	17.12	4.94	120.6	177
Hampshire	148	31.1	1.01	2.1	16.79	5.26	120.5	176
Chester White	100	30.5	1.17	2.0	17.27	4.72	119.9	182
Yorkshire	98	31.5	1.18	1.6	16.66	4.79	114.5	177
Landrace	58	30.9	1.31	1.2	16.46	4.31	107.7	176
Berkshire	76	30.7	1.18	1.7	15.87	4.51	103.8	181
Average		30.9	1.14	1.7	16.85	4.93	117.9	177

*Live show winners 1971 and 1972. All barrows, 1973.

less than 176 days. The other breeds required from 176 to 182 days.

Ratio of Lean to Fat Cuts

In the selection of a breed increased emphasis is being given to the ratio of lean to fat cuts in the carcasses. A high percentage of the weight of the carcass in hams and loin and a low percentage in backfat and trim are desired. Some breeders are now cooperating with the U. S. Department of Agriculture or with their state agricultural extension services and with their breed associations in conducting testing programs. Probes or other analyses are made to determine the amount of backfat on the pigs. Representative animals are slaughtered and careful analyses are made of the carcasses.

CERTIFIED MEAT SIRES AND LITTERS

Breeds differ in number of "certified litters" and "certified sires" produced. To qualify as a "certified litter," eight or more pigs must be raised to a weight of 320 pounds

at 56 days if a sow's litter, or 275 pounds if a gilt's litter. In addition, two pigs from the litter must be slaughtered. Both pigs must weigh 220 pounds in 180 days or the equivalent. The slaughtered pigs must produce carcasses with a maximum backfat of 1.5 inches, a minimum length of 29.5 inches, and a minimum loin eye of 4.5 square inches.

A boar that has sired five "certified litters" from five different sows (only two can be dams and daughters or full sisters) becomes a "certified meat sire." The first boars to meet these standards were King Edward, a Hampshire boar owned by Carr Brothers of McNabb, Illinois, and Flash Chief, a Poland China boar owned by Oscar W. Anderson of Leland, Illinois. Both boars were certified in 1955.

Presented in Table 2-4 is a summary of the numbers of certified meat litters and certified meat sires qualified during 1971 and 1972 by the various breeds.

The Hampshire, Duroc, and Yorkshire breeds led in number of certified litters qualified by December 31, 1972. The Hampshire, Duroc, and Poland China breeds qualified the most certified meat sires.

T A B L E 2-4   CERTIFIED MEAT LITTERS AND CERTIFIED MEAT SIRES BY BREED

	Certified Meat Litters		Total Qualified	Certified Meat Sires		Total Qualified
Breed	1971	1972	12/31/72	1971	1972	12/31/72
Berkshire	145	124	2,189	71	11	176
Chester White	46	53	1,059	3	3	71
Duroc	449	506	6,235	51	60	806
Hampshire	510	363	13,444	48	125	1,329
Landrace	49	28	1,065	5	0	89
Poland China	61	20	2,872	3	3	459
Spot	144	185	1,815	11	16	160
Yorkshire	282	265	5,549	12	12	385

T A B L E 2-5   AVERAGE DATA FOR CERTIFIED LITTERS QUALIFIED IN 1971 AND 1972

Breed	Age at 220 Pounds (Days)	Length (Inches)	Backfat (Inches)	Loin Eye Area (Square Inches)
Berkshire	168	30.8	1.22	5.27
Chester White	163	30.9	1.24	5.48
Duroc	161	30.6	1.21	5.53
Hampshire	165	30.7	1.14	5.68
Landrace	165	30.8	1.25	5.37
Poland China	170	30.8	1.19	5.91
Spot	164	30.5	1.26	5.41
Yorkshire	164	30.7	1.20	5.18

In 1972 there were 506 Duroc, 363 Hampshire, and 265 Yorkshire certified litters qualified. The Hampshire breed led in number of certified sires qualified in 1972. The Duroc breed ranked second and the Spot breed third. More Hampshire boars were certified as meat sires in 1972 than the boars of all other breeds.

Presented in Table 2-5 is a summary of certified litters qualified in 1971 and 1972. Duroc and Chester White litters reached 220 pounds in less than 164 days. Berkshire and Poland China litters required 168 days or more.

All breeds produced carcasses 30.5 inches or more in length. The Chester White breed produced the longest carcasses, 30.9 inches. The Berkshire, Landrace, and Poland China carcasses were 30.8 inches in length. Spot and Duroc carcasses did not average above 30.6 inches. Least backfat was found on Hampshire and Poland China carcasses. Both had less than 1.2 inches. Spot and Landrace carcasses had the most backfat, 1.26 and 1.25 inches respectively.

The high breeds in square inches of loin area were the Poland China, Hampshire, and Duroc. All exceeded 5.50 square inches. The Yorkshire and Berkshire breeds produced carcasses with the smallest loin eyes, 5.18 and 5.27 square inches respectively.

Peak numbers of litters and boars certified were produced in 1967 and 1968. Since that time breeders are devoting more atten-

TABLE 2-6   BOARS TESTED AT 20 CENTRAL TESTING STATIONS, 1972 AND 1973

Breed	Number Tested	Average Daily Gain (Lbs.)	Feed Efficiency (Lbs. Feed per 100 Lbs. Gain)	Backfat Thickness (Inches)
Berkshire	95	2.01	265.31	0.92
Chester White	575	1.89	260.89	0.90
Duroc	2,803	2.08	252.84	0.89
Hampshire	2,170	2.04	258.89	0.80
Landrace	100	1.89	284.01	0.94
Poland China	337	1.98	264.73	0.84
Spot	670	2.03	262.69	0.85
Yorkshire	1,630	2.02	256.93	0.86
Average	8,380	2.04	257.53	0.86

FIGURE 2-2. The Farmland Swine Testing Station at Lisbon, Iowa, is one of the stations assisting swine breeders in comparing bloodlines and breeds. (*Courtesy Farmland Agriservices, Inc.*)

tion to the use of ultrasonic equipment to test individual animals instead of progeny. Some breed associations are establishing new standards for certification. Most seed stock producers rely heavily on testing-station re-

sults and data obtained from the use of ultrasonic equipment.

BOAR TESTING-STATION DATA

Progressive producers of commercial hogs rely heavily on the performance of boars and barrows in swine testing stations in selecting a breed and individuals for their herds. The testing-station data provide the most reliable data concerning the merits of the individual breeds. Approximately 37 stations were in operation in 1974. Presented in Table 2-6 is a summary of the performance of the various breeds in 20 of the stations that were in operation during 1972 and 1973. The Iowa Swine Testing Station at Ames is one of five testing stations in that state. It had the largest number tested among all stations in the nation during the 1972 and 1973 period, 1539 head.

Three breeds contributed two-thirds of the animals tested. Thirty-three percent were Durocs, 25 percent were Hampshires, and 19 percent were Yorkshires. Durocs, Hampshires, and Spots ranked highest in average daily gain. Landrace and Chester White breeds ranked lowest. In feed efficiency (pounds of feed required to produce 100 pounds of gain), the Duroc, Yorkshire, and Hampshire breeds excelled. Landrace and Berkshire boars and barrows required most feed. The Hampshire, Poland China, and Spot breeds produced the least backfat. Landrace and Berkshire breeds had the most backfat.

ORIGIN AND CHARACTERISTICS OF THE OLD ESTABLISHED BREEDS

Seven of the ten old established breeds of swine raised in the United States originated in this country. They are the Chester White, Duroc, Hampshire, Hereford, OIC, Poland China, and Spot. The three breeds of swine that are not native to the United States are the Berkshire, Tamworth, and Yorkshire breeds. They were imported from England.

Berkshire

This English breed is one of the oldest breeds of swine. For many years the Berkshire was considered the best of the meat breeds because of the excellent carcasses that it produced. During the 1939 to 1954 period Berkshire entries were named grand champion 13 times in the carload division at the International Livestock Show.

During recent years the other breeds have been improved, and now there is less difference in the quality of carcasses as indicated by the data shown in Table 2-3 and Table 2-4. Berkshire carcasses at the 1973 National Barrow Show averaged 30.5 inches in length and had 1.19 inches of backfat. Several other breeds had longer carcasses and produced carcasses with less backfat.

FIGURE 2-3. Lynwood Revolt, a Champion Berkshire Boar. (*Moore photograph. Courtesy American Berkshire Association.*)

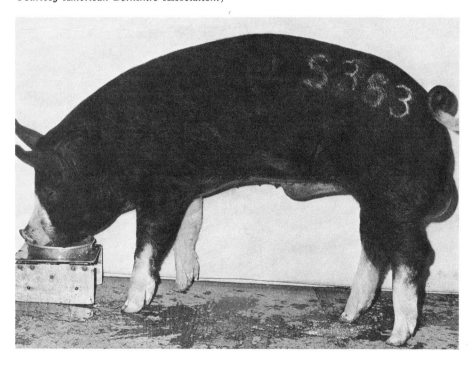

The Berkshire is usually a long animal and in form and fleshing conforms to the ideal meat-type hog. The breed is black, with white markings usually on the feet, head, and tail; it has long been characterized by a short snout and a wide, dished face. The pug nose of the Berkshire of years ago has been well refined through breeding.

Berkshires are slightly smaller at maturity than some of the other breeds. Mature boars weigh 900 pounds or more. In 1973 there were 5,144 Berkshire swine recorded in this country by the American Berkshire Association, Springfield, Illinois. The breed ranked seventh among the breeds in the number of animals recorded in 1973. Farmers have found the Berkshire somewhat less prolific and slower in gaining ability than some of the other breeds but excellent in nicking ability in crossbreeding programs. The 95 boars of the Berkshire breed tested in 20 stations in 1972 and 1973 gained 2.01 pounds per day, used 265.31 pounds of feed to produce a hundred pounds of gain, and had 0.92 inch of backfat.

Chester White

The Chester White had its origin in Chester and Delaware counties in Pennsylvania. The parent stock used to produce the breed included English Yorkshire, Cheshire, and Lincolnshire bloodlines. The Chester White Swine Record Association was established in 1908.

The Chester White has white hair and skin. Small flecks in the skin are not discriminated against, but black or other than white hair is objectionable. The Chester White is intermediate in size and mature boars weigh 900 pounds and over.

FIGURE 2-4. A Champion Berkshire Gilt at the 1974 National Berkshire Type Conference. (*Moore photograph. Courtesy American Berkshire Association.*)

FIGURE 2-5 (*Above*). A Grand Champion Chester White Boar. FIGURE 2-6 (*Below*). A Grand Champion Chester White Gilt. (*Figures 2-5 and 2-6 are Moore photographs. Courtesy Chester White Swine Record Association.*)

Barrows of the Chester White breed have in years past made excellent records at state and national shows. The barrows of this breed which were exhibited at the 1973 National Barrow Show produced carcasses which were 29.9 inches long and had 1.13 inches of backfat. The grand champion barrow at the 1968 National Barrow Show was a Chester White. The breed ranked sixth in ham and loin index.

Sows of this breed produce and raise large litters, and the pigs grow out rapidly. While the white color is objectionable to some breeders, the breed is quite popular with farmers. A total of 18,993 Chester Whites was recorded in 1973.

The record association for the breed is the Chester White Swine Record Association, Rochester, Indiana.

The 575 Chester Whites tested in 20 testing stations in 1972 and 1973 gained 1.89 pounds per day on 2.61 pounds of feed per pound of gain and had 0.9 inch of backfat.

Duroc

The ancestry of this breed is not entirely known, but the Jersey Reds of New Jersey, the red Durocs of New York, and the red Berkshires of Connecticut have contributed to the formation of the breed. The breed was first called the Duroc-Jersey. Standards were established for the breed in 1885.

The Duroc is red in color, with the shades varying from a golden yellow to a very dark red. A medium cherry red is preferred. Black flecks may appear in the skin, but large black spots, black hair, and white hair are objectionable. In type and conformation the Duroc is similar to the Chester White and the Poland China. The breed is prolific, and the sows are good mothers. They have good dispositions and produce large quantities of milk.

More Durocs were recorded in 1973 than the hogs of any other breed. A total of 71,435 animals were recorded by the United Duroc Swine Registry of Peoria, Illinois. This number was larger than the number recorded of the Berkshire, Chester White, Landrace, Poland China, and Tamworth breeds combined.

The Duroc is large and has excellent feeding capacity. Most tests related to rate of gain that have been made by agricultural experiment stations have indicated that the Duroc is a very rapid gainer. The breed is very popular in crossbreeding.

The Duroc barrows that were exhibited at the 1973 National Barrow Show yielded carcasses averaging 30.6 inches long with 1.21 inches of backfat.

The 2,803 Duroc boars tested in 20 stations in 1972 and 1973 gained 2.08 pounds per day on 2.53 pounds of feed per pound of gain and had 0.89 inch of backfat. The breed ranked first in both rate of gain and feed efficiency.

Hampshire

This breed was developed in Boone County, Kentucky, from hogs probably imported from England in the early 1800's. The foundation stock, known as the Thin Rinds and Belted hogs, had been raised in the New England states.

The breed association was organized in 1893, and although the breed is one of the youngest, it has become very popular. In 1973 when 51,135 animals were recorded, Hampshires ranked second to Durocs in the number of purebred animals recorded. The breed association is the Hampshire Swine Registry located at Peoria, Illinois.

The Hampshire is a black hog with a white belt encircling the body and including the front legs. The back legs are usually black, and no white should appear above the hock. The head and tail are black, and the ears are erect. No white should show on the head.

FIGURE 2-7 (*Above*). A Grand Champion Duroc Boar. FIGURE 2-8 (*Below*). A Grand Champion Duroc Gilt. (*Both pictures courtesy United Duroc Swine Registry.*)

The Hampshire is smaller than some of the other breeds. It has been bred for refinement, quality, and prominent eyes. The sows of the breed are very prolific and are good mothers. It is shorter legged than most breeds and sound on its feet and legs in most cases. The breed has been used extensively in crossbreeding because of its quality, fleshing, and prolificness.

The barrows of this breed that were exhibited and slaughtered at the 1973 National Barrow Show yielded carcasses averaging 30.7 inches in length, and had an average backfat thickness of 1.02 inches. Hampshire barrows ranked third in ham and loin index at the 1973 National Barrow Show. More than 50 percent of the hogs in Nebraska and Indiana carry some Hampshire breeding.

The 2,170 Hampshire boars tested in the 20 central testing stations gained 2.04 pounds per day on 2.59 pounds of feed per pound of gain. The boars averaged 0.8 inch of backfat. These tests were made in 1972 and 1973.

A Hampshire litter set a record at the Farmland Swine Testing Station in Iowa. The high boar had an index of 227 with the three boars averaging 220. The littermate barrow had a loin eye of 7.67 square inches.

Hereford

This breed was originated about 1900 by R. U. Weber of La Plata, Missouri. He used Durocs, Chester Whites, and OIC Chester Whites in founding the breed. The first registry association was formed in 1920, and the present National Hereford Hog Association was formed in 1934. It is located at Dwight, Illinois. There were 348 hogs registered in 1972.

The ideal markings for a Hereford are a white head (including the ears), feet, underline, and switch of tail. The remainder of the body is red. The red varies from light to dark, with the darker shades preferred. At least two-thirds of the Hereford must be red.

FIGURE 2-9 (*Left*). A Champion Hampshire Boar at the 1974 Hampshire Type Conference in North Carolina. Bred by A. Ruben Edwards of Missouri. Sold to Robert Meeker for $10,500. FIGURE 2-10 (*Right*). Champion Gilt at the 1974 Hampshire Type Conference in North Carolina. Bred by Jim Foster of Missouri. Sold to Fred Haley of Canton, Georgia, for $2,300. (*Figures 2-9 and 2-10 are Moore photographs. Courtesy Hampshire Swine Registry.*)

FIGURE 2-12 (*Above*). A Landrace boar owned by Harlan Moore of Barrington, Illinois. FIGURE 2-13 (*Below*). A Landrace sow owned by Harlan Moore of Barrington, Illinois. (*Figures 2-12 and 2-13 Courtesy Harlan Moore.*)

FIGURE 2-11. *Above:* A Mulefoot sow owned by R. M. Holiday of Louisiana, Missouri. *Below:* The solid hoof of the Mulefoot hog. (*Courtesy R. M. Holiday.*)

Mulefoot

Mulefoot hogs have been raised in this country since the early 1900's. The breed is referred to in the Bible as the solid-hoofed breed.

Animals are black or black with white feet. The ears are medium in size. They may be erect or with a small droop. The feet have a solid hoof instead of the cloven or two-toe hoof. The National Mulefoot Swine Association, Inc., is located at Knoxville, Iowa.

American Landrace

Landrace hogs originated in Denmark and were first imported to this country in 1934 for experimental crossbreeding purposes. By government agreement we could not at that time produce and release Danish Landrace stock in this country as purebreds. We used them extensively in crossbreeding, and many of the new breeds carry Landrace breeding.

In 1954, 1,912 Landrace hogs were recorded by the American Landrace Association, Inc. This organization was incorporated in December, 1950, with headquarters at Noblesville, Indiana. In addition to the importations from Denmark, a number of Landrace hogs have been brought in from Norway and Sweden.

The Landrace has white hair, and the skin is usually white. Small black spots, however, are common. The breed is extremely long, deep sided, and well hammed. Usually the animals are flat and sometimes low in the back. The ears are very large and cover much of the face. Many of the Landrace breed have weak pasterns. The breed is prolific.

Barrows of the Landrace breed at the 1973 National Barrow Show produced carcasses averaging 30.6 inches long. The carcasses had 1.36 inches of backfat and 4 inches of loin eye.

The 100 boars tested in 1972 and 1973 in the 20 central boar testing stations gained 1.89 pounds per day on 2.84 pounds of feed per pound of gain. The breed ranked lowest among the breeds in both rate of gain and feed efficiency. The boars had 0.94 inch of backfat.

The total number of Landrace animals recorded in 1973 was 6,518.

OIC

This breed was originated by L. B. Silver in Ohio about 1865 from foundation stock obtained in Chester and Delaware counties in Pennsylvania, the same counties that produced the Chester White breed. The breed was first called the Ohio Improved Chesters. The OIC Swine Breeders Association, Inc., was organized in 1897. It is now located at Goshen, Indiana.

The OIC breed resembles the Chester White in color. The sows are good mothers and milkers.

No separate class is provided for this breed at the National Barrow Show.

Poland China

For many years the Poland China was considered the largest of the American breeds. It is today similar in size, type, and conformation to the Duroc, Chester White, and Spot.

The breed was originated between 1800 and 1850 in Warren and Butler counties in Ohio. The white Byfield hog, imported from Russia, and the White Big China hog were used with native hogs in producing the Warren County hog. The use of the Berkshire on the Warren County hogs and later the use of boars imported from Ireland produced the Poland China breed.

The Poland China since about 1875 has been a black hog with six white points: the feet, face, and tip of tail. The typical Poland China has thick, even flesh and is free from wrinkles and flabbiness. The breed has good length and excellent hams. The head is trim, and the ears droop. Farmers have not considered the breed quite as prolific as some of the other breeds, but most bloodlines of the breed produce very satisfactory litters.

Poland Chinas produce excellent carcasses. The barrows of this breed that were exhibited and slaughtered at the 1973 National Barrow Show yielded an average ham and loin index of 115. Their carcasses averaged 29.7 inches in length and had only 1.12 inches of backfat. The loin eyes averaged 4.72 inches.

FIGURE 2-14. A National Champion OIC Senior Boar. (*Courtesy OIC Breeders Association.*)

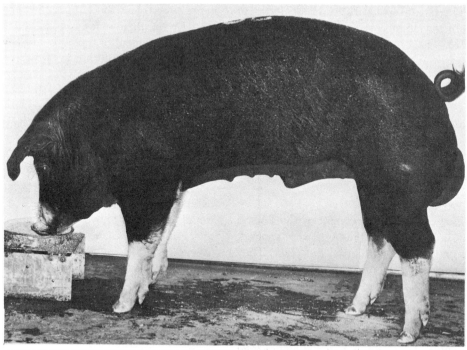

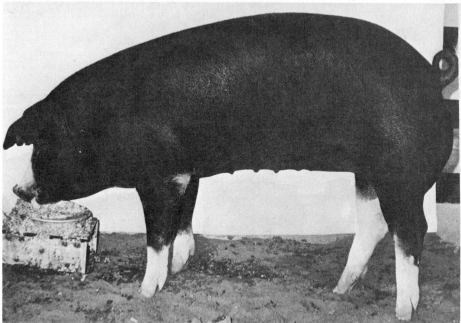

FIGURE 2-15 (*Top*). A Champion Poland China Boar at a recent Poland China Spotlight show. FIGURE 2-16 (*Bottom*). The second prize late February Poland China gilt at the 1973 Illinois State Fair. (*Figures 2-15 and 2-16 courtesy Harlan Moore.*)

The Poland China Record Association of Galesburg, Illinois, recorded 7,958 animals in 1973. The breed ranked sixth among the ten old established breeds in the number of animals recorded in 1973.

The 337 Poland China boars in the 1972 and 1973 swine testing stations gained 1.98 pounds per day on 2.65 pounds of feed per pound of gain. The boars had 0.84 inch of backfat. The breed rated second in backfat thickness but next to last in rate of gain.

Spot

Early in the development of the Poland China breed many of the hogs were spotted, and some breeders preferred this color. They were reluctant to adhere to the color standards set up for the Poland China breed, and many continued to raise spotted hogs. Some of these hogs were crossed with the black Poland China and some with Gloucester Old Spots which had been imported from England.

The formation of the National Spotted Poland China Record Association came about in 1914. The name was changed in 1960 to the Spotted Swine Record.

The Spotted hog resembles the Poland China in type and conformation. It is a large breed, and the animals are good feeders. To be eligible for registration between 80 and 20 percent of the body must be white, but the desired coloring is 50 percent black and 50 percent white.

This comparatively new breed has become quite popular. The breed association located at Indianapolis, Indiana, recorded 17,471 animals in 1973, the fifth largest number of animals recorded by a breed association during that year.

The barrows of this breed which were exhibited and slaughtered at the 1973 National Barrow Show produced carcasses averaging 30.7 inches long. They had an average of 1.19 inches of backfat and produced an average ham and loin index of 116.8. The loin eyes averaged 4.81 inches.

Six hundred and seventy Spot boars were entered in the 20 testing stations in 1972 and 1973. They gained 2.03 pounds per day on 2.63 pounds of feed per pound of gain. The boars had 0.85 inch of backfat.

Tamworth

The Tamworth is red in color, with the shades varying from light to dark. The head is long and narrow, with a long snout and erect ears. The body is long and narrow, and

FIGURE 2-17 (*Left*). A Grand Champion Spotted Boar. FIGURE 2-18 (*Right*). A Grand Champion Spotted Gilt. (*Figures 2-17 and 2-18 courtesy National Spotted Swine Record, Inc.*)

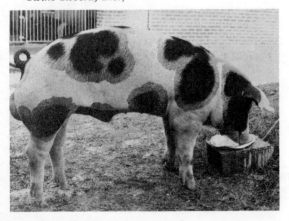 

FIGURE 2-19 (*Left*). A Champion Tamworth Boar. FIGURE 2-20 (*Right*). A Champion Tamworth Sow. (*Figures 2-19 and 2-20 courtesy Tamworth Swine Association.*)

the sides are smooth. Usually the Tamworth has a strong back. Thin shoulders are desired.

The Tamworth is one of the oldest of all breeds of hogs. There is evidence that the breed was well established in England about a hundred years ago. First importations to this country were made about 1881.

Sows of this breed are prolific and are excellent mothers and foragers. Mature boars weigh up to 700 or 800 pounds. The breed was originally developed as a bacon-type hog. Its popularity in this country has been largely due to its use in crossbreeding programs. Tamworth bloodlines were used in producing the Minnesota No. 1 breed of swine.

The Tamworth Swine Association of Evansville, Indiana, recorded 1,415 animals in 1972.

No barrows of this breed were slaughtered at the 1973 National Barrow Show, and no boars of the breed were entered in the 20 central swine testing stations in 1972 and 1973.

Yorkshire

During the time that our American hogs were classified as *bacon* or as *lard* type, the Yorkshire was considered by many as the best bacon-type breed. The breed is raised in large numbers in Canada, England, Scotland, and Ireland. It is a native of northern England, and was imported to this country early in the nineteenth century.

The Yorkshire is white in color but occasionally has black pigment spots in the skin. These spots are objectionable but do not disqualify the animal in the show ring or from recording. The ears are erect. Mature boars weigh from 700 to 1,000 pounds.

The Yorkshire is extremely long and deep, and is firm fleshed. The barrows of this breed that were exhibited and slaughtered at the 1973 National Barrow Show produced carcasses that averaged 31.2 inches in length and had an average of 1.15 inches of backfat. Yorkshire barrows had a ham and loin index of 107.3. The loin eyes averaged 4.36 inches.

The Yorkshire breed ranked third in the number of boars tested in the 20 testing stations during 1972 and 1973. The 1,630 boars tested averaged 2.02 pounds of gain per day on 2.57 pounds of feed per pound of gain. The backfat thickness was 0.86.

The American Yorkshire Club, Inc., of Lafayette, Indiana, recorded 43,433 animals in 1973. The popularity of the breed has

FIGURE 2-21 (*Left*). A Reserve Champion Yorkshire Boar at the Minnesota State Fair. Exhibited by Dick Kuecker, Algona, Iowa. (*Courtesy Dick Kuecker.*) FIGURE 2-22 (*Right*). A Champion Yorkshire Gilt. (*Moore photograph. Courtesy Dick Kuecker.*)

increased greatly since 1952. It has had wide use in crossbreeding.

ORIGIN AND CHARACTERISTICS OF THE NEW BREEDS

A number of new breeds of swine have been developed during the past thirty-five years in an attempt to produce a breed of hogs that is prolific, will gain rapidly, will make efficient use of feed, will produce a carcass high in lean and low in fat cuts, and will have a strong constitution. This has been a big undertaking.

Most of the breeds that have been developed to date have their origin at the U.S.D.A. Agricultural Research Center at Beltsville, Maryland, or at cooperating state agricultural experiment stations. Inbred lines of various existing breeds have also been developed.

In the development of new breeds, an attempt has been made to produce animals that possess the desirable characteristics of two or more parent stocks. The Danish Landrace, Yorkshire, and Tamworth breeds have been used in various combinations with other old established breeds. The goal was to bring together the growing and gaining ability and the ruggedness of some breeds, and the carcass quality and prolificness of others. Following is a description of some of the most promising new breeds.

Beltsville No. 1

This breed was developed by the Bureau of Animal Industry at the Agricultural Research Center, Beltsville, Maryland, from crosses made in 1934 and traces to thirteen

FIGURE 2-23. Beltsville No. 1 boar (Landrace X Poland China). (*Courtesy U.S.D.A. Bureau of Animal Industry.*)

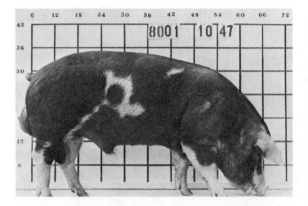

animals of the Danish Landrace breed and three Poland China boars. The breed was admitted to the Inbred Livestock Registry Association, University Farm, St. Paul, Minnesota, in 1951. The breed carries approximately 75 percent Landrace and 25 percent Poland China blood and is about 35 percent inbred.

As shown in Figure 2-23 animals of this breed are black-and-white spotted. They have long bodies, little arch of back, moderate depth of body, smooth sides, and plump hams.

FIGURE 2-24. Maryland No. 1 boar (Landrace X Berkshire). (*Courtesy U.S.D.A. Bureau of Animal Industry.*)

Beltsville No. 2

This breed was also developed at the Agricultural Research Center at Beltsville from crosses begun in 1940. The pigs carry 58 percent Danish Yorkshire, 32 percent Duroc, 5 percent Landrace, and 5 percent Hampshire blood.

Beltsville No. 2 hogs are usually solid red in color and have white underlines. The ears are usually short and erect. The head is intermediate in length and has a moderately trim jowl. Hogs of this breed have the length of the Yorkshire. The back is of a medium width and has little arch.

Carcass data obtained on 72 hogs of this breed averaging 213 pounds indicated a cold dressing percentage of 79.3 and an average thickness of backfat of 1.68 inches.

The breed was accepted by the Inbred Livestock Registry Association in 1952.

Maryland No. 1

This breed of swine was produced cooperatively by the Bureau of Animal Industry of the U. S. Department of Agriculture, and the Maryland Agricultural Experiment Station. The Maryland No. 1 was admitted to the Inbred Livestock Registry Association in 1951.

The Maryland No. 1 line was established in 1941 and carries approximately 62 percent of Landrace and 38 percent of Berkshire blood.

Hogs of this breed are black-and-white spotted and are intermediate in conformation between the Landrace and Berkshire. The back is slightly arched and medium in width. The head is long and the jowl is somewhat heavier than that of the Landrace. The ears are medium in size and are usually erect.

Minnesota No. 1

This breed was developed by the Minnesota Agricultural Experiment Station in cooperation with the Regional Swine Breeding Laboratory of the U. S. Department of Agriculture. The breed traces in origin to the pigs produced by crossing six Tamworth females with a Landrace boar in 1936.

Minnesota No. 1 is an inbred line. Breeding stock has been selected from within the herd on the basis of performance; prolificness, survival, growth rate, economy of gain, and carcass quality were the criteria considered.

Hogs of this breed are red in color. Occasionally black spots may appear in the skin. The body is long, about two inches longer than most American breeds. The back usually has no arch. The jowl is refined, and the neck is thin. The snout is usually long and trim. The ears are fine textured and vary from erect to drooping.

FIGURE 2-25 (*Above*). Minnesota No. 1 boar (Landrace X Tamworth). (*Courtesy George Slater, Augusta, Illinois.*)   FIGURE 2-26 (*Below*). Minnesota No. 2 boar (Yorkshire X Poland China). (*Courtesy Inbred Livestock Registry Association.*)

Minnesota No. 2

The Minnesota No. 2 hog was developed by the Minnesota Agricultural Experiment Station in cooperation with the Regional Swine Breeding Laboratory of the U.S.D.A.

An inbred Canadian Yorkshire boar was mated to 13 inbred Poland China gilts in 1941. Animals of the first and second generation were intermated, and some backcrossing was done. The breed contains 40 percent Yorkshire and 60 percent Poland China blood.

Animals of this breed are black and white in color, having long bodies, well muscled loins, full and deep hams, and have heads with shorter snouts than the Minnesota No. 1. The ears are medium in size and are erect.

The breed was recognized in 1948 by the Inbred Livestock Registry Association.

Minnesota No. 3

The Minnesota No. 3 breed is an inbred line developed from eight other breeds. The foundation for the line was established in 1950. The following breeds were used:

1. Gloucester Old Spot from England
2. Welsh pig from England
3. English Large White from England
4. Beltsville No. 2 from Beltsville, Maryland
5. San Pierre from Indiana
6. Inbred Poland China
7. Minnesota No. 1
8. Minnesota No. 2

Most of the pigs are black and red spotted. The color is not completely fixed.

Montana No. 1 or Hamprace

This breed is solid black in color and was developed from crosses made of inbred Landrace and Hampshire bloodlines by the

Montana Agricultural Experiment Station and the Bureau of Animal Industry of the U. S. Department of Agriculture. The breed is about 32 percent inbred and contains approximately 55 percent Landrace and 45 percent Hampshire blood.

Production records based on a nine-year period show that the Montana No. 1 averages about 10.6 pigs per litter at birth, 8.1 pigs per litter at weaning, and 248 pounds in weight of litter at 56 days. The average daily gain from weaning until about 225 pounds varies from 1.13 to 1.69 pounds.

The Hamprace Breeders Association, which was formed in 1949, was discontinued in 1953.

Palouse

The Palouse breed of swine was developed by Washington State University by crossing in 1945 three Landrace boars with 18 Chester White gilts and sows. No blood from outside the line has been introduced since 1947.

In producing the breed, effort was made to produce a hog that at 210 pounds would have at least 50 percent of live weight in ham, loin, picnic, and butt, and would have a backfat thickness of 1.3 to 1.6 inches. The hogs are solid white in color and resemble the Landrace in body type and conformation.

San Pierre

The San Pierre hog originated at the Inbred Swine Farm, San Pierre, Indiana, then owned by Gerald Johnson. The foundation stock were Canadian Berkshire and Chester White. No outside blood has been introduced since the original crosses were made. This is the only new breed which has been developed by a private producer.

The Johnson farm was sold a few years after the line was started. A few breeders acquired foundation stock, and the breed has been maintained.

San Pierre hogs are black and white in color, have length similar to the Berkshire, and the growth vigor of the Chester White.

FIGURE 2-27. A Five-Star Montana No. 1 Sow (Landrace X Hampshire). (*Courtesy George Slater, Augusta, Illinois.*)

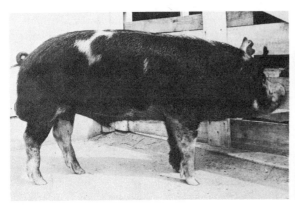

FIGURE 2-28. A San Pierre boar (Berkshire X Chester White). (*Courtesy Inbred Livestock Registry Association.*)

Their ears are erect. They are a meat-type hog and have been used in crossbreeding programs.

Lacombe

This breed of hogs was developed by the Canadian Department of Agriculture by combining several breeds. Bloodlines are now stabilized at 55 percent Landrace, 23 percent Berkshire, and 22 percent Chester White.

The breed has been refined during the past 20 years in Canada. The first imports to the U. S. were made in 1962.

Lacombe hogs are white with drooping ears of medium length. Their general appearance resembles the Landrace breed. The American Lacombe Swine Association is located at Grand Mound, Iowa.

HYBRID HOGS

Hybrid Defined

A hybrid is produced by crossing two or more inbred lines. Some commercial hybrid seed corn is a result of a double cross process involving four inbred lines. Line A is crossed with line B (A x B), and line C is crossed with line D (C x D) during one growing

FIGURE 2-29. A large and healthy litter of hybrid pigs. (*Courtesy Farmers Hybrid Hogs.*)

season. The next year the progenies of the two crosses are crossed (AB x CD). The corn produced from the latter cross is sold as commercial seed corn.

Hybrid hogs are produced in much the same manner as is hybrid corn. Lines of hogs are inbred for several generations and then crosses are made of the inbred lines. The extent that the hybrid hog is more productive than the parent stock is dependent upon the genetic make-up of the various lines, and how well they supplement each other when they are brought together.

Hybrid Vigor or Heterosis

The crossing of two breeds of hogs has been a popular procedure in commercial hog production. The offspring usually make rapid and economical gains and have excellent constitutions. The crossbred gilts farrow and raise larger litters than do noncrossbred sows. This increase in vigor, growth rate, and productiveness is called *hybrid vigor* or *heterosis*.

It has been found that when animals of two different breeds of hogs are mated, an increase in performance of approximately 6 to 8 percent is realized in the offspring. When an animal from a third breed is crossed with the offspring of the original cross, another 2 to 4 percent increase in performance can be expected. The use of three or four breeds in a crossbreeding program further increases the hybrid vigor, providing the lines are carefully selected.

The use of inbred lines in the crossbreeding program further increases the hybrid vigor. Inbreeding increases the degree of genetic difference among the breeds. The greater the genetic difference, the greater the

hybrid vigor. Results reported in Oklahoma show that a cross of three inbred lines improved the five months' litter weights as much as 15 percent over a crossing of two outbred breeds, and 48 percent over outbred matings within a breed.

Hybrid Hog Production

There is no set pattern in the production of hybrid hogs in this country. The hybrid is produced by crossing two or more inbred lines, usually from different breeds. Many farmers and breeders do not understand the true meaning of the term "hybrid," and as a result some hogs are called hybrids incorrectly.

Hybrid hog production is a complicated procedure and one that involves careful planning and management. Producers of

FIGURE 2-30. Reserve Champion Truckload of Market Pigs at the 1972 National Barrow Show. These crossbreds were exhibited by Cindy Smith, Leland, Illinois. (*Courtesy Geo. A. Hormel & Co.*)

hybrid hogs must understand genetic laws and be able to apply them. Usually there is quite an outlay in breeding stock and equipment. The number of producers of hybrid hogs is small. However, some of them do produce large numbers of hogs.

The following are producers of large numbers of seed stock. Some producers market purebred, crossbred, and incrossbred lines as well as true hybrids.

American Hog Farm, Wiggins, Colorado
Conner Prairie Swine, Inc., Noblesville, Indiana
DeKalb AgResearch, DeKalb, Illinois
Farmers Hybrid Companies, Hampton, Iowa
Lubbock Swine Breeders, Lubbock, Texas
Midwestern Swine Breeders, Rochester, Minnesota

Following is a brief summary of the production programs of some of the larger producers of hybrid hogs.

Farmers Hybrid Hogs

This organization, located at Hampton, Iowa, has been producing hybrid boars since 1945 and every year markets more than 10,000 boars to commercial hog producers.

The boars are not the progeny resulting from the crossing of two inbred lines. The boars and gilts marketed are the result of nearly 30 years of swine breeding. The organization has developed synthetic pure strains of hogs. Each strain combines the desirable genetic qualities of two or more of the existing swine breeds. Most of the foundation germ plasm for the hybrid hogs was contributed by seed stock of the Duroc, Hampshire, Spot, Poland China, Yorkshire, and Chester White breeds.

The organization maintains an "elite" foundation herd of each of the five strains. More than 1,000 litters are farrowed and the pigs are performance tested. The very top ones are selected to be retained in the elite herd. A second selection is made to determine boars and gilts to be placed in multiplier herds.

Approximately 160 multiplier herds are maintained. These herds are scattered throughout Iowa, Illinois, Indiana, and South Dakota. The boars and gilts marketed com-

FIGURE 2-31. Commercial hogs sired by hybrid boars. (*Courtesy Farmers Hybrid Hogs.*)

FIGURE 2-32. Crossbred pigs on an Iowa farm. (*Courtesy Land O Lakes, Felco Division.*)

mercially come from these herds. Several distribution centers are provided.

Customers are encouraged to follow a recommended breeding program using the various strains of breeding stock in rotation.

No tests comparing hybrid hogs from this source with other types and breeds of hogs have been made by experiment stations. Commercial producers who used hybrid boars from this source, however, have been represented in the Iowa Master Swine Producers' selections since about 1949.

A 212-pound Farmers Hybrid hog set a new record for meatiness in the 1969 Ft. Dodge, Iowa, carcass show. The carcass scored 195.2 points based on the amount of carcass in ham and loin cuts. The carcass had an 8.06 square inch loin eye.

Conner Prairie Swine, Inc.

This farm, located at Noblesville, Indiana, produces three separate hybrid lines of breeding animals. They began their breed-

FIGURE 2-33. Boars produced by Conner Prairie Swine, Inc. of Noblesville, Indiana. (*Courtesy Conner Prairie Swine, Inc.*)

ing operations in 1949 and each year sell approximately 2,000 boars and gilts.

The original breeding program on the Conner Prairie Swine farm involved six breeds of inbred swine and the crossing of these breeds to produce incross boars. The breeds involved were the Minnesota No. 1, Minnesota No. 2, Montana No. 1, Maryland No. 1, San Pierre No. 1, and Beltsville No. 2.

Their breeding operation followed the pattern of the hybrid corn procedure in that related inbred lines were crossed. The hybrid boars are more active, have greater disease resistance, and adapt themselves more readily to new surroundings than do the inbred boars. Since they genetically possess the growth factors of two or more lines of breeding, they have more hybrid vigor.

This organization recommends that the commercial producer use the three lines in rotation just as he rotates corn, oats, and clover. The three lines which they produce

FIGURE 2-34 (*Above*). Weighing and blood testing pigs on Connor Prairie Farms. FIGURE 2-35 (*Below*). An F-Line boar bred by Connor Prairier Swine, Inc., of Noblesville, Indiana. (*Figures 2-34 and 2-35 courtesy Connor Prairie Swine, Inc.*)

are (1) the C Line, a cross of the Montana No. 1 and the Minnesota No. 1; (2) the E Line, a cross of the San Pierre No. 1 and the Maryland No. 1; and (3) the F Line, a cross between the Minnesota No. 2 and the Belts-

FIGURE 2-36. A shipment of meat-type commercial hybrids. (*Courtesy Lucie Hybrid Hog Farms.*)

ville No. 2. New blood has been introduced in each line during recent years.

The inbred lines are produced largely by Conner Prairie Swine, Inc. The boars sold to commercial hog producers are raised by approximately 30 producer farmers.

Boars have been sold to swine producers in a large number of states from Maryland to California and from Texas to Minnesota. Eight service men are maintained to assist farmers in making effective use of the breeding stock.

Since the breeds of hogs used in their breeding program are high in Landrace, Berkshire, and Yorkshire bloodlines, the hogs produced usually have excellent carcasses. Carcasses from 203- to 206-pound hogs usually have an average of about 1.25 inches of

backfat thickness and 31 inches of carcass length.

Lucie Hybrid Hogs

A large producer of hybrid boars is the Lucie Hybrid Hog Farms of Augusta, Illinois. This firm produced and sold approximately 1,500 boars in one year to commercial hog producers, largely in Illinois, Indiana, Iowa, Missouri, and Wisconsin.

Bloodlines used in developing the original stock included Landrace, Hampshire, Large English White, Welsh, Yorkshire, Poland China, Berkshire, and Old Spot. Carefully selected lines of these inbreds are maintained through performance testing. Records of performance are kept on (1) rate of gain, (2) prolificness, (3) feed efficiency,

(4) mothering ability, (5) carcass quality, (6) type, (7) number of teats, and (8) survival of pigs farrowed.

High-scoring animals of one inbred line are crossed with high scoring animals of another inbred line to produce the hybrid boars.

This firm recommends the use of a continuous four-boar rotation breeding system in order to maintain the maximum of hybrid vigor.

The decision as to the bloodlines and the order in which they are to be used must be determined on the basis of the kind of breeding in the sow herd at the time the rotation is begun.

SUMMARY

There is no one best breed of hogs. The differences among the animals within a breed may be greater than those between breeds. The following factors should be considered in selecting a breed: availability of good breeding stock, prolificness, growth rate, temperament, carcass quality, nicking ability, market demand, disease resistance, feeds available, and personal likes and dislikes.

Breeds are no longer classified as lard or bacon type. Attention is being focused upon meat-type hogs, but breeds differ somewhat in the quality of carcass that they produce.

Of the ten old established breeds, the Duroc, Yorkshire, and Hampshire are most popular among farmers.

The Berkshire, Hampshire, Tamworth, and Yorkshire have erect ears. The Chester White, Duroc, Hereford, OIC, Poland China, and Spotted hogs have drooping ears.

The Beltsville No. 1, Beltsville No. 2, Maryland No. 1, Minnesota No. 1, Minnesota

FIGURE 2-37. A Minnesota No. 3 gilt. (*Courtesy Harold Beane.*)

FIGURE 2-38. The Grand Champion Barrow at the 1973 National Barrow Show. A crossbred exhibited by Roy B. Keppy, Davenport, Iowa. (*Courtesy Geo. A. Hormel & Co.*)

No. 2, Montana No. 1, Palouse, and San Pierre are new breeds developed in this country by crossing two or more of our existing breeds or by crossing Danish Landrace hogs with one or more of our old established breeds.

The Landrace breed is an outgrowth of importations of Landrace hogs from Denmark, Norway, and Sweden. It has been used extensively in the production of new breeds.

At the 1973 National Barrow Show the crossbred, Duroc, and Poland China hogs had the highest ham and loin indexes.

The Yorkshire and crossbred barrows had the most length. The Hampshire, Poland China, and Chester White had the least backfat.

Chester White, Spot, and Poland China pigs had the highest percentages of ham. The crossbred, Hampshire, and Spot hogs had the largest loin eyes. The crossbred and Spot barrows reached 220 pounds at the earliest age.

Hybrid hogs are produced by crossing two or more inbred lines. The extent that the hybrid hog is more productive than the parent stock is dependent upon the genetic make-up of the various lines, and how well they supplement each other when they are brought together.

Not all hogs that are called hybrids are true hybrids. The number of producers of hybrid hogs is small, but some produce them in large numbers.

QUESTIONS

1 What breeds of hogs are most common in your community?

2 Why do you raise the breed of hogs that you have on your farm?

3 What factors should be considered in selecting a breed?

4 What use can be made of the new breeds of hogs in your community?

5 Are there differences in the established breeds in rate of gain? Explain.

6 Which breeds of swine do you consider most prolific?

7 Which breeds of swine have been producing the best carcasses at the barrow and carcass shows?

8 Make a chart in which you list the old established breeds and show their rank in regard to each of the factors listed under Question 3.

9 Make a chart listing the new breeds, where they were developed, the parent stock used, and their physical characteristics.

10 In what ways are hybrid hogs different from the old established breeds and from the new breeds? Explain the methods used in producing hybrid hogs.

11 Which breed, breeds, or kind of hogs do you think are best suited to your home farm enterprise? Why?

REFERENCES

Anderson, A. L., and J. J. Kiser, *Introductory Animal Science.* New York, N.Y.: The Macmillan Company, 1963.

Briggs, H. M., *Modern Breeds of Livestock,* 3rd ed. New York, N.Y.: The Macmillan Company, 1969.

Ensminger, M. E., *Swine Science,* 4th ed. Danville, Ill.: The Interstate Printers and Publishers, 1970.

Krider, J. L., and W. E. Carroll, *Swine Production,* 4th ed. New York, N.Y.: McGraw-Hill Book Company, 1971.

Zeller, John H., *Breeds of Swine,* Farmers' Bulletin No. 1263. Washington, D.C.: U. S. Department of Agriculture, 1946.

———, *New Breeds of Meat-Type Hogs Developed by Research,* A.H.D. No. 144. Washington, D.C.: U. S. Department of Agriculture Bureau of Animal Industry, 1952.

3 FACTORS IN SELECTING THE BREEDING AND FEEDING STOCK

Good breeding stock is essential for the most profitable hog production enterprise. It is sometimes possible for a hog producer to overcome in part the lack of good breeding stock by efficient feeding, good management, and disease control. For maximum returns, however, it is necessary to start out with the right kind of breeding animals.

The selection of a good breed, or of good breeds, is not enough. The wide differences among the animals of a breed in type, body conformation, rate of gain, prolificness, and carcass quality make it imperative that these factors be given careful consideration in selecting breeding and feeding stock.

The factors to be considered in selecting, breeding, and feeding stock are enumerated in this chapter, and suggestions are offered to aid the reader in making wise decisions when choosing animals for the home farm enterprise.

MEAT-TYPE HOGS

The type of hog produced in this country has varied from time to time during the past century. Originally there were two types. The lard type, as the name implies, was a thick-bodied hog that, when in market condition, carried a large amount of fat. The bacon-type hog was imported from England and the Scandinavian countries, where it was developed to meet consumer demand for "Wiltshire Sides" of bacon.

The competition of vegetable fats on the market has de-emphasized the value of lard, and hog breeders are now trying to produce a hog that will yield a carcass high in lean meat, but comparatively low in lard. Regardless of the breed of hogs that is being raised, the average farmer is trying to produce a *meat-type* hog.

Meat-type Hog Defined

While there is not complete agreement among swine producers as to what is meant

FIGURE 3-1 (*Above*). A Grand Champion Hampshire Boar in 1929. We have made great strides in swine breeding since then. (*Smith-Morton photograph.*) FIGURE 3-2 (*Right*). A Champion Poland China Boar at the Iowa State Fair about 1909. (*Figures 3-1 and 3-2 courtesy* Wallaces' Farmer.)

FIGURE 3-3 (*Left*). Second prize light weight Chester White barrow at the 1972 National Barrow Show. FIGURE 3-4 (*Right*). Reserve Grand Champion Barrow at the 1973 National Barrow Show. (*Figures 3-3 and 3-4 courtesy Dave Huinker, Ames, Iowa.*)

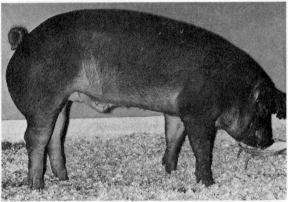

by the term meat-type hog, they generally will agree to the following specifications:

1. The hog will yield, when slaughtered, 45 to 50 percent of its carcass weight in ham and loin.
2. The animal will weigh about 210 to 230 pounds at 5 months of age.
3. The 210 to 230 pound carcass will be 29½ to 31 inches long at about 5 months of age.
4. The backfat thickness will not exceed 1.1 inches.

5. The ideal carcass should yield about 19 percent ham, 15.2 percent bacon, 17 percent loin, 9.7 percent picnic, and 6 percent Boston butt.
6. The loin eye should exceed 5 square inches.
7. The animal will have a feed conversion ability of 1 pound of gain from 3 pounds of feed.
8. The females will produce litters of 8 to 9 pigs raised to market weight.

Comparison of Meaty and Average Hogs

Two pigs were slaughtered and the dressed carcasses frozen in upright position by Iowa State University staff members in order to demonstrate differences in carcasses and carcass values. Hog A (Figure 3-5) was a meaty gilt weighing 233 pounds. Hog B (Figure 3-6) was considered to be an average barrow weighing 225 pounds. The gilt (Hog A) shows much more trimness of jowl, neck, shoulder, and middle. She shows more muscling throughout, but especially so over the loin, rump, and ham. Hog B shows a heavy shoulder and a short flat ham.

FIGURE 3-5 (*Above*). A meaty gilt weighing 233 pounds. FIGURE 3-6 (*Below*). An average barrow weighing 225 pounds. (*Figures 3-5 and 3-6 courtesy Drs. E. A. Kline and L. L. Christian, Iowa State University.*)

Presented in Figures 3-7 and 3-8 are cross sections of the hams of these two hogs. Note the high percentage of muscle and absence of fat in the ham of Hog A. There is little cushion fat above the hocks. The ham is 81.3 percent lean and 17 percent fat. The ham of Hog B is 51 percent fat and only 47.4 percent muscle. Note the thick layer of fat around the entire ham and the cushion fat carrying down to the hock.

Figures 3-9 and 3-10 show the cross sections of the carcasses of the two hogs at the tenth rib. The meaty hog has only 0.77 inch of backfat. Hog B has 1.73 inches of backfat. Note the 6.5 square inches of loin eye on Hog A as compared to the 3.9 square inches of loin eye on Hog B.

The shape of the loin eye is important. Note the large, rounded loin eye on Hog A as compared to the flat, kidney-shaped loin

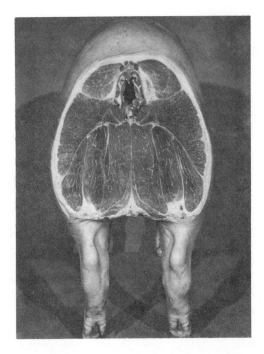 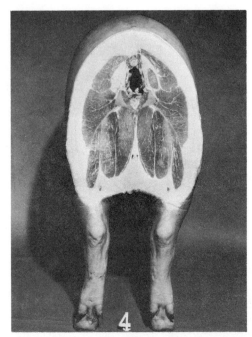

FIGURE 3-7 (*Left*). The cross-section of the ham of the gilt shown in Figure 3-5. Note the muscle—81.3 percent lean, 17 percent fat.   FIGURE 3-8 (*Right*). The cross-section of the ham of the barrow shown in Figure 3-6. It contained 51 percent fat and 47.4 percent muscle. (*Figures 3-7 and 3-8 courtesy Drs. E. A. Kline and L. L. Christian, Iowa State University.*)

eye of Hog B. The cross sections also show the superior bacon on the sides of Hog A as compared with those of Hog B. Hog B was considered to be an average market animal. Think what the differences in these cuts would have been had they selected a poorly muscled or a lardy animal.

The cross sections of hogs A and B made at the first rib are presented in Figures 3-11 and 3-12. A high percentage of muscle and a low percentage of fat covering on Hog A are evident. Excessive fat over the shoulder and back and seam fat within the muscling of Hog B greatly reduce the retail value of the cuts from this part of the carcass.

Consumption of pork can be greatly increased with increased production of market hogs like Hog A. Plump hams with little fat, large chops and lean, meaty roasts with a minimum of fat are desired by housewives and restaurant operators. Improvement in quality of carcass can greatly increase the profits from swine enterprises.

Swine breeders must select breeding stock that will produce carcasses like that of Hog A. Improvement in carcass and production traits of swine is dependent largely on the selection of superior sires and replacement females. Carcass characteristics are highly heritable. Much of the superiority or inferiority of an animal is transferred to its offspring. Length of body and percentage of ham are considered to be 60 percent inherited; loin eye area and backfat thickness are each 50 percent inherited. These are the traits that affect carcass quality. Hog producers can, through proper selection of breeding stock, greatly improve the quality of carcasses produced by the hogs that they market.

Swine Testing Stations

Boar testing and other swine testing stations are doing much to provide hog producers with information helpful in selection of breeding stock. There are nearly 37 testing stations in 24 states. More than 10,000 animals are tested each year. Most of these are boars. Illinois has seven stations; Iowa has five. Many states have one or two.

Commercial and purebred swine breeders rely heavily on swine testing station results. The following information is usually available: (1) daily rate of gain, (2) pounds of feed per pound of gain, (3) probed inches of backfat, (4) length of littermate barrow carcass, (5) loin eye area of barrow carcass, (6) percentage of ham and loin in barrow carcass, and (7) index score.

Great improvement has come about in swine breeding stock since testing began. In Table 3-1 (page 55) is a summary of improvements in Iowa tests since tests were started in 1956. It should be noted that (1) rate of gain increased from 1.92 to 2.15 pounds per day; (2) feed efficiency improved from 291 pounds required to produce 100 pounds of gain in 1956 to 251 pounds in 1972; (3) backfat probe decreased from 1.39 inches to 0.85 inch; (4) index increased from 105 to 189; (5) carcass length increased from

FIGURE 3-9 (*Left*). The cross-section at the tenth rib of the gilt shown in Figure 3-5. Note the 0.77 inch of backfat and the 6.5-inch loin eye. FIGURE 3-10 (*Right*). The cross-section at the tenth rib of the carcass of the barrow shown in Figure 3-6. There were 1.73 inches of backfat and a 3.9-inch loin eye. (*Figures 3-9 and 3-10 courtesy Drs. E. A. Kline and L. L. Christian, Iowa State University.*)

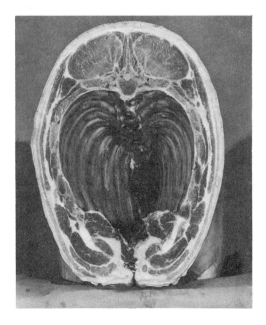
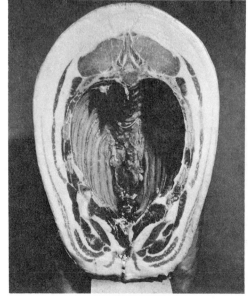

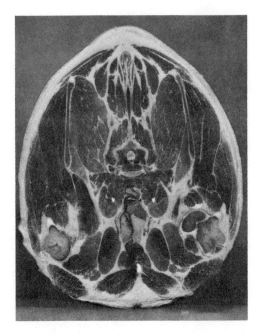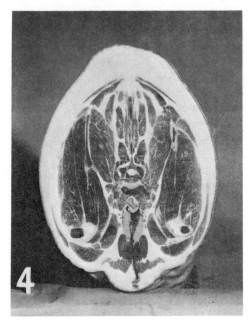

FIGURE 3-11 (*Left*). The cross-section at the first rib of the gilt shown in Figure 3-5. Note the muscling. FIGURE 3-12 (*Right*). The cross-section at the first rib of the barrow shown in Figure 3-6. Note the excessive fat over the shoulder. (*Figures 3-11 and 3-12 courtesy of Drs. E. A. Kline and L. L. Christian, Iowa State University.*)

29 to 30.5 inches; (6) backfat on barrow decreased from 1.63 to 1.24 inches; (7) ham and loin percentage increased from 32.8 to 44.4 percent; and (8) the loin eye increased from 3.36 to 5.35 square inches.

Meat-type Hogs Make Rapid and Economical Gains

Some hog producers, remembering their experiences with the leggy, "hamless wonder" hogs raised in the 1920's, have questioned the rate and economy of gain of meat-type hogs. However several tests have been made and it has been proven that meat-type hogs can be produced as rapidly and as economically as fat-type hogs.

Shown in Table 3-2 is a summary of tests conducted by the Iowa Agricultural Experiment Station. This information shows rather clearly that there is no pronounced relationship between the leanness of carcass and the growth rate to five months of age or to the feed requirements of the pigs.

Tests conducted at the Wisconsin Agricultural Experiment Station also showed that fast gain is not associated with a fat carcass. The reverse was found to be true. A summary of the Wisconsin tests is shown in Table 3-3.

A comparison of the growth rate and economy of gain of pigs sired by chuffy and meat-type boars was made at the Ohio Agricultural Experiment Station by Prof. W. L. Robison. The pigs sired by the chuffy boar averaged 206 pounds at 5 months and 19 days. They gained 1.51 pounds daily and used 399 pounds of feed to produce 100 pounds of gain.

The pigs sired by the meat-type boar averaged 210 pounds at 5 months and 12

T A B L E 3-1   IOWA SWINE TESTING STATION RESULTS, 1956-1973

	PEN AVERAGE		BOAR AVERAGE		BARROW CUTOUT			
Year	Gain (Pounds per Day)	Feed Efficiency (Lbs. per 100 Lbs. Gain)	Backfat Probe (Inches)	Index	Length (Inches)	Backfat (Inches)	Ham and Loin (%)	Loin Eye (Square Inches)
1956	1.92	291	1.39	105	29.0	1.63	32.8	3.36
1958	1.78	299	1.21	112	29.3	1.51	34.6	3.72
1960	1.86	282	1.20	129	29.3	1.52	35.9	3.87
1962	1.94	268	1.06	152	29.7	1.37	37.8	4.09
1964	1.98	271	.97	160	29.6	1.35	40.0	4.16
1966	2.18	260	.99	177	29.5	1.34	40.5	4.41
1968	2.11	253	.91	188	29.6	1.23	42.9	4.92
1970	2.05	250	.75	190	29.8	1.20	44.1	5.11
1972	2.15	251	.85	189	30.5	1.24	44.4	5.35
1973	2.09	252	.89	188	30.3	1.24	45.0	5.45

T A B L E 3-2   FEED ECONOMY AND GROWTH RATE AS RELATED TO LEAN CUTS

Lean Cuts (Percent of Live Weight)	Percent of Fat Cuts	Feed per 100 Lbs. of Gain	Weight at 5 Months of Age
Less than 36.9	26.4	340	229
37.0-37.9	25.4	341	226
38.0-38.9	25.1	345	222
39.0-39.9	24.1	343	244
40.0-40.9	25.3	323	230
Over 41.0	22.9	333	227

Source: Dr. L. N. Hazel, Iowa Agricultural Experiment Station.

T A B L E 3-3   FEED ECONOMY AND GROWTH RATE AS RELATED TO TYPE

	Short Chuffy	Type Intermediate	Long Rangy
Av. daily gain (lbs.)	1.25	1.30	1.34
Feed per cwt. gain (lbs.)	387	390	367
Av. backfat thickness (in.)	1.88	1.78	1.48
Percent of live weight in primal cuts	46.70	47.44	48.62

Source: Dr. R. H. Grummer, Wisconsin Agricultural Experiment Station.

FIGURE 3-13. A fat or lardy-type market hog. (*Courtesy Wilson & Co., Inc.*)

Since it does not cost any more to produce meat-type hogs than it does to produce fat hogs, any increase in selling price received for the improved carcass quality is extra profit. Three hundred pigs weighing an average of 230 pounds might earn their farmer a bonus of $1.50 per hundredweight because they are meat-type hogs. This would be $1,185 more than they would earn if they were lard-type hogs. If the premium should amount to $2.00 per hundredweight, the additional income would be $1,380.

If the pigs were sold to a packer who paid an extra $2.50 per hundredweight for meat-type hogs, the farmer would receive $1,725 in premiums.

days. They gained 1.63 pounds daily and required 385 pounds of feed for 100 pounds of gain. The pigs sired by the meat-type boar made more rapid gains and used less feed per pound of gain.

Packers Pay More for Meat-type Hogs

Most packer buyers are now bidding from $1.00 to $3.00 per hundredweight above the current market price for uniform loads of meat-type hogs. Many interior markets are buying hogs on the basis of carcass grade and yield. Some farmers have received as much as $2.50 more per hundredweight by selling meat-type hogs at their carcass value.

Meat-type hogs yield higher than other hogs in the four lean cuts: ham, loin, picnic, and Boston butt. These cuts represent nearly 50 percent of the chilled carcass weight of a 200- to 210-pound hog. The four cuts in 1974 were worth about 85 percent of the value of the entire carcass on the retail market. These cuts represented about 66 percent of the retail value of the carcass in 1951, and 55 percent of the value of the carcass in 1940. It pays in dollars to produce meat-type hogs.

BREEDING ANIMALS

The factors to be considered in selecting boars and sows for the breeding herd are much the same regardless of the breeds and breeding programs involved. It is very important that hog producers select carefully the animals that are to be used in the breeding herd because it is difficult to grow hogs profitably when inferior breeding stock is being used.

Parts of a Hog

Farmers, breeders and packers use much the same terms in describing hogs. Since these terms will be used repeatedly in the following paragraphs, the reader should become familiar with them. Figure 3-14 shows the various parts of the body of a Yorkshire barrow.

The Ideal Type and Conformation

Breeders of purebreds and commercial pork producers usually have some ideal in mind in selecting breeding and feeding stock. Usually they do not find animals that possess all the characteristics that they are looking for, and must select those that are nearest their ideal. Not all breeders and feeders agree as to what makes up the ideal type and conformation of a hog.

The ideal type changes from time to time with changes in market demands. More attention is usually given to quality of carcass when hogs are plentiful than when hogs are grown in smaller numbers. Factors associated with prolificness may be given more emphasis when the price of pork is high than when pork is cheap.

Producers select animals to be mated with other animals. Quite often the animal selected is not ideal in several respects but is sufficiently outstanding to mate well.

Ideal type of market hog. The ultimate goal in all hog production is to produce efficiently and profitably a hog that will yield, when slaughtered, a carcass high in the cuts

of pork desired by the consumer. Many think that the barrow shown in Figure 3-15 (page 58) comes close to the present-day concept of the ideal meat-type hog. For illustrative purposes a similar animal could have been selected from each of the other meat-type breeds. It is the goal of the farmer and hog breeder to select breeding stock that will produce market hogs with the muscling, conformation, and carcass quality possessed by this barrow.

This barrow has good length and width of body. He carries his depth uniformly from front to rear, and he is deep and smooth sided. He is especially good in the ham. Note the plumpness and depth of ham. This barrow gives evidence of being firm fleshed. There are no signs of flabbiness; he is well muscled and has a trim head and jowl.

This barrow should produce an excellent carcass and yield a 73 percent dressing per-

FIGURE 3-14. The parts of a hog. (*Moore photograph. Courtesy Dave Huinker, Ames, Iowa.*)

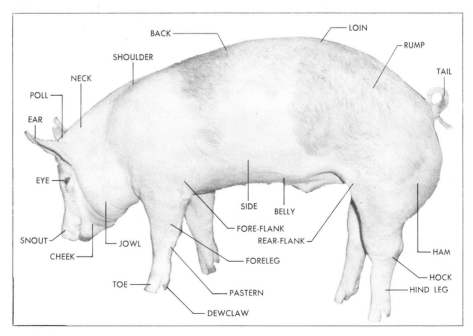

centage or better. *Dressing percentages* refer to the percentage of the live weight of the hog found in the carcass after the animal has been slaughtered. Lard-type or exceedingly fat hogs may yield dressing percentages of 75 percent or more. In the meat-type hog a carcass with a high dressing percentage is desired, but it must also be a carcass high in the choice cuts: ham, loin, Boston butt, and picnic shoulder. This barrow should produce a high percentage of lean cuts and a minimum of backfat and lard.

FIGURE 3-15 (*Above*). The Grand Champion Barrow at the 1972 National Barrow Show with the National Pork Queen. Exhibited by Johnny Peugh of Stanton, Texas. (*Courtesy Geo. A. Hormel & Co.*) FIGURE 3-16 (*Below*). Carcasses of hogs: intermediate, chuffy, and rangy. Note the differences in length, muscling, and backfat thickness. (*Courtesy Wilson & Co., Inc.*)

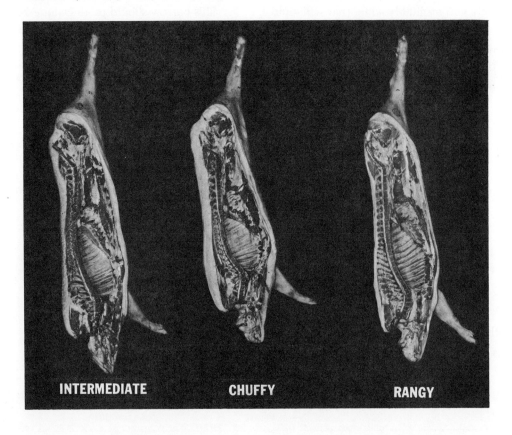

INTERMEDIATE   CHUFFY   RANGY

FIGURE 3-17. Grand Champion Barrow at the Cedar Rapids Spring Barrow Show. A Chester White. (*Courtesy Dave Huinker, Ames, Iowa.*)

The carcasses of three barrows of different types are shown in Figure 3-16. The carcass on the left is from the intermediate or meat-type hog. It is the type of carcass which we would expect to get from the barrow pictured in Figure 3-17. Note the thickness of the layer of fat over the back and compare it with the backfat on the carcass in the center, which is from a chuffy hog. There is nearly twice as much backfat on the carcass from the chuffy hog as there is on the carcass from the intermediate hog. It has been found that the percentage of lean cuts in a carcass is closely related to the backfat thickness.

The carcass on the right is from a rangy barrow. It has a very thin layer of backfat, but the ham and loin are undeveloped. A higher percentage of the carcass is in the form of bone and skin. Usually, rangy hogs require more time in being finished for market, and they are marketed at heavier weights.

Today the ideal market hog is the intermediate, middle-of-the-road, meat-type hog. Breeders and packers shy away from both the chuffy and the rangy animals. The chuffy hog produces too much backfat and lard; the rangy hog requires too much time to be properly finished and produces a heavy carcass.

Swine specialists have set the following goals for producers to attain by 1985:

Pigs raised per litter	10
Market weight (pounds)	240
Feed per 100 pounds gain	270
Days to 240 pounds	150
Carcass weight (pounds)	178
Chilled dressing (percent)	74.2

Carcass length (inches)	30–34
Backfat (inches)	1.0–1.3
Loin eye (square inches)	5.0–7.0
Belly with rind (inches)	1.25
Belly rindless (inches)	1.10

These goals should serve as guides in selecting breeding animals.

Ideal type of breeding animal. Now that we know which type of market hog we want to produce, we can determine the type, conformation, and other qualities desired in the boars and sows which will produce this kind of pig. The kinds of feeds used and the methods of feeding also influence the growth rate, the economy of gain, and the quality of the carcass produced. Good feeding practices are of little value unless we have the right kind of breeding stock to begin with.

The mature sow shown in Figure 3-18 possesses many of the characteristics desired in a brood sow. She has a medium-long body and a strong, well-arched back. She is deep sided and has the capacity of the chest and middle that insures good feeding quality and vigor. Note her well-developed, deep, and

full hams. This sow is good on her feet and legs. She has ample bone, strong but medium-length legs, and short, straight pasterns. A brood sow must have good feet and legs; an inactive or clumsy sow usually is unable to raise a good litter.

This sow has a trim head and jowl. She is feminine and has a prominent eye. Coarse-headed and heavy-jowled sows should not be kept in the herd. Femininity and a reasonable amount of refinement are desirable in brood sows.

This brood sow is smooth in the shoulders, has a wide, well-muscled loin, and has an excellent hair coat. The refinement of the head and ear, shoulder, and hair coat are important items in selecting female herd material.

The udder of the sow should be well developed with 12 or 14 sound teats. The teats should be prominent and well spaced. The sow shown in Figure 3-18 has an excellent udder.

Most hog producers select young animals as replacements in their herds. This is especially true in making replacements in the sow

FIGURE 3-18. All-American and Grand Champion Hampshire Junior Yearling Sow at the Indiana State Fair. Owned by Ralph Bishop, Tipton, Indiana. (*Courtesy Hampshire Swine Registry.*)

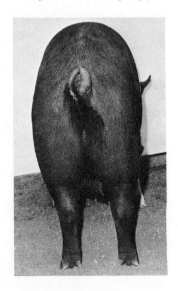

FIGURE 3-19. An outstanding Duroc gilt. She is well muscled, firm, and has a deep wide ham. (*Courtesy United Duroc Swine Registry.*)

FIGURE 3-20. *Above:* The Grand Champion Yorkshire Boar at the 1973 National Barrow Show. Exhibited by Dick Kuecker, Algona, Iowa. (*Courtesy Geo. A. Hormel & Co.*) *Below:* Tex, a junior yearling boar owned by Klein and Hartl of Iowa Falls, Iowa. A masculine, massive, well-muscled boar with exceptional feet and legs, he was Grand Champion Boar at the 1974 Iowa State Fair. (*Courtesy United Duroc Swine Registry.*)

herd. Most farmers select gilts that are six to eight months of age. Shown in Figure 3-19 is a gilt that possesses many of the characteristics desired in a future brood sow. She has excellent length, uniform width and depth, and shows firmness of fleshing. She has a feminine head and a clean-cut jowl. She is sound on her feet and legs, is well hammed, and has a good underline.

The selection of the herd boar is a major undertaking for most farmers. The boar genetically represents one-half of the herd. It has been through the use of good boars that much of the progress has been made in swine improvement since one boar can be mated to a large number of gilts.

Shown in Figure 3-20 are two boars of different breeds with much the same body conformation. Boars should show masculinity and breed character. They should show cleanness and firmness of jowl, a wide-open eye, and a neck that blends in well with the shoulders. The shoulders should be smooth and the back strong. The width should be uniform from front to rear, and there should be no tendency to be narrow at the loin or ham.

The tail setting should be high, and there should be no fat around the tail setting. The hams should be deep and full. The lower ham should be smooth and firm. The side should be smooth, and the middle trim, but there should be good depth at both the fore and rear flank.

Boars should be big for their age, medium long, and rugged. They should have

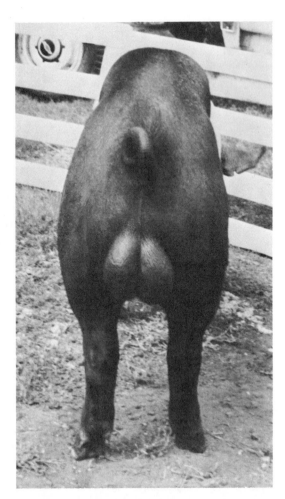 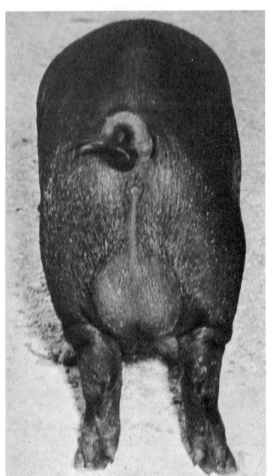

FIGURE 3-21. *Left:* A junior spring boar with sound, straight hind legs. He is firm, meaty and nicely turned over the back. *Right:* A senior spring boar with poor hind legs and pasterns. He is flabby and poorly turned over the back. (*Courtesy Hampshire Swine Registry.*)   FIGURE 3-22 (*Below*). There are 5 normal and 7 inverted nipples on this underline. (*Courtesy Iowa State University.*)

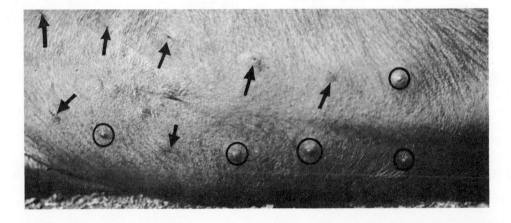

strong bone and stand squarely on all four legs. The pasterns should be short, and the legs should be set out on the corners. Shown in Figure 3-21 is a comparison of good and poor feet and legs.

A boar must have a good constitution. He should have width and depth in the heart area. He should have well-developed sex organs. The testicles should be of equal size and prominent.

Bidders on gilts and boars in purebred sales usually inspect the teats or rudimentaries on the animal being sold. The teats on boars are called *rudimentaries* or rudimentary teats. Sometimes a teat on a sow is not connected with a milk gland and will not produce milk. This is called a *blind teat*. A teat which has an inverted nipple is more serious. These teats usually do not come down and produce milk. A sow with 12 teats can suckle only 9 pigs if she has 3 blind teats or inverted nipples. Boars and sows will pass this characteristic on to their offspring. Figure 3-22 shows normal and inverted nipples on an underline.

Fifty percent of the genetic potential of gilts comes from their sire. That sire and the one to which they are bred will account for 75 percent of the genetic makeup in the next pig crop. The most recent four boars used in a herd contribute more than 90 percent of the genetic potential of the herd. It is important that more than visual factors be considered in selecting herd sires.

Tests for Carcass Lean and Fat

It is possible to take much of the guesswork out of selecting meat-type breeding stock by using tests which have been developed by swine research specialists.

Backfat probe. Dr. L. N. Hazel of Iowa State University developed a method of probing the back to determine the thickness of backfat. A small incision is made in the skin of the animal with a sharp knife. The cut is made 1½ to 2 inches to one side of the exact center of the back and just behind the shoulders.

A small metal ruler is carefully pushed through the incision into and through the fat until it reaches the back muscle. The depth of backfat can be read on the ruler. By repeating the process at the last rib and halfway between the last rib and the base of the tail, it is possible to get a good idea of the

FIGURE 3-23. Weighing the 1973 National Barrow Show barrows on completion of tests at St. Ansgar, Iowa. All production-tested barrows must gain 1.3 or more pounds per day and pass a soundness test. (*Courtesy Geo. A. Hormel & Co.*)

thickness of fat carried by the animal. The probe does not injure the animal and there is little danger of infection.

Purdue electric lean-fat meter. Dr. R. M. Whaley and Dr. Fred Andrews of Purdue University developed an electronic device which will tell the difference in the proportion of lean and fat inside the loin, ham, shoulder, or bacon side of live hogs. See Figure 3-25. There is no injury or pain more than that caused by a needle's prick.

The device functions because lean is two to five times better as a conductor of electricity than fat. The needle on the instrument is part of an electrical circuit. When flesh completes the circuit, a high current in lean and low current in fat results. The needle is set for depth, then pushed into the flesh. The meter, which is a part of the instrument, then shows whether the needle is in fat or lean.

Sonoray or Ultrasonics. The use of ultrasonic or high-frequency sound (Figure 3-26)

has been proven effective in measuring backfat thickness and loin eye areas in hogs. Sonoray service is available in some states from swine specialists associated with state colleges, or from technicians employed by breeding associations or commercial organizations. The equipment and technology necessary will prevent most commercial pork producers from making their own tests.

Sonoray estimates are made from readings on the oscilloscope. The loin eye is determined by plotting the reading on graph paper. Persons with experience and a thorough knowledge of the anatomy of the area can plot rather accurately the loin eye area. A planometer is used to measure the estimated square inches of loin eye area (Figure 3-27, page 66).

Ultrasonic estimates made at the University of Missouri have been found to be quite accurate when compared with actual measurements of slaughtered animals.

FIGURE 3-24. *Top:* Senior spring gilt with a good underline and good feet and legs. *Bottom:* Senior spring gilt with a poor underline and poor feet and legs. (*Courtesy* American Hampshire Herdsman.)

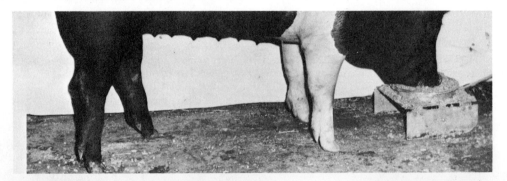

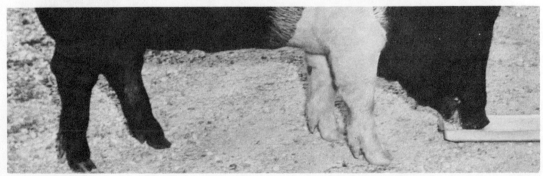

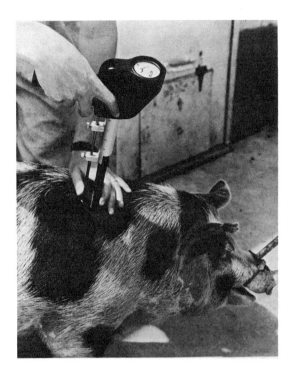

FIGURE 3-25. A Lean-Meter being used in making a backfat measurement. (*Courtesy Chas. Pfizer and Co., Inc.*)

K-40 Counter. This instrument measures accurately the leanness of live animals. It operates on the principle that most of the potassium in an animal is contained in the muscle tissue. A sensitive detection system measures the amounts of potassium in the body. This provides a reliable index of the amount of muscle tissue.

The minute amounts of gamma radiation emitted by the potassium in the various parts of the animal's body are measured electronically. The machine was developed by the Nuclear Electronics Counter Corporation and has had much use in Oklahoma.

The equipment is very expensive and complex. Pigs are cleaned carefully and given no feed during the 24 hours before the test. Feeds high in potassium and mud or dirt could affect the accuracy of the test.

EMME (Electronic Meat Measuring Equipment). This recent development makes it possible to obtain almost instantly the leanness of 200 or more hogs an hour with about 99 percent accuracy. It operates similarly to the K-40 on the principle that lean tissue conducts electricity better than fat does.

Although the machine costs about $10,-000, it can be operated by a nontechnically trained person. It stands about 4 feet high, is 8 feet long and 5 feet wide. It weighs about 1,000 pounds and is portable. It is manufactured by the EMME Company of Glendale, Arizona.

Swine specialists are recommending that EMME data be obtained on the animals tested in the various boar testing stations. Data on live animals is much more valuable than data on littermate barrows.

The grand Champion Duroc boar at the 1973 National Barrow Show was tested by the EMME machine two months before the show. The test determined that the hog contained 60.8 percent lean cuts. In addition to becoming grand champion, the boar sold for $38,000.

FIGURE 3-26. Measuring backfat thickness and loin area by use of sonoray (high-frequency sound). (*Courtesy Conner Prairie Swine, Inc., Noblesville, Indiana.*)

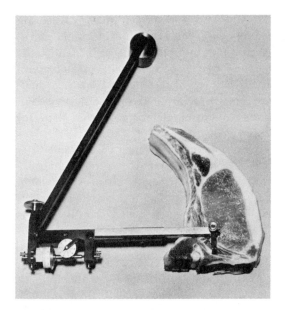

FIGURE 3-27. A planometer used to measure the loin-eye muscle. (*Courtesy John Morrell & Co.*)

State pork producers' associations, purebred swine breeders' associations, packing companies, and other commercial concerns are acquiring EMME equipment. Many of the more progressive producers of swine seed stock now provide EMME data on all breeding stock marketed.

Swine producers should make use of EMME data as a supplement to Sonoray, backfat probes, testing-station performance, and other data. Many breeders are contracting with the International Livestock Services Corporation (ILIS) of Ames, Iowa, or with other similar commercial organizations, for assistance in obtaining data on their breeding stock.

Other Factors in Selecting Breeding Stock

Although type and carcass quality are important, they are just two of a number of factors which should be considered in selecting breeding stock. In selecting both boars and gilts, age, pedigree, performance or production records, health, and disposition should be considered.

Age. Mature sows have a higher conception rate and usually wean one or two more pigs per litter than gilts. They may have built up some immunity to diseases which affect the newborn pigs. Based on averages, sows will farrow about 1.7 litters per year as compared with 1.3 litters farrowed by gilts. Usually sows are better milkers and are more reliable than gilts in suckling their litters.

Gilts are less clumsy than sows and require less space in housing. From an income tax standpoint, there may be some advantage to gilts, for they may qualify for long-term capital gains at selling time. These factors need to be considered by producers in deciding whether to keep tried sows or to select gilt replacements.

Mature boars usually settle more sows and are more dependable during the breeding season. They do not get the "flu" as easily as do young boars, and they are usually in better health during the breeding season.

Most farmers and breeders, however, use young boars and gilts in their breeding operations. The young animals require a smaller investment, need less feed to maintain themselves, and need less space in housing. When a young animal dies, there is less loss than there is when a mature animal is lost. The young animal gains in weight during the production period, and can be sold at a good price after it is no longer useful in the herd.

Pedigree. Every animal has a pedigree, but only the pedigrees of purebred animals can be recorded by a breed association. A pedigree is merely a record of the bloodlines of the ancestors of the animal. The pedigree is very important in the breeding of purebreds, but the breeder of commercial hogs is more concerned with the production record of the sire and dam of the litter than with their pedigrees.

Performance or production records. It is possible to judge the performance of a mature animal by the size of the litters produced, by the conformation of the pigs, and by the weight of the pigs at 21, 35, 56, 150, or 180 days of age. Most breeders prefer breeding stock from litters of eight or more pigs raised. They want the pigs to average 40 to 50 pounds at 56 days, and 220 to 230 at 5 months of age. By ear-marking the pigs at birth and weighing them at 21, 35, 56, 150, or 180 days, it is possible to select breeding animals from the most productive litters.

Several of the breeding associations have had production registry programs. The pro-duction of the ancestors is recorded on the pedigree in the same way as production records of dairy cows are reported on the pedigrees of cows. The National Association of Swine Records developed a production record plan based upon the number of pigs farrowed in the litter and the weight of the litter at 56 days of age. A litter qualifies for production registry when there are 8 or more pigs farrowed and raised by a mature sow to a 56-day weight of at least 320 pounds. A first litter gilt must raise the same number of pigs to a 56-day weight of at least 275 pounds.

Sows qualify for production registry after producing 2 production registry litters. Boars must sire 5 qualified daughters, or 15

FIGURE 3-28. This Yorkshire litter had a combined weight of 947 pounds at 56 days of age. Raised by Eugene Wagner, Newton, Illinois. (*Courtesy Paul Walker.*)

daughters which have produced one production registry litter, to qualify as production registry sires.

The production registry program has much merit but is not currently being encouraged by the various breed associations. Many swine specialists are urging that increased emphasis be given the program.

The *Meat Certification Program* inaugurated in 1956 by most swine breed associations has provided a valuable aid for use in selecting breeding stock. Through this program it has been possible to find the strains of hogs that are producing acceptable meat hogs at desired market weights. Carcass quality is determined with the testing of carcass length, backfat thickness, and size of loin eye.

Swine breed certification is a three-point program based on production registry, rate of gain, and carcass quality. Two pigs of the litter that has met Production Registry requirements must be slaughtered. The pigs must weigh 220 pounds in 180 days or less. They cannot weigh more than 242 pounds when slaughtered. If overaged at slaughter time, equivalent weight at 220 pounds is adjusted on the basis of a 2-pound-per-day gain beyond 180 days.

The carcass standards for the two slaughtered pigs are as follows:

Slaughter weight	220.0 pounds
Minimum loin area	4.5 square inches
Carcass length (minimum)	29.5 inches
Maximum backfat thickness	1.5 inches

A feature of the certification program has been its emphasis on both production in terms of weight of pigs per litter and carcass quality. The use of backfat probes, Sonoray, and EMME in evaluating the carcass quality of live animals has been found to be much more effective than the certification program in evaluating the meatiness of breeding animals. As a result, the certification program is being de-emphasized. The standards for both production registry and certification programs perhaps should be increased. They are excellent means of evaluating performance of genetic bloodlines.

Many swine producers have developed on-farm testing programs. Litters are kept in individual pens. Pigs are weighed at birth and at 21, 35, and 150 days. Feed records are very often kept, and costs of gains are calculated. Backfat probes, Sonoray and/or EMME tests are made to determine carcass quality of each pig. These tests are very valuable to both the breeder and the prospective buyer of breeding stock. Since the tests are not supervised by a disinterestd person, they are only as valuable as the breeder is honest and technically efficient.

Many breeders of commercial as well as purebred swine are relying heavily upon swine-testing-station results in the selection of breeding stock.

A Yorkshire boar tested at Midland Co-op Swine Testing Station at Belleville, Wisconsin, is the highest indexing boar tested to date. The boar had an index of 237. A Duroc barrow on test at the Northeast Iowa Swine Testing Station produced a carcass with 51.24 percent ham and loin. This was the highest ham-loin percentage reported up to 1974. Determination of index is explained in Chapter 13.

Boar-testing-station data provide pork producers the best source of information in selecting boars and replacement gilts. Since most producers select replacement gilts from the home herd, it is imperative that they develop performance and carcass testing programs involving the home herd. Gilts should be selected on the basis of their genetic potential, and sows retained according to their performance.

Health. Breeding animals that are large and heavy for their age usually are in good health, but the best policy is to buy only animals that have been tested for brucellosis and leptospirosis. In some areas it may be desirable to buy animals that have also been vaccinated for erysipelas at weaning and at eight months. The herd from which the animals come should be inspected, and breeding and feeding stock should be purchased only from disease-free herds.

Atrophic rhinitis, necrotic enteritis, and various forms of scours have caused losses of hogs in some areas during recent years. Careful hog producers inspect the herds and the farms on which hogs are raised before buying breeding stock. Many breeders will not buy breeding or feeding animals that have been marketed through public sales barns or stockyards. The policy of buying only disease-free stock and keeping them away from other hogs for two or three weeks is a sound one.

An advancement in swine production is the development of Specific Pathogen Free (disease free) hogs. The process is described in Chapter 11. While still somewhat experimental, the process has proven to have merit. SPF hogs, kept isolated from other hogs and in clean surroundings, tend to eliminate disease losses due to virus pneumonia and atrophic rhinitis. SPF hogs are no better than the production bred into them. SPF lines have been developed from only a few of the highest testing lines in the Swine Testing programs. Some breeders presently must choose between SPF lines and production proven testing station lines.

Disposition. The disposition of a sow may materially affect the number of pigs she will save at farrowing time. A sow that is nervous and easily disturbed is more likely to lie or step on pigs than is a sow with a quiet disposition. Quiet animals and those that like to be scratched usually have good dispositions and are good gainers. A good brood sow should always let you enter the pen.

The disposition of a boar is equally important because the boar will be the sire of the next year's sow herd. An extremely nervous boar is undesirable, as is an extremely quiet or inactive individual. A boar should be friendly in disposition, active, and a good rustler. Inactive boars are usually slow breeders.

"Code of Fair Practice for Buyers and Sellers." The following code of fair practice for buyers and sellers of purebred registered breeding stock is followed universally in the exchange of breeding animals. Buyers should check with the seller and make certain that if there are variations in the guarantee by the seller, they are understood by both parties. It is very important that the buyer understand his responsibilities as well as those of the seller.

A CODE OF FAIR PRACTICE FOR BUYERS AND SELLERS OF PUREBRED REGISTERED SWINE

Adopted and Recommended by
National Association of Swine Records
April 17, 1973

Buyers of purebred, registered boars and gilts buy them to be breeders. Many factors may affect an animal's breeding capabilities. Many of these are not visible at time of purchase. Some problems may be the result of management before sale. Some the result of handling and management by the buyer after purchase. Some may be hereditary. Because of this, adjustments need to be a sharing of responsibilities.

All Adjustments Are a Matter Between Buyer and Seller

The National Association of Swine Records or individual Registries assume no responsibilities for enforcement of these recommendations.

Standard Warranty

All purebred, registered hogs, over four months of age, (Not used for breeding under 7 months of age) sold as breeding animals for breeding purposes are sold with a warranty that they are capable of and will breed.

If and when any said animal does prove to be a non-breeder the seller shall make an adjustment to the satisfaction of the buyer, provided the buyer informs the seller of the situation within a reasonable time after purchase in keeping with normal management practice characteristic of buyer's type of operation.

In all purebred transactions the registration certificate is an integral part of the transaction and shall be delivered to the buyer properly transferred on the Association records at the expense of the seller.

Suggested Warranty Adjustments

(These are only suggestions considered as generally acceptable within the industry. Other adjustments may be made if satisfactory to both buyer and seller.)

Boars Failing to Serve or Settle Sows

1. Refund the purchase price or difference between purchase price and market value as shown by sales receipt if boar sold on the market.

2. Make a replacement of another boar satisfactory to buyer.

3. Give buyer credit (amount to be agreed on by both buyer and seller) on the purchase of another animal or animals.

Gilts Sold as Open

A—If proven to be bred . . . 1—Refund of purchase price upon return to seller. 2—Refund difference, if any, between price paid for gilt and value of a commercial bred sow.

B—If proven to be a non-breeder . . . 1—Make a replacement of another gilt satisfactory to buyer. 2—Refund purchase price or difference between purchase price and market value of gilt as shown by sale receipt. 3—Give buyer credit, (amount to be agreed upon by both buyer and seller) on the purchase of another gilt in the future.

Bred Sows

A—Are expected to be bred to a designated boar and date of service. When proven otherwise . . . 1—Replace the sow with another satisfactory to the buyer. 2—Refund the difference between the purchase price and market value of sow as shown by sale receipt. 3—If buyer desires to keep sow, refund one-half difference

between purchase price and market value of gilt at time of purchase.

Handling the New Boar

1—Buy boars two months before needed for breeding. 2—Keep boar in isolation from other hogs for 30 days. 3—Don't use a young boar until seven months or more of age. 4—Give boar fence line contact with females for three weeks. 5—Hand mate one to three gilts thoroughly in heat to start boar. 6—Keep boars cool. 7—When pen mating allow one young boar to 12 to 15 gilts.

Handling Gilts Before Breeding

1—Don't breed gilts until at least seven months old. 2—Breed gilts on third heat period or later. 3—Expose gilts to boar by fence line contact. 4—Limit feed 200-pound gilts until just before breeding.

FEEDER PIGS

The number of farmers who buy feeder pigs has increased in recent years. It is estimated that 40 percent of all hogs slaughtered start as feeder pigs. Although many are farrowed and sold as feeder pigs in the state where they are slaughtered, many are shipped in from other states. Several million feeder pigs are purchased by farmers in Iowa and Illinois each year. Some are produced by cooperative feeder pig corporations, some by neighboring farmers. Many are imported from southern states, some from Wisconsin. About one-third of the Tennessee pig crop is sold as graded feeder pigs.

It has been estimated that any farmer who cannot raise an average of 7 pigs per litter on his farm will find it to his advantage to buy pigs rather than to try to produce them himself. The average number of pigs weaned per litter in this country is rarely much above 7 pigs. There is considerable interest among commercial pork producers in the purchase of feeder pigs.

Housing and equipment needed in farrowing and nursery quarters are expensive, and therefore it is important for those who have these facilities to use them continuously. In doing so they may find it profitable to sell a part or all of the pigs produced as feeder pigs.

Most of the losses of pigs occur at or shortly after farrowing. Many swine producers do not have the facilities or skill necessary to manage the sows and litters effectively. These men find it profitable to buy feeder pigs.

Feeder-Pig Production

Many farmers are now producing pigs to be sold at weaning age to farmers and

FIGURE 3-29 (*Left*). Nearly a million feeder pigs are brought into Iowa each year. Wisconsin and Missouri are the major sources. (*Courtesy American Cyanamid Co.*)   FIGURE 3-30 (*Below*). A prize winning group of feeder pigs marketed by a Tennessee producer. (*Courtesy Tennessee Livestock Association.*)

feeders. The greater share of these producers are in the southern states, Wisconsin, Iowa Illinois, Missouri, and Nebraska. The production and marketing of feeder pigs have expanded greatly since 1960. Large numbers of feeder pigs are being raised in southern states by small producers. The pigs are shipped to northern states, where they are fed out for market.

Feeder-pig production has become a profitable business, especially in areas where feed, housing, and capital are not available to grow out the pigs to market. It is estimated that more than a million pigs are sold annually in Wisconsin. Some 495,000 pigs were imported from other states into Indiana in 1972. Feeder-pig inshipments to Ohio in 1966 totaled 336,000 head. Illinois and Iowa import about 1.5 million pigs each year.

Selection of Feeder Pigs

Regardless of the source, extreme care must be taken in the selection and purchase of feeder pigs. The reputation of the breeder,

the breeding stock used in producing the pigs, the type and conformation, the health of the animals, and the purchase price must be carefully considered.

Reputation of the producer. In buying an automobile, a tractor, or a life insurance policy, we check upon the reputation of the manufacturer or the company. The same method should be practiced in buying feeder pigs. Learn from the neighbors, the vocational agricultural instructor, or the agricul-

FIGURE 3-31 (*Above*). High quality pigs marketed through the cooperative feeder pig sales in North Carolina. (*Courtesy Farmers Cooperative Service, New Bern, N.C.*)
FIGURE 3-32 (*Below*). A thrifty group of feeder pigs on an Iowa farm. (*Courtesy Kent Feeds.*)

tural extension director about the experiences of others who have purchased pigs from the various sources. If possible, visit the producer and evaluate the man and his producduction program.

Breeding stock. Quite often we buy feeder pigs because the producer of the pigs was able to use better breeding stock than we could provide for a small breeding operation on our own farms. We are anxious to get as much hybrid vigor as possible, so it is important that we study carefully the quality and conformation of the breeding stock and the methods used in breeding.

Type and conformation. We desire in the feeder pig the type and conformation necessary to produce a high-quality market hog. The pigs must be growthy and large for their age. They must have good length of body and carry down deep and wide in the ham. We like them to have good width throughout, especially over the loin and in the chest.

The U. S. Department of Agriculture announced the adoption of revised standards for grades of feeder pigs effective April 1, 1969. The revised standards provide for four numerical grades. U. S. No. 1, U. S. No. 2, U. S. No. 3, and U. S. No. 4, and one lower grade, U. S. Utility. Thriftiness and slaughter potential are the basis for grading. The U. S. No. 1 feeder pig must be sufficiently muscled and thrifty to reach market weight at 220 pounds and produce a U. S. No. 1 carcass. The other numerical grades are parallel to the carcass grades expected from the pigs at 220 pounds. Pigs lacking general constitution and thriftiness will be graded U. S. Utility. The U. S. grades of feeder pigs are shown in Figure 3-33.

Most feeder pigs sold in major markets are graded according to U.S.D.A. standards. Some states or markets have attached their own name to the top grades. The Wisconsin Feeder-Pig Marketing Cooperative grades pigs as Northland Select, Northland Choice,

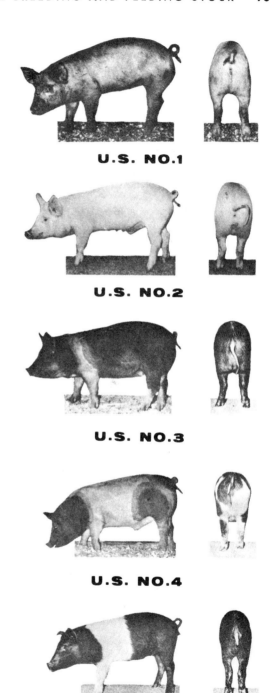

U.S. NO.1

U.S. NO.2

U.S. NO.3

U.S. NO.4

U.S. UTILITY

FIGURE 3-33. U.S. Grades of Feeder Pigs. (*Courtesy U.S.D.A.*) .

and Northland Good. The grading of pigs in many Missouri feeder-pig markets is handled jointly by the Missouri Department of Agriculture and the U.S.D.A. In Virginia all U.S. No. 1 and No. 2 pigs are penned and sold together by weight classifications. The No. 1 pigs are marked with two blue dots on the shoulder; one blue dot is painted on the shoulder of each No. 2 pig.

Nearly 40 percent of the 483,000 feeder pigs sold in 384 cooperative sales in Tennessee in 1972 were graded No. 1 or No. 2. Feeder-pig buyers will profit from the purchase of graded pigs assuming that they meet the other criteria described in this chapter.

Sound underpinning and quality of bone are preferred. The muscling or meatiness of the pig is very important; a pig which is flabby will not usually grow and fatten out well. Trimness of head, jowl, and shoulder are necessary if we are to produce carcasses of high quality.

Health. You cannot afford to take chances in buying feeder pigs. Make certain the pigs are from a disease-free herd and that they have not had an opportunity to become infected while being transported or marketed. Purchase only pigs which have been vaccinated for erysipelas. Avoid pigs with heavy skin or rough hair coats.

It is a good policy to observe pigs carefully in buying feeder pigs. Make certain that there are no signs of diarrhea, coughing, or sneezing. Watch also for thumping or labored breathing indicating pneumonia or anemia. The pigs should be growthy, have smooth hair coats, and be active.

Each state has health laws concerning the sale and transportation of livestock. Some have specific laws concerning feeder pigs. The health certificates should be checked for completeness and accuracy. A veterinarian's inspection at source and at location of the sale is a must. It is desirable to buy pigs from the producer, but this is not always possible.

Most states have laws requiring the ear tagging of feeder pigs. This will permit identification of the pigs and of the producer. The producer's reputation is very important.

Price. How much is a feeder pig worth? There is no one answer to this question. The price varies with the needs of the prospective buyer, the availability of the pigs, the price of farm grains, the market outlook for pork products, and the crop outlook. In 1974, 40-pound feeder pigs that were wormed and castrated were selling for $34 per head delivered to the farm. The top price on No. 1 market hogs at that time was $36 per hundredweight. Efficient producers could then raise their own feeder pigs for less money. The inefficient producers, and those without breeding stock and facilities, purchased the pigs at $34 per head and felt that they had made a good buy.

There is no advantage in buying SPF (specific pathogen free) feeder pigs unless all non-SPF hogs have been removed from the premises for at least a two-month period, and the premises have been thoroughly cleaned and disinfected.

Feeders should consider the quality of the pigs, their weight, and the current market price for commercial hogs in estimating the price of feeder pigs. One feeder calculates the value of the pigs on a sliding scale based upon the weight of the pigs and the price of No. 2 hogs at the nearest packing plant. The price of feeder pigs changes as the market price for finished hogs fluctuates. A 40-pound pig is priced at one and one-half times the price per pound of a No. 2 hog plus a $5 premium. With No. 2 hogs at 30 cents, a 40-pound pig would be valued at $23.

With market hogs at $30 per hundredweight, feeder pigs with superior breeding are worth about $4 more than pigs with average breeding. Pigs with poor breeding are usually not a good investment, even if purchased at a cost of $4 per head less than the price of pigs with superior breeding.

TABLE 3-4   PRICES YOU CAN PAY FOR 40-POUND FEEDER PIGS

Price of Corn (Bushel)	Cost of Gain (Hundred-weight)	Necessary Break-even Price at Various Expected Markets			
		$30.00	$34.00	$38.00	$42.00
$1.20	$18.50	$32.70	$41.50	$50.30	$59.10
1.40	19.78	30.40	39.20	48.00	56.80
1.60	21.06	28.09	36.89	45.69	54.49
1.80	22.34	25.79	34.59	43.39	52.19
2.00	23.62	23.49	32.29	41.09	49.89
2.20	24.90	21.19	29.99	38.79	47.59
2.40	26.18	18.89	27.69	36.49	45.29

Source: Purdue University.

Presented in Table 3-4 is a guide for use in deciding the price of feeder pigs.

SUMMARY

Good breeding stock is essential for the most profitable hog enterprise. Lard-type hogs are out of the picture. We must grow meat-type hogs.

A meat-type hog is one that will yield 45 to 50 percent of its carcass in the ham and loin. It will weigh 210 to 230 pounds in 5 months and have a carcass which is 29½ to 31 inches long with a maximum of 1.1 inches of backfat. It will produce a pound of pork on 3 pounds of feed. The litters will average from 8 to 9 pigs raised.

Tests indicate that meat-type hogs make as rapid and as economical gains as the fat- or chubby-type of hog.

Packers are paying premiums of from $1 to $3 per hundredweight for uniform loads of meaty hogs. Packers who buy hogs on a carcass cutout and yield basis may pay as much as $2.50 a hundredweight above market price for outstanding meat-type animals.

The ideal meat-type hog may be found in all breeds. The goal of hog breeders is to bring together breeding stock that will produce meat-type hogs. In the selection of breeding stock, emphasis must be placed upon the productiveness of the animals in terms of prolificness, rate and economy of gain, and the quality of carcasses produced. We need a better muscled carcass, one low in backfat and lard and high in lean cuts.

We need more length in many of our common breeds of hogs. At the same time we need to develop hogs with trimmer jowls and heads. Boars and gilts should be selected from large uniform litters. Breeding animals should be large for their age, have good length, uniform width and depth, and be firm fleshed. There should be no signs of wasteness or wrinkling. The feet and legs should be set out on the corners and be straight. Sound underlines with 12 or more nipples are preferred. Good eyesight and a good disposition are musts.

Swine testing-station data provide the best information concerning desirable sources of breeding stock. Boars tested in the Iowa Swine Testing Station in 1972 gained 2.15 pounds a day on 2.51 pounds of feed per pound of gain. They had an average backfat probe of only 0.85 inch. Littermate barrows had 44.4 percent ham and loin and 5.35 square inches of loin eye.

About 40 percent of the hogs marketed for slaughter were not farrowed on the farm where they were grown to market. Iowa, Missouri, Wisconsin, and Illinois produce the largest numbers of feeder pigs. From 25 to 40 percent of the pigs farrowed in some southern states are sold as feeder pigs.

QUESTIONS

1   What factors have brought about the changes in type of hog produced in this country?

2   How does the carcass of a meat-type hog differ from that of a fat-type hog in percentage of ham and loin, backfat thickness, and yield?

3   Which type of hog will make the most rapid and economical gain, the meat type or the fat type?

4   How much additional income could you get from the hog enterprise on your home farm if you were to raise meat-type hogs?

5   Are the barrows which have been winning the barrow classes at the fairs the type of hogs which should be raised on our commercial hog production farms? Explain.

6   Describe the factors you would consider in selecting open gilts.

7   What additional factors would you consider in selecting bred sows?

8   Outline the qualities you would seek in selecting a young boar for use in your home herd.

9   What are the requirements for a Certified Meat Litter and a Certified Meat Sire?

10   Of what value are swine testing station data to you in selecting breeding stock?

11   What is the difference between a blind and an inverted nipple?

12   In selecting breeding and feeding stock, how can you be certain that the animals are from a disease-free herd?

13   Under what conditions would it be best for you to purchase feeder pigs rather than produce them on your farm?

14   Outline precautions that should be taken in buying feeder pigs.

REFERENCES

Bundy, Clarence E., R. V. Diggins and Virgil W. Christensen, *Livestock and Poultry Production,* 4th ed. Englewood Cliffs, N.J.: Prentice-Hall, Inc., 1974.

Christians, Charles J., *Livestock Judging,* Extension Bulletin 340. St. Paul, Minn.: University of Minnesota, 1967.

Ensminger, M. E., *Swine Science,* 4th ed. Danville, Ill.: The Interstate Printers and Publishers, 1970.

Hunsley, R. E., W. M. Beeson, and J. E. Nordby, *Livestock Judging and Evaluation.* Danville, Ill.: The Interstate Printers and Publishers, 1970.

Krider, J. L., and W. E. Carroll, *Swine Production,* 4th ed. New York, N.Y.: McGraw-Hill Book Company, 1971.

Tanksley, T. D., Jr., *Selecting Meatier Hogs,* B-9222. College Station, Tex.: Texas A & M University, 1967.

4 FEED SELECTION IN RATION FORMULATION

Feed costs represent from 60 to 70 percent of the total cost of producing 100 pounds of pork. Consequently, the profit from the swine enterprise is directly affected by the extent to which efficient and economical use of feeds has been made. Hogs like and do better on some feeds than others. Feeds differ in their nutritive value and in palatability. Price is also a factor. Some feeds will provide the nutritional needs of hogs more cheaply than others.

The nutritional needs of pigs vary with age. The needs of breeding animals are affected by their condition at breeding time and by the stage of gestation or suckling period.

Disease is also a factor in swine nutrition. Malnutrition and disease usually go hand-in-hand. Diseased pigs make very inefficient use of feeds. Therefore, one of the best ways of obtaining efficient use of feeds is to maintain a healthy herd of hogs and feed adequate feeds in proper balance.

We present in this chapter (1) principles in swine nutrition, (2) the nutritional needs of pigs, and hogs of various ages and productivity, (3) a classification of feeds and the composition of each, and (4) suggestions for effective selection of feeds in ration formulation.

PRINCIPLES OF SWINE NUTRITION

To become an efficient hog producer, one must first know what foods swine require in order to grow and reproduce. To understand the requirements of hogs, one must first become familiar with certain terms.

Nutrient

The term nutrient means any single class of food or group of like foods that aid in the support of life and make it possible for animals to produce the things expected of them.

Classes and Functions of Nutrients

Nutrients are divided into five classes: carbohydrates, fats, proteins, minerals, and vitamins.

Function of carbohydrates and fats. Carbohydrates and fats furnish the heat and energy for animals and provide the material necessary for fattening. Fats furnish 2.25 times as much heat and energy as do carbohydrates.

Composition of carbohydrates and fats. Carbohydrates are made up largely of sugars and starches. Fats consist of the same chemical compounds, but in different combinations.

Function of proteins. Proteins are essential in livestock feeding because they help to form the greater part of muscles, internal organs, skin, hair, and hoofs. Milk also contains protein.

FIGURE 4-1. Protein made the difference. The pig in the back was fed a 14-percent protein ration on alfalfa-red clover pasture. The pig in front was the same age but was fed an 8 percent protein ration on similar pasture (*Iowa State University test. Courtesy* Wallaces' Farmer.)

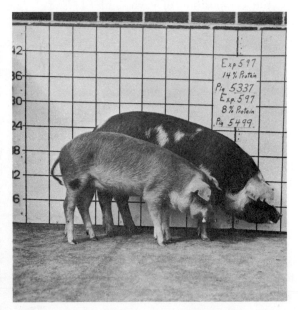

Composition of proteins. Proteins are made up of a group of acids known as *amino acids*. Twenty-five or more amino acids have been identified; at least ten of these acids are needed by animals. Cattle and sheep, with four compartments to the stomach, actually manufacture their own amino acids from nitrogen compounds. Hogs, however, are *simple-stomach animals* and must be fed all the essential amino acids. Since no single feed contains all the proteins, several sources of protein should be provided for this class of animals.

Function of minerals. Minerals are used primarily in the bones and in the teeth, and make up an important part of the blood. Even the heart depends upon mineral balance to maintain its regular beating.

Essential minerals. Minerals are generally divided into two groups—major minerals and trace minerals. The major minerals, salt, calcium, and phosphorus, are needed in the greatest quantity and are most likely to be lacking in the feed supplied. The trace minerals are those needed in very small amounts and yet are essential to the health of the animal. They include iron, copper, manganese, iodine, cobalt, sulfur, magnesium, zinc, potassium, boron, and selenium.

Vitamins. All the vitamins are essential in swine feeding. They are vitamin A, B complex, B_{12}, C, D, E, and K. The B complex group includes thiamine, riboflavin, niacin, pyridoxine, pantothenic acid, choline, biotin (vitamin H), pyracin, tara-amino benzoic acid, inosital, and folic acid.

Vitamin functions. The purpose or function of many of these vitamins is not clear, but we do know that serious results occur when feeds are fed which do not provide the necessary vitamins. If vitamin A is lacking, animals fail to reproduce, eyesight is impaired, and growth slows down. If the B complex group is not provided, appetite fails and disease may become a problem. B_{12} improves the protein in grains and other feeds.

A lack of vitamin E may cause a failure of the reproductive system.

In general we can say that vitamins provide a defense against disease, promote growth and reproduction, and contribute to the general health of the animal.

Function of water. Water is not generally considered a nutrient, but its importance to the animal cannot be overestimated. The animal body is made up of about 45 to 70 percent water, and nutrients must be in liquid form before they can be absorbed by the body. Water is important in controlling body temperature. Hogs must have an abundance of fresh, clean water if the best results are to be expected.

Antibiotics

In recent years we have learned a great deal about antibiotics and their value when supplied in small amounts in the daily ration of certain kinds of livestock. Antibiotics are

FIGURE 4-2. Implanting an antibiotic pellet behind the ear of a pig. (*Courtesy* Wallaces' Farmer.)

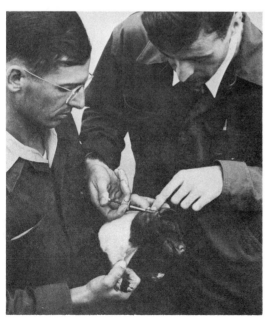

considered a form of medication rather than a food nutrient.

Function of antibiotics. The function of antibiotics is not clear. Some authorities believe that they help control undesirable microorganisms in the digestive system and thereby increase the rate of growth. We do know that there is a generally healthier condition and an increased growth rate in most animals when they are fed daily small amounts of certain beneficial antibiotics. The known antibiotics that have shown value when fed are Aureomycin, Terramycin, bacitracin, neomycin, Tylosin, streptomycin, oleandomycin, penicillin, and Virginiamycin.

Digestible Nutrients

Nearly all feeds contain a certain percentage of material that is not digestible in the animal body. The nondigestible portions are eliminated by the animal as waste. The parts which are absorbed by the body are referred to as *digestible nutrients*.

Fiber. The portion of carbohydrates not easily digested is known as *fiber* or *crude fiber*. The amount of fiber which a feed contains is important in determining its value.

Nitrogen-free extract. The term *nitrogen-free extract* often appears on the label listing the ingredients of a feed. It refers to the more digestible part of the carbohydrates rather than to the fiber, which is much less digestible.

Balanced Rations. A *ration* is the amount of feed which is given to an animal during a 24-hour period. A *balanced ration* is one which supplies in their correct proportion all the food nutrients necessary to properly nourish the animal during a 24-hour period.

NUTRITIONAL NEEDS OF PIGS AND HOGS IN VARIOUS STAGES OF PRODUCTION

It has been pointed out that the nutritional needs of swine vary with the age and

TABLE 4-1 RECOMMENDED NUTRIENT ALLOWANCES FOR SWINE[1]

Stage of Life Cycle	Weight of Pig, lbs.	Percent of Ration						Units/Lb. of Ration		Mcg./Lb. Ration	Mg./Lb. of Ration			Feed Additive[5] Grams/ton of Ration
		Protein[2]	Lysine	Methionine and Cystine	Tryptophan	Ca	P	Vitamin A	D	Vitamin B_{12}	Riboflavin	Pantothenic Acid	Niacin	
SOWS, GILTS AND BOARS														
Pregestation, breeding and gestation														
Feed intake per day														
3 pounds	—	13.0	0.55	0.30	0.13	1.00	0.75	3000	400	10	2	8	15	0-300[5]
4 pounds	—	12.0	0.45	0.25	0.11	0.80	0.55	2250	300	7.5	1.5	6	11	0-300[5]
5 pounds	—	11.0	0.35	0.20	0.08	0.60	0.45	1500	200	5	1	4	7.5	0-300[5]
Lactation	—	13.0	0.60	0.35	0.12	0.60	0.50	2000	200	10	2	8	15	0-300[5]
YOUNG PIGS														
Milk replacer (dry feed)	- to 12	20-24	1.00	0.60	0.18	0.70	0.60	2000	200	10	2	8	15	100-300
Starter	creep to 40	18-20	0.95	0.50	0.15	0.70	0.60	2000	200	10	2	8	15	100-300
G-F HOGS														
Grower	40-120	14-16	0.65-0.75	0.40	0.12	0.60	0.50	1000	100	5	1	4	7.5	0-100[7]
Finisher	120-240	12-14	0.55-0.65	0.30	0.10	0.60[8]	0.40[8]	1000	100	5	1	4	7.5	0-100[7]

SUGGESTED LEVEL TO BE ADDED[4]

[1] The nutrient allowances are suggested for maximum performance, not as minimum requirements. They are based on research work with natural feedstuffs and have been found to give satisfactory results.

[2] Sow protein recommendations are based on corn-soybean meal rations, other feedstuffs may require more protein to meet the amino acid requirements.

[3] Cystine can satisfy 40 percent of the total methionine need.

[4] The vitamin levels listed should be added to the ration in addition to that occuring in natural feedstuffs. Most natural feedstuffs contain very little vitamin D or B_{12}. The amount of provitamin A (carotene) in feedstuffs will depend on processing and storing. The niacin in most grains, moreover, is relatively unavailable to swine. Thus, of the vitamins that need to be added to the ration, only riboflavin and pantothenic acid can be supplied, in part, by natural foodstuffs. The level of the other vitamins not shown above should be adequate in natural feedstuffs except for vitamins E and K, which may require supplementation in special cases.

[5] The feed additives may be antibiotics, arsenicals or other chemotherapeutics or combinations of these. Levels and combinations used and stage of production for which they are used must comply with the current Food and Drug Administration regulations.

[6] High levels of feed additives may be beneficial just prior to and at breeding and farrowing time but are not recommended during the entire gestation-lactation period unless specific disease problems are present.

[7] The feed additive and the level used during the growing-finishing phase should be primarily for growth promotion and improvement of feed efficiency.

[8] Replacement gilts should receive 0.60 percent calcium and 0.50 percent phosphorus.

Source: *Life Cycle Swine Nutrition*, Iowa State University, 1974.

stage of production. Specialists in swine nutrition indicate that there are at least nine distinct stages in hog production. These stages are: (1) the unbred gilts, (2) the pregnant sow, (3) the sow at farrowing time, (4) the sow suckling a litter, (5) pigs weighing 5 to 12 pounds, (6) pigs weighing 12 to 25 pounds, (7) pigs weighing 25 to 50 pounds, (8) the growing pigs weighing 50 to 125 pounds, and (9) pigs during the finishing period weighing 125 to 240 pounds.

In Table 4-1 are shown those nutritional allowances recommended for optimum performance of sows and pigs in the various stages of production. These recommendations are based upon research at Iowa State University. Recommendations may be obtained from other state agricultural experiment stations, from the Bureau of Animal Industry of the U. S. Department of Agriculture, and from the National Research Council, Washington, D.C.

Sows

Pregnant and lactating sows require rations that are fairly high in mineral and vitamin content and contain 13 percent protein. Pregnant sows must have rations containing an abundance of minerals in order to produce large litters of healthy pigs. Sows which are suckling litters need proteins, minerals, and vitamins in order to produce an abundance of milk. While the quantities of feeds needed by the two groups are different, both need feeds with the same nutritive balance. Sows suckling litters need more feed than do pregnant sows.

The lactating sow's ration should contain 0.6 percent lysine, 0.35 percent methionine and cystine, 0.12 percent tryptophan, 0.6 percent calcium, and 0.5 percent phosphorus. It should also contain 2,000 units of vitamin A per pound, 200 units of vitamin D, 10 micrograms of vitamin B_{12}, 2 milligrams of ribo-

FIGURE 4-3. Sows suckling large litters require rations high in protein, minerals, and vitamins. (*Courtesy Land O Lakes, Felco Division.*)

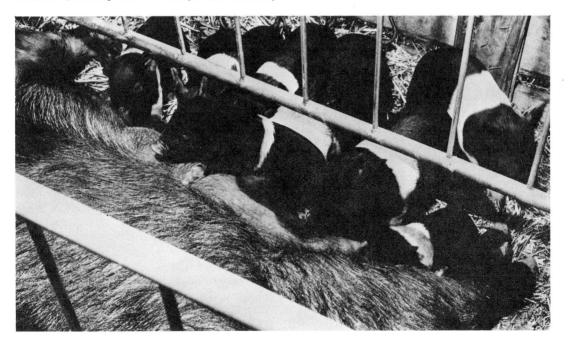

flavin, 8 milligrams of pantothenic acid, and 15 milligrams of niacin per pound of ration.

Pigs (Up to 40 Pounds)

As shown in Table 4-1, during the first few weeks of their lives pigs receive rations high in protein, calcium, phosophorus, vitamins, and antibiotics. The milk replacer or prestarter ration should contain 20 to 24 percent protein, 0.7 percent calcium, and 0.6 percent phosphorus. It should contain 2,000 units of vitamin A and 200 units of vitamin D per pound. It should also contain 2 milligrams of riboflavin, 15 milligrams of niacin, 8 milligrams of pantothenic acid, and 10 micrograms of vitamin B_{12} per pound.

The starter ration recommended for pigs weighing from 12 to 40 pounds is much the same as the milk replacer ration with the exception that the protein content is reduced to 18 to 20 percent, and a small decrease is made in the percentages of lysine, tryptophan, methionine, and cystine in the ration.

Pigs (Up to 240 Pounds)

With the exception of the protein requirements, the nutritional allowances for pigs weighing 40 to 120 pounds are about the same as for pigs weighing 120 to 240 pounds. The grower ration recommended for pigs weighing 40 to 120 pounds should contain 14 to 16 percent protein, 0.65 to 0.75 percent lysine, 0.40 percent methionine and cystine, 0.12 percent tryptophan, 0.60 percent calcium, and 0.50 percent phosphorus. The ration should also contain 1,000 units of vitamin A per pound of ration, 100 units of vitamin D, 5 micrograms of vitamin B_{12}, 1 milligram of riboflavin, 4 milligrams of pantothenic acid, and about 7.5 milligrams of nacin per pound of ration.

As shown in Table 4-1, the nutritional requirements for hogs weighing 120 to 240 pounds permit reduction of protein content to 12 to 14 percent, and slight reduction in percentages of lysine, methionine and cystine, tryptophan, calcium, and phosphorus.

Life Cycle Swine Feeding

The underlying principles of swine nutrition have been described in previous paragraphs. The life cycle of swine has been divided into the various stages according to nutritional needs, and the needs of the animals in the various stages were enumerated in Table 4-1. From a nutritional standpoint, some of the stages are more critical than others. The latter stages of the gestation, the period of lactation, the prestarting, starting, and growing periods have been classified as nutritionally critical periods. It is during these periods that the best and most costly rations are needed.

Shown in Table 4-2 is a summary of life-cycle swine feeding indicating the various stages in the production cycle, the length of the feeding program, the protein content, and amount of ration recommended to be fed daily. Note that full feeding is recommended for gilts and sows during lactation and for pigs up to 125 pounds in weight. Pigs over 125 pounds in weight may be fed limited rations when this system appears desirable.

COMPOSITION AND CLASSIFICATION OF FEEDS

Feeds for hogs are generally classified according to the amount of total digestible nutrients that they provide, or according to the amount of a specific nutrient that they furnish in the ration. On the basis of total digestible nutrients they are classified as *roughages* and *concentrates*.

Roughages

Feeds containing relatively large amounts of fiber and nondigestible material

TABLE 4-2   LIFE CYCLE SWINE FEEDING PROGRAM

Stage of Cycle	Length of Feeding Program	Season	Complete Ration % Protein	Complete Ration Lbs./day	Lbs. Corn or Grain/ day	Lbs. Supplement/ day
BOARS	From time purchased at 5-6 months of age	Summer Winter	12-16 11-14	4-6 5-7	3-5 4-6	0.8-1.0
	Increase intake 1-2 lbs. during the heavy breeding season.					
GILTS						
Pregestation	From time selected at 5-6 months until breeding at 7-9 months of age	Summer Winter	12-16 11-14	4-6 5-7	3-5 4-6	0.8-1.0
Flushing and breeding	For 3 weeks prior to breeding (do not continue after breeding)		11-14	6-9	5-8	1.0-1.2
	Increase corn intake approximately 2 lbs./day.					
Gestation		Summer Winter	12-16 11-14	4-5 5-6	3-4 4-5	0.8-1.0
	Increase intake 1-2 lbs. last 3 to 5 weeks of gestation if gilts appear to be too thin.					
Lactation	Wean at 3-5 weeks after farrowing.		13-16	10-14	Full feed complete ration.	
SOWS						
Breeding and gestation	Breed back at first heat period after weaning (flushing is not beneficial with sows)	Summer Winter	13-18 12-16	3-4 4-5	2-3 3-4	0.8-1.0
	Increase intake 1-2 lbs. last 3 to 5 weeks of gestation if sows appear to be too thin.					
Lactation	Wean 3-5 weeks after farrowing.		13-16	12-16	Full feed complete ration.	
PIGS						
Milk replacer	Use only if weaned before 3 weeks and feed until pig weighs 12 lbs.		20-24		Full feed complete ration.	
Starter	Use as a creep feed and continue after weaning until pigs are 8 weeks of age or 40 lbs.		18-20		Full feed complete ration.	
Grower-finisher	From 8 weeks of age or 40 lbs. until market weight.		12-16	Full feed (may limit feed after 125 lbs.)	With free choice, corn consumption varies depending on weight of pig. Supplement intake should be approx. 0.75 lbs. regardless of the weight of the pig.	

Source: *Life Cycle Swine Nutrition*, Iowa State University.

are called *roughages*. In swine feeding rough-ages are fed only to mature breeding stock.

Concentrates

Concentrates are feeds which have a comparatively high digestibility. They are relatively low in fiber and include all grains and many by-products of grains and animals.

Protein Concentrates

Protein concentrates refer to that group of concentrates which furnish a relatively high percentage of protein. We usually con-sider any feed which contains 20 percent or more protein a protein concentrate.

Animal proteins. Proteins that are de-rived from animals or animal by-products, such as tankage and dried skim milk, are called animal proteins. In addition to prob-ably containing amino acids not found in other proteins, animal proteins also contain growth factors which make them especially valuable in the feeding of swine.

Vegetable proteins. Vegetable proteins are those found in plants or in the by-prod-ucts of plants. The plant or vegetable pro-teins, as a class, are not nearly as complete nor as high in quality as are the animal proteins. In swine rations, one or more animal

FIGURE 4-4. This litter of Landrace pigs weighed 810 pounds at 56 days. The sow owned by Howard Cowden of Missouri farrowed 49 pigs in three litters. (*Courtesy* National Hog Farmer.)

FIGURE 4-5. These pigs are fed in confinement on a slatted floor a complete pelleted ration. (*Courtesy American Cyanamid Co.*)

proteins are generally considered necessary in order to give the amino acid balance essential for best results.

Sources of Carbohydrates

Of the common feeds that are fed to hogs, the grains provide the best source of energy and have the highest value especially during the finishing program.

Grains

Corn. Yellow corn is considered the best grain for swine. It ranks high in total digestible nutrients (about 80 percent for No. 2 grade), low in fiber, and higher in fat than any cereal grain except oats. Corn is very palatable. It is deficient, however, in proteins and minerals.

High-lysine corn is lower in carbohydrate content than normal corn but contains 1 to 3 percent more crude protein. The lysine content of normal corn averages about 0.28 percent, whereas the percentage in high-lysine varieties ranges from 0.25 to 0.50 percent.

Oats. As a swine feed, oats rank second to corn. They have about the same general nutritional value as corn but contain more fiber and are less digestible. Their protein content is a little higher than that of corn. Oats from which the hulls have been removed by mechanical means or hull-less oats have a food value equal to or higher than corn.

Barley. Barley is a very good substitute for corn. It is less digestible than corn and has more fiber but has the advantage of containing more protein. Barley is higher in food value per hundred pounds than oats and is considered better than oats as a finishing feed. Barley infected with scab, however, cannot be fed to swine.

Wheat. Wheat resembles the other cereal grains in nutritional value. It has a higher carbohydrate content and a higher protein content than corn.

Rye. Rye has about the same general food value as other grains but is not palatable to livestock and, when fed in large quantities, may cause trouble. It should be mixed with other grains and fed only as a part of the ration.

Grain sorghums. Grain sorghums resemble corn in all respects, except that they are lower in fat. The sorghums make very good feed and are well liked by hogs. Some sorghum grain is rather hard, and grinding is necessary.

Triticale. Triticale is a cross between durum wheat and rye. It is somewhat higher in protein than corn. It is more palatable as a feed than rye, but it is often contaminated with ergot, as is rye. Tests indicate that it can be used in place of corn or sorghum in growing-pig rations. Supplementing high triticale rations with lysine is recommended.

Rice bran. Rice bran is fed in rice growing areas and may be used in replacing up to about 40 percent of the corn or milo in pig rations. Since it is higher in protein than corn, it also replaces some soybean meal in the ration.

Molasses. Molasses has a fairly high nitrogen-free extract content but is fed largely because of its palatability. It is usually used as a supplement to other feeds in order to increase consumption.

Sources of Animal Proteins

It has been pointed out that animal protein feeds come from animal by-products, dairy products, and from dairy by-products.

Animal By-products

Tankage and meat scraps. Tankage and meat scraps are very much alike in composition. The digestible protein content will vary considerably, depending upon the contents of the feed. Since it is impossible for the user to determine what amount of nondigestible protein a feed contains, it is wise to buy only from a company known to be reliable.

Meat and bone scraps. The combination of meat and bone scraps contains a higher percentage of bone. Since bone is lower in protein than are meat scraps, the total protein content of the feed is generally lower.

Blood meal. Blood meal is high in total protein, but its quality and digestibility are lower than that of good grade tankage or meat scraps.

Fish meal. Fish meal is one of the best feeds for swine because of its high percentage of good quality protein. As in tankage and meat scraps, the protein content varies, depending upon the materials used and the methods of manufacture.

Dairy Products and By-products

Skim milk and buttermilk. Both of these products contain over 90 percent water, but the dry matter is about one-third protein. Nearly all the protein is digestible, and when these products are available, they furnish one of the best sources of protein.

Dried skim milk and buttermilk. A heating process drives off most of the water in skim milk and buttermilk, leaving the milk solids in a dry condition. These dried milk solids furnish a high protein food of good quality.

Sources of Vegetable Proteins

All seeds, roughages, and grain by-products furnish various amounts of proteins, but the chief sources of vegetable proteins are the seed by-products and legume roughages.

Seed By-products

Soybean oil meal. Soybean oil meal varies in protein content from 41 to 46 percent, depending upon the method of manufacture. The protein is of very high quality and may be used to furnish a major part of the protein concentrate in rations for swine.

Soybeans. Soybeans have the highest protein content of any seeds commonly used for feed. Soybeans will average about 37 percent total protein, but the quality is low. The protein in soybeans is improved by the heating process used in the manufacture of soybean oil meal. Soybeans can be fed only in small quantities because of the high oil content.

Cottonseed meal and cake. The protein content of cottonseed meal varies from 38 to

FIGURE 4-6. High quality gilts on an Indiana farm making effective use of pasture. (*Agricultural Associates photograph. Courtesy Starcraft Livestock Equipment.*)

47 percent, depending upon the variety of cotton from which the seed is taken and upon the method of processing. Cottonseed meal is a good source of protein for cattle but should be used in limited amounts as a protein concentrate for swine.

Linseed meal. Linseed oil meal has been a widely used protein feed. Its protein content ranges from 31 to 36 percent. Linseed meal, however, should be used as only a part of the protein concentrate in the ration for swine.

Peanut oil meal. Peanut oil meal can be used as protein feed for swine. When fed as the only protein, it may be too laxative.

Corn gluten meal. The protein content ranges from 40 to 44 percent but the quality of protein is not high. Best results are obtained when it is fed in combination with other proteins.

High-lysine corn. A part of the protein needs of swine may be provided by high-lysine corn. The protein content of this type of corn varies widely from field to field; it averages about 0.4 percent lysine as compared to 0.28 percent in normal corn. It also contains nearly twice as much tryptophan and some additional methionine and cystine. The overall protein content may exceed that of normal corn by 3 percent. Since the lysine content varies widely, it is recommended that laboratory tests be made to determine actual content before using it in a ration.

Synthetic lysine. Sources of synthetic lysine are available. It can be used to replace 100 pounds of 44 percent soybean meal in a ton of the ration. Three pounds of a synthetic lysine product containing 98 percent lysine and 97 pounds of corn will replace 100 pounds of soybean meal in the ration. With

FIGURE 4-7. Mechanized feed handling permits the feeding of complete rations in large swine enterprises. (*Courtesy Land O Lakes, Felco Division.*)

soybean meal at $200 per ton, a pound of lysine would be worth $2.69.

Legume Roughages

Legumes, when fed as dry roughages, silage, or pasture, furnish an excellent source of high-quality protein. Under some conditions legumes will furnish all the needed protein.

Dehydrated alfalfa meal is widely used in swine rations. It contains 17 percent protein and is high in vitamins. It is especially desirable in the rations of the breeding stock, both boars and sows. Sun-cured alfalfa meal contains only 13 percent protein, some less essential vitamins, but 6 percent more fiber than dehydrated alfalfa meal.

Sources of Vitamins

Swine, when fed a well-balanced ration consisting of grains, protein concentrates, and good forage, generally get enough of the needed vitamins. However, under certain conditions, it is necessary to select feeds high in essential vitamins. Vitamins most likely to be deficient in the feed are A or carotene, the B-complex group, D, and E.

Vitamin A

Vitamin A functions in both skeleton and tissue growth, improves vision, and contributes to the reproductive process. It is also related to disease resistance.

Vitamin A in forage. Vitamin A is found in plants and in plant products as carotene, which is converted into vitamin A in the digestive system of animals. Green plants have an abundance of carotene, and there is seldom any need to give any special attention to the vitamin A content of a ration for swine on pasture. Roughages that retain their green color and most of the leaves when cured will still be high in carotene content.

Vitamin A in grain. Carotene is found in yellow corn and is an important source of vitamin A.

Commercial sources of vitamin A. The fish liver oils are one of the best sources of concentrated amounts of vitamin A. For feed-mixing purposes vitamin A may also be purchased in a supplement containing a mixture of several other vitamins.

Vitamin B

The vitamin B complex consists of nine water soluble vitamins. They are thiamine,

FIGURE 4-8. The result of a ration low in Vitamin A. (*Courtesy U.S.D.A. Bureau of Animal Industry.*)

riboflavin, niacin, pantothenic acid, pyridoxine, B_{12}, biotin, choline, and folic acid. Biotin, folic acid, thiamine, and pyridoxine are not generally a concern in ration formulation. Biotin and folic acid are synthesized by intestinal bacteria. There is an abundance of thiamine and pyridoxine in grains and protein supplements. The other five B vitamins should be supplied in the ration.

A deficiency in riboflavin causes poor growth, red exudate around the eyes, and the birth of weak or dead pigs. Niacin also affects growth and is very important in the control of necrotic enteritis. A lack in the amount of pantothenic acid results in poor growth, posterior paralysis, goose stepping, and the birth of weak pigs.

Vitamin B_{12} deficiences cause poor growth, nervous and irritable movement, and the birth of small, weak pigs. Cholin has been found to be related to litter size and possibly to the birth of pigs with spraddled legs. It is closely related to methionine and cystine in cellular metabolism.

Vitamin B complex group in forages. With the exception of vitamin B_{12}, which will be discussed separately, forages provide an excellent source of the B complex group of vitamins.

Vitamin B complex group in animal products. The animal products, especially milk or milk products, furnish substantial amounts of vitamin B complex when included in swine rations.

Vitamin B complex group in grains. Grain crops provide, to some extent, most of the B vitamins, but they are not high enough in these essential vitamins to be relied upon entirely.

Vitamin B group in brewer's yeast and distillers' solubles. Small amounts of yeast or distillers' solubles are often used to boost the B vitamin content of the ration.

B_{12} in animal products. With the exception of commercial sources, B_{12} is found only in animal products, such as milk, tankage, meat scraps, fish meal, and similar feeds. Swine following cattle are supplied B_{12} to some extent through the feces of the cattle. When B_{12} is supplied in other forms, the amount of animal protein fed swine can be considerably reduced without affecting the growth rate of the animal.

Commercial sources of vitamin B_{12}. Many laboratories and feed companies manufacture concentrated forms of vitamin B_{12} that may be purchased for feed-mixing purposes.

Vitamin D

Vitamin D in sunlight. Animals exposed to direct sunlight seldom suffer from a vitamin D deficiency. It is reflected in poor utilization of calcium and phosphorus in skeletal development. Poor growth, weak legs, and a ricket condition may result. Pigs fed in confinement must have vitamin D in their rations.

Vitamin D in forage crops. Forage crops, especially sun-cured legume hays, are very good sources of vitamin D.

Commercial sources of vitamin D. Concentrated amounts of vitamin D may be secured in fish liver oils and irradiated yeast.

Vitamin E

Vitamin E serves as a biological antioxidant in the body, preventing the formation of peroxide which may cause liver necrosis, pale muscles, edema, jaundice, or death. It functions with selenium in maintaining membrane integrity. Vitamin E and selenium have been found to be deficient in the grain produced in a number of states.

Vitamin E in farm feeds. Vitamin E is found in nearly all forages, feed grains, and protein concentrates. It is essential for fertility. Boars used heavily in breeding programs may show some benefit from additional amounts of vitamin E.

Wheat germ oil as a source of vitamin E. When amounts of vitamin E greater than those supplied by the ordinary ration are desired, wheat germ oil is an excellent source. The alpha tocopherol additive is perhaps the best source because of its high vitamin E nutritive value and its ease in mixing.

Vitamin K

Vitamin K assists the blood clotting mechanism and is especially important in rations for pigs being fed moldy corn in confinement. Moldy feed may cause internal hemorrhaging resulting in the death of the pig. In the past, pigs could synthesize the needed vitamin K by using forages and the feces of animals. The 1973 swine nutrient requirements publication of the National Research Council recommends 2.2 grams of menadione (vitamin K) in each ton of feed.

Sources of Minerals

Most of the essential minerals are present in common feeds, but the amounts are generally not great enough to provide for the mineral needs of swine.

Minerals in farm-produced feeds. All farm-grown feeds contain minerals, but the legume forages are the best sources. Legumes are high in both phosphorus and calcium. The mineral content of all farm-grown feeds depends upon the soil on which they are grown. The trace mineral content of grains and roughages also depends largely upon the types of soil on which they were grown.

Minerals in protein concentrates. Animal protein feeds, especially meat and bone scraps, tankage, and fish meal, are excellent sources of minerals. The mineral content in these feeds varies from 15 to 25 percent. Vegetable protein meals are much lower in

FIGURE 4-9. This pig was fed a ration deficient in calcium. Note the bone weakness. (*Courtesy U.S.D.A. Bureau of Animal Industry.*)

mineral content than are the animal proteins, but higher than are the farm grains.

Chief sources of mineral supplements. High-quality ground limestone is one of the best sources of calcium for mixing mineral supplements. The calcium-phosphorus ratio in a mineral should be no greater than 1.5 parts of calcium to every one of phosphorus.

Steamed bonemeal is an excellent source of phosphorus and contains much calcium. It is, however, expensive, and quantities are not available to meet the needs. It contains 24 percent calcium and 12 percent phosphorus. Dicalcium phosphate contains 26 percent calcium and 18.5 percent phosphorus. Defluorinated rock phosphate and defluorinated treble phosphate may also be used.

Salt is needed by swine and is easily obtainable. Mineralized salt, which contains a mixture of trace minerals, is also available.

Trace minerals can best be secured by purchasing a *trace mineral mixture* or a complete mineral mixture.

Following is a trace mineral premix that contains 7 percent iron (Fe), 0.45 percent copper (Cu), 5.5 percent manganese (mn), 8 percent zinc (Zn), and 0.005 percent selenium:

	Pounds
Ferrous sulfate	35.0
Copper sulfate	2.0
Manganese sulfate	24.0
Zinc sulfate	36.0
Potassium iodide	0.05
Sodium selenite	0.011
Carrier	2.939
Total	100.00

Sources of Antibiotics

Because no farm-produced feeds contain antibiotics, they must be added to the ration. Otherwise a complete ration containing antibiotics should be purchased.

Nearly 70 percent of swine producers mix their own rations or have them mixed by the feed dealer. Since separate rations are fed during the various stages of production, the antibiotic most suitable for feeding at the specific stage is added. Mixers and feed dealers can provide a wide choice of antibiotics.

In some areas antibiotic-vitamin premixes are available. In most cases producers use separate vitamin, trace mineral, and mineral premixes. The antibiotic, or combination of antibiotics, is provided separately.

There is evidence that a mixture of certain antibiotics may be better than any single antibiotic. Therefore, a premix containing two or more antibiotics is used. Sometimes the antibiotic or antibiotics are used in premixes with sulfa and other drugs.

Antibiotics used in swine feeding are:

1. **Aureomycin**
2. **Bacitracin**
3. **Neomycin**
4. **Oleandomycin**
5. **Penicillin**
6. **Streptomycin**
7. **Terramycin**
8. **Tylosin**
9. **Virginiamycin**

Combinations of antibiotics and of drugs available are:

1. **Aureomycin, sulfamethazine and penicillin**
2. **Penicillin and streptomycin**
3. **Neomycin and Terramycin**
4. **Tylosin and sulfamethazine**

Following is an antibiotic feeding program recommended to be used as a guide in selecting feeds and antibiotics for home mixing, or for use in purchase of complete rations:

Stage of Production	Suggested Level gms/ton
Prestarter (early weaning)	200
Starter (12–25 lbs.)	100–200
Grower (25–75 lbs.)	50–100
Finisher (75 lbs. to market)	20– 50

Nutrient Content of Common Feeds

Shown in Table 4-3 are analyses of the food nutrients in the carbohydrate feeds most commonly fed to hogs. The analyses of swine protein feeds are shown in Table 4-4. The compositions of hays and silage commonly fed to hogs are shown in Table 4-5. The mineral and vitamin content of feeds commonly fed to hogs is shown in Table 4-6.

FIGURE 4-10. These Yorkshire pigs like the aroma of the pelleted prestarter feed. (*Courtesy Land O Lakes, Felco Division.*)

TABLE 4-3   AVERAGE COMPOSITION OF COMMON CONCENTRATES HIGH IN ENERGY AND FAT-PRODUCING VALUE
(Expressed in Percent)

Feed	Total Dry Matter	Total Digestible Nutrient	Total Protein	Digestible Protein	N-Free Extracts or Carbo-hydrates	Fat	Mineral Matter	Fiber
Barley	89.4	77.7	12.7	10.0	66.6	1.9	2.8	5.4
Corn (No. 2)	85.0	80.1	8.7	6.7	69.2	3.9	1.2	2.0
Ground ear corn	86.1	73.2	7.3	5.3	66.3	3.2	1.3	8.0
Feterita (grain)	89.4	79.8	12.2	9.5	70.1	3.2	1.7	2.2
Kafir (grain)	89.8	81.6	10.9	8.8	72.7	2.9	1.6	1.7
Milo (grain)	89.0	79.4	10.9	8.5	70.7	3.0	2.1	2.3
Blackstrap molasses	73.4	53.7	3.0	0	61.7	0	8.6	0
Oats (hulled)	90.4	91.9	16.2	14.6	63.7	6.1	2.2	2.2
Oats	90.2	70.1	12.0	9.4	58.6	4.6	4.0	11.0
Rye	89.5	76.1	12.6	10.0	70.9	1.7	1.9	2.4
Wheat	89.5	80.0	13.2	11.1	69.9	1.9	1.9	2.6
Wheat bran	90.1	66.9	16.9	13.3	53.1	4.5	6.1	10.0
Wheat middlings	90.1	79.2	17.5	15.4	60.0	4.5	3.8	4.3

Source: Adapted by special permission of the Morrison Publishing Company, Ithaca, N.Y., from *Feeds and Feeding*, 22nd ed., by F. B. Morrison.

TABLE 4-4   AVERAGE COMPOSITION OF COMMON PROTEIN CONCENTRATES
(Expressed in Percent)

Feed	Total Dry Matter	Total Digestible Nutrient	Total Protein	Digestible Protein	N-Free Extracts or Carbo-hydrates	Fat	Mineral Matter	Fiber
Blood meal	91.6	60.4	82.2	60.4	0.9	1.9	5.7	0.9
Buttermilk (dried)	92.0	83.1	31.8	28.6	43.6	6.1	10.0	0.5
Corn gluten meal	91.6	79.7	43.2	36.7	38.9	2.2	3.5	3.8
Cottonseed meal	92.7	72.6	43.3	35.9	27.4	5.1	6.0	11.0
Fish meal	92.0	70.8	60.9	53.6	5.4	6.9	18.3	0.9
Linseed meal, solvent process	91.0	70.3	36.6	30.7	38.3	1.0	5.8	9.3
Meat scraps	94.2	66.7	54.9	45.0	2.5	9.4	24.9	2.5
Meat and bone scraps	93.7	65.3	49.7	40.8	3.1	10.6	28.1	2.2
Milk (dried skim)	94.0	80.7	34.7	31.2	50.3	1.2	7.8	0.2
Peanut oil meal, solvent process	93.0	77.3	52.3	47.6	26.3	1.6	5.9	6.9
Soybean seed	90.0	87.6	37.9	33.7	24.5	18.0	4.6	5.0
Soybean oil meal, solvent process	90.4	78.1	45.7	42.0	31.4	1.3	6.1	5.9
Tankage (high-grade)	92.8	65.8	59.4	50.5	2.6	7.5	21.4	1.9

Source: Adapted by special permission of the Morrison Publishing Company, Ithaca, N.Y., from *Feeds and Feeding*, 22nd ed., by F. B. Morrison.

TABLE 4-5 AVERAGE COMPOSITION OF HAYS AND SILAGES
(Expressed in Percent)

Feed	Total Dry Matter	Total Digestible Nutrient	Total Protein	Digestible Protein	N-Free Extracts or Carbo-hydrates	Fat	Mineral Matter	Fiber
HAY								
Alfalfa hay (average)	90.5	50.7	15.3	10.9	36.7	1.9	8.0	28.6
Alfalfa hay (high-grade)	90.5	52.7	17.5	12.8	39.5	2.4	8.4	22.7
Alfalfa meal (dehydrated)	92.7	54.4	17.7	12.4	38.4	2.5	10.1	24.0
Clover hay—red (good)	88.3	51.8	12.0	7.2	40.3	2.5	6.4	27.1
Clover hay—red (second cutting)	88.1	54.1	13.4	8.4	40.4	2.9	6.9	24.5
SILAGE								
Alfalfa	36.0	21.3	6.0	4.1	13.7	1.4	3.2	11.7
Corn	27.6	18.3	2.3	1.2	16.2	0.8	1.6	6.7
Grass-legume mixture	33.3	19.1	5.2	2.9	14.2	1.3	3.8	8.8
Grain sorghum	30.0	17.1	2.6	1.4	18.6	0.7	2.1	6.0

Source: Adapted by special permission of the Morrison Publishing Company, Ithaca, N.Y., from *Feeds and Feeding*, 22nd ed., by F. B. Morrison.

FEED SELECTION IN RATION FORMULATION

The rations recommended for use during the various stages of the life cycle of swine will be discussed in later chapters. In selecting feeds to include in the rations it is necessary to consider the following factors: (1) availability, (2) cost, (3) nutritional value, (4) palatability, and (5) ease of feeding.

It is usually necessary to calculate the cost of a pound of total digestible nutrients (TDN) and of digestible protein. When all other factors are equal, the cost per pound of TDN or of protein will be the determining factor. When comparing carbohydrate feeds the cost of one pound of TDN is important. If we are comparing protein feeds, the cost per pound of digestible protein is important.

Feed Grain Substitution Scale

Shown in Figure 4-11 is a scale which can be used to determine the cheapest grain for swine feeding according to its price per bushel and feeding value. It is assumed that grains will be of average quality and will be properly prepared and fed in a balanced ration.

To use the scale, place a ruler or piece of paper vertical to the line and at the point showing the price of one of the feeds. The points where the paper or ruler cross the lines for the other feeds indicate the values of these feeds in relation to the price of the feed with which they are being compared. The scale will show that when corn is priced at $2.00 per bushel, wheat would be worth $2.25, and oats about 96 cents per bushel.

Ingredient (Air dry)	Metabolizable Energy Kcal/lb.	Protein %	Calcium %	Phosphorus %	Fat %	Fiber %	Lysine %	Methionine %	Cystine %	Tryptophan %
Alfalfa meal (dehydrated)	543	17.0	1.30	0.23	2.5	27.0	0.80	0.20	0.34	0.40
Alfalfa meal (sun-cured)	460	13.0	1.20	0.20	1.5	33.0	0.50	0.20	0.26	0.23
Animal fat, stabilized	3,590	—	—	—	100.0	—	—	—	—	—
Barley	1,200	11.5	0.06	0.36	1.8	7.0	0.40	0.22	0.26	0.17
Beet pulp, dried	1,020	8.0	0.60	0.10	0.5	21.0	0.60	—	—	0.10
Blood meal	1,329	80.0	0.28	0.22	1.0	1.0	6.90	1.00	1.40	1.05
Corn (yellow)	1,390	8.9	0.01	0.25	3.8	2.9	0.28	0.18	0.16	0.07
Corn and cob meal (yellow)	1,200	7.5	0.04	0.20	3.0	10.0	0.23	0.15	0.13	0.06
Cottonseed meal (solvent)	1,060	41.0	0.15	1.10	1.5	13.0	1.60	0.60	0.80	0.50
Distillers dried grain with solubles	1,180	27.0	0.12	0.68	7.5	9.0	0.80	0.45	0.32	0.20
Fish meal, anchovia	1,150	63.5	3.60	2.40	4.0	1.0	4.70	1.88	0.56	0.69
Fish meal, menhaden	1,165	61.0	4.90	2.80	9.4	1.0	4.50	1.67	0.52	0.62
Fish solubles (50 percent solids)	740	31.0	0.10	0.50	4.0	0.5	1.50	0.90	0.21	0.12
Linseed meal (solvent)	820	33.0	0.35	0.75	0.5	9.5	1.20	0.63	0.66	0.48
Meat and bone meal	800	50.0	8.10	4.10	8.6	2.8	2.60	0.65	0.60	0.26
Milo, grain sorghum	1,357	9.0	0.02	0.27	2.5	2.7	0.20	0.16	0.19	0.12
Molasses, cane	1,060	3.0	0.50	0.05	—	—	—	—	—	—
Oats	1,064	12.0	0.10	0.33	4.0	12.0	0.50	0.19	0.25	0.15
Oat groats, rolled	1,429	15.0	0.07	0.44	5.5	4.5	0.60	0.20	0.20	0.18
Skim milk, dried	1,545	33.0	1.25	1.00	0.5	—	2.70	0.80	0.40	0.45
Soybean meal (solvent, hulled)	1,380	48.5	0.20	0.65	0.5	3.0	3.30	0.68	0.73	0.68
Soybean meal (solvent)	1,224	44.0	0.25	0.60	0.5	7.0	3.00	0.63	0.67	0.63
Soybeans (whole, cooked)	1,570	37.0	0.25	0.58	17.5	5.0	2.40	0.51	0.54	0.55
Sugar	1,383	—	—	—	—	—	—	—	—	—
Tankage	990	60.0	4.60	2.50	6.4	2.0	4.20	0.66	0.38	0.65
Wheat (hard)	1,416	12.5	0.05	0.35	1.5	2.4	0.32	0.20	0.32	0.15
Wheat bran	893	14.5	0.10	1.15	3.0	11.0	0.58	0.20	0.35	0.27
Wheat midds	1,130	15.5	0.05	0.80	4.0	7.0	0.80	0.27	0.38	0.23
Whey (dried)	1,446	12.0	0.90	0.70	0.5	—	0.90	0.18	0.27	0.14
Yeast (brewers, dried)	1,050	45.0	0.10	1.40	1.0	3.0	3.40	0.70	0.50	0.50

Dicalcium phosphate—26% Ca, 18.5% P; steamed bone meal—24% Ca, 12.0% P; ground limestone—38% Ca, 0.0% P; defluorinated rock phosphate—34% Ca, 14.5% P; defluorinated treble phosphate—30% Ca, 18% P; monocalcium phosphate—25% Ca, 21% P.

Source: *Life Cycle Swine Nutrition,* Iowa State University, 1974.

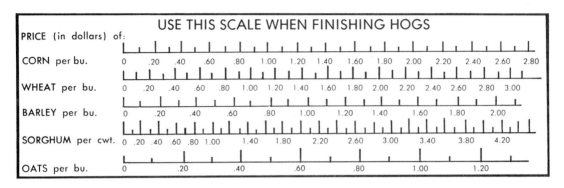

FIGURE 4-11. Feed grain substitution scale (*Scale copyrighted by Leonard W. Schruben, Kansas State University.*)

Protein Feed Price Comparisons

The costs of crude protein in protein feeds commonly fed to hogs at various prices are shown in Table 4-7.

Suggested Range in Amounts of Feeds in Rations

The amounts of various feeds that have been used successfully in swine rations during the several stages of swine production are shown in Table 4-8 (pages 98 and 99). These suggestions may be very helpful in deciding upon the kinds and amounts of various feeds to include in the ration. Note the differences in the recommendations for the various stages in the life cycle.

Commercial Mixed Feeds

There are many companies engaged in the business of producing hog feeds. When questions arise as to the value of one company's product as compared to that of another, we must consider the reliability of the manufacturer and the success that feeders have had with his products. A study of the analysis given on the container may be helpful. Most states permit feed companies to sell under what is known as the *closed formula, the open formula*, or both.

SUMMARY

Feed costs represent about 60 to 70 percent of the total cost of producing pork. The ration fed determines to a large extent the health of the animals, their rate of gain, their productivity in breeding, their feed efficiency, the type of carcasses produced, and the profit from the swine enterprise.

Hogs need proteins to develop muscles, body tissues, and offspring. They need carbohydrates and fats to provide heat and energy, and to produce lard. Minerals are needed to develop bone, muscle, and teeth. Minerals are also needed in the blood. Vitamins and antibiotics are needed for the hogs to use feed effectively.

The nutritional needs of swine vary with the age and stage of production. Sows need rations that contain from 11 to 13 percent protein and are fairly high in minerals and vitamins. Prestarter rations for pigs should contain about 20 to 24 percent protein, 0.7 percent calcium, 0.6 percent phosphorus, 2,000 units of vitamin A, and 200 units of vitamin D per pound, and large amounts of the B vitamins and antibiotics.

T A B L E 4-7 COST PER POUND OF CRUDE PROTEIN AT VARIOUS PRICES

Price Per Ton	Soybean Meal 44% (Cents)	Soybean Meal 41% (Cents)	Cottonseed Meal 42% (Cents)	Linseed Meal 35% (Cents)	Linseed Meal 31% (Cents)	Peanut Oil Meal 45% (Cents)	Corn Gluten Meal 43% (Cents)	Tankage 60% (Cents)	Meat Scraps 50% (Cents)	Fish Meal 63% (Cents)
$ 50	5.7	6.1	6.0	7.1	8.1	5.6	5.8	4.2	5.0	4.0
60	6.8	7.3	7.1	8.6	9.7	6.7	7.0	5.0	6.0	4.8
70	8.0	8.5	8.3	10.0	11.3	7.8	8.1	5.8	7.0	5.6
80	9.1	9.8	9.5	11.4	12.9	8.9	9.3	6.7	8.0	6.3
90	10.2	11.0	10.7	12.9	14.5	10.0	10.5	7.5	9.0	7.1
100	11.4	12.2	11.9	14.3	16.1	11.1	11.6	8.3	10.0	7.9
110	12.5	13.4	13.1	15.7	17.7	12.2	12.8	9.2	11.0	8.7
120	13.6	14.6	14.3	17.1	19.4	13.3	14.0	10.0	12.0	9.5
130	14.8	15.8	15.4	18.6	21.0	14.4	15.2	10.8	13.0	10.4
140	16.0	17.0	16.6	20.0	22.6	15.6	16.2	11.6	14.0	11.2
150	17.1	18.3	17.8	21.4	24.2	16.7	17.4	12.5	15.0	11.9
160	18.2	19.6	19.0	22.8	25.8	17.8	18.6	13.4	16.0	12.6
170	19.3	20.8	20.2	24.3	27.4	18.9	19.8	14.2	17.0	13.4
180	20.4	22.0	21.4	25.8	29.0	20.0	21.0	15.0	18.0	14.2
190	21.6	23.2	22.6	27.2	30.6	21.1	22.1	15.8	19.0	15.0
200	22.8	24.4	23.8	28.6	32.2	22.2	23.2	16.6	20.0	15.8
210	24.9	25.6	25.0	30.0	33.8	23.3	24.4	17.5	21.0	16.6
220	25.0	26.8	26.2	31.4	35.4	24.4	25.6	18.4	22.0	17.4
230	26.1	28.0	27.4	32.8	37.1	25.5	26.8	19.2	23.0	18.2
240	27.2	29.2	28.6	34.2	38.8	26.6	28.0	20.0	24.0	19.0

T A B L E   4-8   SUGGESTED RANGE OF COMMONLY USED INGREDIENTS THAT HAVE PRODUCED SATISFACTORY RESULTS

Percent Protein	Ingredients	Percent of Complete Ration					Percent of Supplement	Remarks
		Gestation	Lactation	Starter	Growing	Finishing		
13-17	Alfalfa meal	10-50	5-10	0	0-5	0-5	5-25	Good source of carotene and B vit., unpalatable to baby pigs
11.5	Barley (48 lb./bu.)	25-70	50-70	5-25	50-70	60-90	—	Corn substitute
80	Blood meal	1-3	1-3	0	1-3	1-3	—	Availability
8.8	Corn, yellow (56 lb./bu.)	25-70	35-80	10-60	60-80	75-95	—	High-energy, palatable, available
7	Corn and cob meal	20-40	0	0	0	10	—	Bulky, low energy
41	Cottonseed meal (solv.)	0-10	0-5	0	0-5	0-5	5-20	Gossypol toxicity problem. Low in lysine
27	Distillers' dried solubles (corn)	0-5	0-5	2-5	0-5	0-5	10-20	High in B vit. Source of "unidentified fermentation factors"
57-70	Fish meal	2-10	2-10	0-5	2-10	2-5	5-30	Availability, cost limitations, good amino acid make-up
32	Fish solubles (50 percent solids)	2-3	2-3	2-3	2-3	2-3	2-5	Excellent source of "unidentified fish factor"
32	Linseed meal	0-5	0-5	2-5	2-5	2-5	5-20	Low in lysine
50	Meat and bone scraps	0-10	0-10	0-5	0-5	0-5	5-30	Low in tryptophan
8.8	Grain sorghum	25-70	50-70	5-35	50-70	60-90	—	Corn substitute. Lacks vitamin A

Ingredient								Remarks
Molasses (cane, 11.7 lb./gal.)	3	0-20	0-10	0-5	0-10	5-10	0-10	Used for energy, better physical appearance and harder pellets
Oats (32 lb./bu.)	12	10-40	10-15	0	5-20	5-20	—	—
Oats, rolled or oat groats	16.0	—	—	0-40	0-10	—	—	Palatable, cost limitations
Skim milk, dried	33	0	0	0-40	0	0	—	Excellent protein quality, palatable but cost limitations
Soybean meal	44-50	10-22	10-22	10-25	10-20	5-16	50-80	Good amino acid makeup. Low in methionine, palatable
Tankage	60	0-10	0-10	—	0-5	0-5	5-30	Low in tryptophan and isoleucine
Wheat (60 lb./bu.)	12.7	25-90	25-90	5-35	60-80	70-90	—	Corn subs., cost, no vitamin A
Wheat bran	15	5-30	5-20	0	0-5	0-5	20 (max.)	Bulky, high fiber, laxative
Wheat midds	16	5-20	5-20	0-5	5-30	5-30	—	Low energy, partial substitute for grain
Whey, dried whole cheese	12	0-5	0-10	0-20	0-5	0-5	5	Lactose is carbohydrate "of choice" for baby pigs. Unidentified whey factor

Note: In using the above figures, consideration must be given to quantity of ingredients used in rations of a similar nature and composition.
Source: *Life Cycle Swine Nutrition*, AS-90 Iowa State University, 1968.

Pig starters should contain 18 to 20 percent protein, and the same amounts of minerals, vitamins A and D, antibiotics, and nearly the same amounts of B vitamins recommended for prestarter or milk replacer rations.

The protein content of rations for pigs over 40 pounds in weight is decreased gradually as the pigs grow heavier, from 14 percent at 40 pounds to 12 to 14 percent for 120- to 240-pound pigs.

Feeds are divided into two groups: roughages and concentrates. Roughages are those feeds relatively high in fiber. Concentrates are the low-fiber feeds.

Protein concentrates are feeds high in protein; they are classified as vegetable or animal proteins depending upon the material from which they are made.

Tankage, meat scraps, meat and bone scraps, fish meal, milk, and milk products are sources of animal protein.

The vegetable proteins are supplied in soybean oil meal, linseed meal, cottonseed meal, and other feeds derived from plants. Legumes when used as feed for older animals will replace much of the protein available from other sources.

Three pounds of synthetic lysine containing 98 percent lysine and 97 pounds of corn will replace 100 pounds of soybean meal in a ton of ration. High-lysine corn contains 1 to 3 percent more protein and 0.03 to 0.22 percent more lysine than does normal corn.

Older animals receiving rations consisting of legumes, proteins, and grains seldom need added vitamins, except possibly B_{12}. However, vitamin concentrates are usually essential for best results in feedng young animals. The vitamin concentrates are found in such feeds as fish liver oil (vitamins A and D) and wheat germ oil (vitamin E). The B complex group may be purchased in a vitamin premix available in most localities.

Antibiotics have an important place in livestock feeding and are provided separately or in commercial ready-mixed feeds.

In selecting feeds to include in swine rations it is necessary to consider the following factors: (1) availability, (2) cost, (3) nutritive value, (4) palatability, and (5) ease of feeding.

QUESTIONS

1   What is meant by the term *nutrient?*

2   List the classes of nutrients.

3   What is the main function of each nutrient class?

4   How do fats and carbohydrates differ in their ability to produce energy and fatty tissue?

5   What are the major minerals?

6   Why are trace minerals so called?

7   Why have antibiotics become important in certain livestock feeds?

8   What is meant by *digestible nutrient?*

9   To what part of a feed does the term *nitrogen-free extract* refer?

10   What is the difference between a ration and a balanced ration?

11   How do roughages and concentrates differ?

12   What are protein concentrates?

13 What is the chief difference between animal and vegetable proteins?

14 List the feeds commonly fed to hogs for their carbohydrate content.

15 List some of the common animal protein feeds.

16 What are some popular vegetable protein feeds?

17 What vitamins are most likely to be lacking in swine rations?

18 Explain how vitamins may be added to a ration when the ordinary feeds fail to provide essential amounts.

19 What materials may be used in mixing a mineral supplement for hogs?

20 Give two common methods of supplying antibiotics in the ration.

21 What are the differences in the nutritive requirements of sows during gestation, 5- to 12-pound pigs, 12- to 40-pound pigs, 40- to 120-pound pigs, and 120- to 240-pound pigs?

22 How may triticale be used in swine growing rations?

23 Of what value is high-lysine corn in swine rations?

24 Explain how synthetic forms of lysine might be used in swine rations?

25 What factors must be considered in selecting feeds in ration formulation? Explain.

REFERENCES

Ackerman, C. W., and D. L. Handlin, *Swine Feeding Suggestions,* Circular 509. Clemson, S.C.: Clemson University, 1972.

Holden, Palmer, V. C. Speer, and E. J. Stevermer, *Life Cycle Swine Nutrition,* Pm. 489. Ames, Ia.: Iowa State University, 1974.

Krider, J. L., and W. E. Carroll, *Swine Production,* 4th ed. New York, N.Y.: McGraw-Hill Book Company, 1971.

Miller, E. C., E. R. Miller, and D. E. Ulrey, *Swine Feeds and Feeding,* Ext. Bul. 537. East Lansing, Mich.: Michigan State University, 1970.

Morrison, Frank B., *Feeds and Feeding,* 23rd ed. Clinton, Ia.: The Morrison Publishing Company, 1967.

Moyer, Wendell, and B. A. Koch, *Swine Nutrition,* C-333 Revised. Manhattan, Kan.: Kansas State University, 1967.

National Academy of Sciences, National Research Council, *Nutrient Requirements of Domestic Animals, Number 2; Nutrient Requirements of Swine,* 1973.

Stevermer, E. J., and Palmer Holden, *Feeding High-Lysine Corn to Swine,* As. 385. Ames, Ia.: Iowa State University, 1973.

Stevens, Vernon, and Wm. G. Luce, *Formulating Swine Rations,* No. 3501. Stillwater, Okla.: Oklahoma State University, 1972.

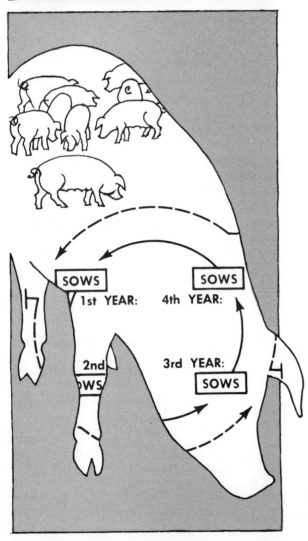

5 REPRODUCTION AND BREEDING MANAGEMENT

Much of the improvement in prolificness, body conformation, carcass quality, feed efficiency, and rate of gain in swine has come about through the use of carefully planned breeding programs. Some animals will fit into certain breeding programs but not into others. The "nicking ability" of breeds and lines in swine breeding is very important. The pedigree and ancestry of each animal must also be considered. The breeding of swine requires careful planning based upon sound principles of reproduction and inheritance.

The goal in swine breeding is to bring together and mate individual animals which will produce large litters of strong, healthy pigs, which in turn will grow rapidly on a minimum of feed and produce desirable breeding stock and carcasses.

We present in this chapter brief descriptions of (1) reproduction in swine, (2) laws of inheritance, (3) systems of breeding, (4) multiple farrowing, (5) ovulation control, (6) artificial insemination, and (7) management of the breeding herd.

REPRODUCTION IN SWINE

Reproduction is the process by which new individuals are produced in plant and animal life. The process begins when the female germ cell is fertilized by the male germ cell. The female cell is called the ovum, or egg, and the male cell is called the sperm. This fertilization process follows closely the breeding of the female by the male.

The egg or ovum contains the hereditary materials of the dam, and the sperm contains the materials of inheritance contributed by the sire. The contributions of the sire and dam, as far as inheritance is concerned, are equal.

Female Reproductive Organs

As shown in Figure 5-1, the reproductive system of the sow consists of the ovaries, the

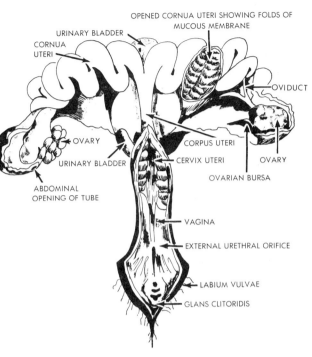

Figure labels:
CORNUA UTERI
URINARY BLADDER
OPENED CORNUA UTERI SHOWING FOLDS OF MUCOUS MEMBRANE
OVIDUCT
OVARY
URINARY BLADDER
CORPUS UTERI
CERVIX UTERI
OVARY
OVARIAN BURSA
ABDOMINAL OPENING OF TUBE
VAGINA
EXTERNAL URETHRAL ORIFICE
LABIUM VULVAE
GLANS CLITORIDIS

FIGURE 5-1. Reproductive organs of a sow. (*After a drawing by David C. Opheim.*)

oviducts, the uterus, the cervix, the vagina, and the vulva.

Ovaries. These two glandular organs are located in the sublumbar region and produce the eggs. As the eggs mature, they are dropped into the oviduct. The process is called ovulation.

The heat period usually occurs every 21 days, but the interval may vary from 18 to 24 days. Ovulation—the liberation of the eggs—usually occurs at about the end of the first day of the heat period. Each follicle in the ovary contains one egg. Gilts will produce from 10 to 15 and sows an average of 17 eggs during each heat period.

Breeding should be timed so that the sperm is in the oviduct when ovulation occurs. It normally requires several hours for the sperm to reach the ovum. It is thought that sperms can live in the female for about 15 hours.

Oviducts. These tubes lead from the ovaries to the horns of the uterus. The fertilization of the eggs usually takes place near the upper end of the oviduct. Several billion sperms are deposited by the boar in one service. Only one sperm can fertilize each egg.

Uterus. The fertilized eggs move from the oviduct into the uterus and become attached to the wall. The fertilized eggs develop in the uterus.

Cervix. The cervix connects the uterus with the vagina. At the time of mating, the boar deposits semen in the cervix. The sperms then move from the cervix through the uterus to the oviducts.

Vagina. The vagina connects the vulva and the cervix.

Vulva. Both the urinary and reproductive organs of the female terminate in the vulva.

Male Reproductive Organs

The reproductive system of the male consists of (1) the testicles, (2) the sperm ducts, (3) the seminal vesicles, (4) the prostate, (5) the Cowper's glands, (6) the urethra, and (7) the penis.

Testicles. Sperm cells are produced in the two testicles, which are suspended in the scrotum. The testicles also produce a hormone called testosterone. This hormone is released into the blood stream, which distributes it to other parts of the body. Testosterone is the hormone that causes body changes making the males different from the females.

Sperm ducts. These tubes connect the testicles with the urethra. Sperms pass through and may be stored at the upper end of these tubes.

Seminal vesicles. These glands open to the urethra and secrete a fluid.

Prostate. The prostate gland is located near the bladder and the urethra. It produces a secretion that becomes a part of the seminal fluid.

Cowper's glands. These glands secrete a fluid that precedes the passage of the sperm cells down the urethra.

Urethra. This long tube extends from the bladder to the penis and carries both urine and semen.

Penis. This organ deposits the sperm cells within the female reproductive system.

LAWS OF INHERITANCE

Chromosomes and Genes

Each germ cell (sperm and egg) contains chromosomes, which in turn carry genes. The genes determine the characteristics to be found in the individual.

The number of chromosomes in the nucleus of a cell is constant. They occur in pairs. In swine there are 19 pairs.

Dominant and Recessive Characteristics

When a Yorkshire boar is mated to a Poland China sow, the offspring are white. Certain characteristics are dominant, whereas others are recessive. When dominant and recessive characteristics are brought together, the progeny will possess the dominant characteristics but will produce in the next generation some animals that show the dominant characteristics and others that possess the appearance of the recessive ones.

White hair color in hogs is dominant to black and to red. Many of the factors which positively affect vigor, rate of gain, feed efficiency, and carcass quality are thought to be dominant, while those which affect these characteristics in a negative manner are thought to be recessive. Navel hernias appear when two recessive genes are brought together.

FIGURE 5-2. A growthy lot of crossbred pigs being finished for market. A high percentage of all pigs marketed are crossbred. (*Courtesy Kent Feeds.*)

Heritability in Swine

Through selection and breeding, hog breeders try to improve prolificness, rate of gain, feed efficiency, and carcass quality. The traits affecting these factors are not inherited equally. Carcass characteristics are more highly hereditary than other economically important traits. Litter size and weaning weight of pigs have low heritabilities. Improvement in these traits can be brought about more readily through improved feeding and management than through selection and breeding. Carcass length, percentage of ham, backfat thickness, and loin eye area are medium to high in heritability and can be improved fairly rapidly through selection and breeding.

Presented in Table 5-1 is a summary of approximate heritability figures for certain traits in hogs.

Hybrid

An animal is considered a hybrid for any one characteristic when it possesses one dominant and recessive gene. When a hybrid

TABLE 5-1   APPROXIMATE HERITABILITY OF SWINE TRAITS

Level of Heritability	Characteristic	Heritability (%)
High	Carcass length	61
	Percent ham (based on carcass weight)	58
	Percent fat cuts (based on carcass weight)	60
	Backfat thickness	50
	Loin eye area	50
	Percent ham and loin (based on carcass weight)	65
Medium	Meat tenderness	30
	Meat color	30
	Marbling in loin	30
	Firmness of meat	30
	Feed efficiency	38
	Growth rate (weaning to market)	30
	Five-month weight	25
	Conformation score	26
Low	Weaning weight	17
	Number farrowed	10
	Number weaned	19
	Birth weight	5

FIGURE 5-3. These pigs are the result of several generations of the use of hybrid boars. (*Courtesy Farmers Hybrid Hogs.*)

is crossed with another hybrid, about 75 percent of the progeny will show the characteristics of the dominant, and 25 percent will show recessive characteristics. Of 75 percent of the progeny that show dominance, only about one-third are pure dominant for the one characteristic. The other two-thirds have the appearance of the dominant but are hybrid. Those that show recessive characteristics are pure recessives.

The following example illustrates the inheritance in crossing a purebred Poland China boar and a purebred Yorkshire sow. The Yorkshire carries two genes, one on each chromosome, for the dominant white color. The Poland China boar carries two genes for the black color, which is recessive.

In the reproductive processes which take place in the testicles of the male and in the ovaries of the female, a reduction process occurs. The chromosomes line up in pairs, and only one of each pair is included in the reproductive cell, the sperm, or the female egg cell. When the two cells unite, the fertilized egg has 19 pairs of chromosomes with one or more genes on each chromosome affecting color of hair.

In this case all offspring of this cross will be white, for each will carry one dominant gene (W) in the reproductive cell. The other gene, however, will be for the black color, which is recessive (w). The combination would appear in a diagram as Ww. In mating progeny, each hybrid for the color characteristic, only one of each pair of chromosomes is included in the reproductive cell due to the reduction process. In this case, when the reduction process takes place, there is one chance in four of the W gene uniting with the other W gene, one chance in four of the w gene uniting with the w gene, and two chances in four that the W gene will unite with the w gene. One-fourth of the progeny will be dominant white (WW), one-half of the progeny will be hybrid white (Ww), and one-fourth will be recessive black (ww).

Hybrid Vigor

The crossing of two superior animals of different breeds usually results in increased growth rate, increased efficiency of fertilization, improvement in body conformation, and increased production.

Hereditary Defects in Swine

Two types of defects occur in swine. *Lethal* defects result in the death of pigs within a few hours or days. Pigs with cleft palates cannot suck. They starve. Pigs with no anus cannot eliminate solid wastes. They die. Some pigs have paralysis of hind quarters. They usually die. These defects cannot be transmitted by the animals affected because they do not live. It is possible, however, for some of these defects to be transmitted by littermates that carry the defect.

Several hereditary defects in swine do not cause death but destroy or impair the economic value of the animal. These include hernia (rupture), cryptochid (ridgling), swirl, and inverted nipple. These defects are transmitted by recessive genes. Thus the defect appears only when both the sire and dam carry the recessive gene responsible for the defect.

There are four ways of ridding the herd of these undesirable genes:

1. Eliminate from the breeding herd boars and sows exhibiting any of the defects.
2. Market boars or sows that have produced progeny with defects.
3. Do not keep for breeding purposes offspring from boars or sows that have transmitted these defects.
4. Market all close relatives of animals having or transmitting the defects.

Heterosis, or hybrid vigor, is the term applied to the increase in vigor and performance resulting when two animals of unrelated breeds are crossed. The genes producing vigor are dominant to those producing a lack of vigor. By crossing breeds, a larger number of dominant genes are brought together in the progeny than are involved in breeding animals of the same breed. The problem in crossbreeding is to determine the crosses that produce the most hybrid vigor.

BREEDING SYSTEMS

Most growers produce hogs for slaughter. It is estimated that less than 2 percent of all growers are raising purebred hogs to be sold for breeding purposes. More than 98 percent of our producers produce only market hogs.

The breeding programs followed by producers of purebred hogs are usually quite different from those followed by producers of market hogs. These breeders mate purebred boars and sows of the same breed. Some breeders do inbreeding or line breeding. Most breeders of commercial hogs follow upgrading, crossbreeding, or crisscrossing methods, which may involve two or more breeds. A description of each of the various swine breeding systems follows.

Upgrading

The system of mating purebred boars with grade sows of the same breed is called *upgrading*. A *purebred* is an animal that is registered or eligible for registration by a breed association. A *grade* animal is one whose sire is a purebred but whose dam or mother is of the same breed but is not eligible for registration.

Purebred Breeding

In the breeding of purebreds, purebred boars are mated to purebred sows of the same breed. The mating of purebred Hampshire boars with purebred Hampshire sows is an example of purebred breeding. Breed-

ers of purebreds must produce boars for use by their neighbors who raise market hogs; they must also produce purebreds for use by other breeders.

According to the National Society of Livestock Record Associations, there were 16,795 active breeders of swine in the nation in 1973 who registered purebred animals.

Crossbreeding

The mating of purebred or inbred boars of one breed with purebred or grade sows of another breed is called *crossbreeding*. About 90 percent of commercial hog producers are crossbreeding or are crisscrossing their hogs. An example of this system is the mating of a Duroc boar with Chester White sows.

In tests conducted at Iowa State University over a ten-year period, it was found

FIGURE 5-4. LeRoy Faint of Lisbon, Iowa with an outstanding Berkshire X Duroc crossbred litter. (*Courtesy American Berkshire Association.*)

T A B L E 5-2   SUPERIORITY OF SUBSEQUENT CROSSES, EXPRESSED AS A PERCENT-
AGE OF AVERAGE PUREBRED PERFORMANCE

Trait	Pure-bred	First Cross	Three-breed Crosses	Four-breed Crosses
Litter size at birth	100	101	111	113
Litter size at 8 weeks	100	107	125	126
Weaning weight	100	108	110	109
154-day weight	100	114	113	111
Litter production	100	122	141	140

Source: *Your Hog Business*, University of Illinois, 1969.

that fewer pigs were born dead among the crossbreds and slightly more of the crossbred pigs lived to weaning age than did the purebred pigs. The crossbred pigs weighed an average of nearly four pounds heavier at weaning time than did the purebreds. The crossbreds gained more rapidly from weaning to market and reached market weight of 225 pounds about ten days earlier than did the purebreds. Twenty-five to 30 more pounds of feed were required to bring the purebreds up to 225 pounds than were required for the crossbreds.

The crossbred sows were good mothers when bred back to a boar of either of the parent breeds or to a boar of a third breed. The pigs produced compared favorably with the first-cross pigs in rate of gain and in economy of gain.

Presented in Table 5-2 is a summary of Illinois tests indicating the superiority of subsequent crosses in using purebred stock. Note that little was gained by using a four-breed cross over a three-breed cross.

In a 1,700 litter study of crossbreeding in Oklahoma, it was found that there was an 11 percent increase in litter size from crossbred sows. The crossbred pigs were 10 percent heavier at birth, and crossbred sows farrowed 3 percent heavier pigs than pure-

bred sows. There was a 13 percent lower death loss of pigs between birth and weaning from the crossbred pigs. At weaning time, the crossbred litters weighed 6 percent more, and the litters from crossbred sows weighed 15 percent heavier than those from purebred sows. There was little difference between the purebred and crossbred pigs in pounds of feed required to produce a pound of gain. The crossbred pigs, however, reached market weight 7 days earlier than the purebreds. Meatiness was not greatly improved in the Oklahoma tests through crossbreeding. Hampshire, Duroc, and Beltsville No. 1 lines were used in the experiment.

Presented in Table 5-3 is a summary of the advantages of crossbreeding in commercial hog production in Oklahoma.

It has been found that using three breeds in a crossing program results in greater hybrid vigor than when two breeds are used. Recent tests in Oklahoma involving the use of purebred boars of one breed on purebred sows of another breed in one group, and purebred boars mated to crossbred sows representing two other breeds in the second group, indicated much improved litter size at farrowing time, weight of litter at birth, and litter weight at weaning time by using crossbred gilts. There was no advantage of

T A B L E 5-3   COMPARISON OF TWO- AND THREE-LINE CROSSES WITH PUREBREDS

	Two-line Crosses (Duroc & Beltsville)	Purebred (Duroc & Beltsville)	Three-line Crosses	
			Crossbred Dam	Purebred Dam
Litter size	9.2	10.2	10.4	9.4
Birth weight, lbs.	3.1	2.8	3.1	3.0
Pigs weaned/litter	7.3	7.1	7.8	7.4
Pig weaning weight, lbs.	43.7	42.2	41.6	38.4
Litter weaning weight, lbs.	319	300	325	284
Percent survival	80	`67	75	76

Source: Oklahoma State University, 1968.

T A B L E 5-4   APPROXIMATE ADVANTAGE OF CROSSBREDS OVER PUREBREDS

Trait	Percent Advantage over Purebreds	
	2-Breed Cross (Both boars and sows purebred)	3-Breed Cross (Boars purebred and crossbred sows)
PRE-WEANING TRAITS		
Litter size at farrowing	0	10
Pig birth weight	10	10
Litter birth weight	5	15
Survival rate to weaning	5	5
Pig weaning weight	negligible	5
Litter weaning weight	5	15
POST-WEANING TRAITS		
Daily gain	5	5
Feed efficiency	0	0
Carcass merit	negligible	negligible

Source: Oklahoma State University, 1971.

the three-breed cross in feed efficiency, carcass quality, or daily gain.

A summary of the Oklahoma tests is presented in Table 5-4.

Crisscrossing

In this system purebred boars of one breed are mated with grade or purebred sows of another breed (as in crossbreeding); then the gilts produced from the first mating are mated with a boar of the same breed as the original sows. The gilts produced from this cross are mated the next year with a boar of the same breed as the boar used in making the first cross. Boars of the parent breeds are then used in alternate years thereafter. A farmer who breeds a group of Spot gilts to a Duroc boar and then alternates the use of boars of these two breeds in the following years is using the crisscross method of breeding.

TABLE 5-5   PERFORMANCE OF PUREBREDS, FIRST-CROSSBREDS, CRISSCROSS-
BREDS, AND ROTATIONAL-CROSSBREDS AT MINNESOTA

Breeding Group	No. of Litters	No. Pigs Farrowed	No. Pigs Weaned	Survival %	Wt./Pig Weaned	Wt./Litter Weaned
Purebreds*	76	8.3	5.6	67	28	159
First-cross	45	9.2	6.0	65	33	196
Crisscross	16	8.1	6.3	78	36	225
Rotational-cross	24	9.9	7.7	78	33	254

*Poland China, Chester White, Duroc, and Yorkshire breeds.
Source: Minnesota Agricultural Experiment Station Bulletin No. 320.

Rotation Breeding

Swine specialists in several states are now recommending the use of boars of three or four breeds of hogs in rotation. This method is especially recommended when inbred boars are used in producing commercial hogs. The use of three or four carefully selected breeds maintains the hybrid vigor desired in hog production. The boars used in this system of breeding may be purebreds, inbreds, or crosslines. Crosslines are the progeny resulting from the crossing of two inbred lines of the same breed.

Some breeders prefer to use just three breeds in the four years. A rotation involving Duroc, Poland China, Duroc, and Hampshire boars has produced excellent results. A diagram of rotation breeding is shown in Figure 5-5.

The value of rotation is shown in Table 5-5. The first crossbreds from this system

FIGURE 5-5. An example of a rotation system of breeding using Poland China, Yorkshire, Duroc, and Hampshire boars. (*After a drawing by David C. Opheim.*)

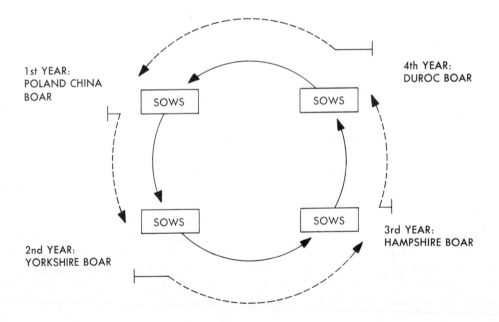

1st YEAR: POLAND CHINA BOAR

4th YEAR: DUROC BOAR

2nd YEAR: YORKSHIRE BOAR

3rd YEAR: HAMPSHIRE BOAR

FIGURE 5-6. A healthy litter resulting from a four-breed rotation breeding program. (*Courtesy Walnut Grove Products Co., Inc.*)

exceeded the purebreds in number of pigs farrowed and weaned, and in individual and litter weights at weaning time. The crisscrossbred and rotation-crossbred pigs—all produced by crossbred sows—exceeded both the

T A B L E 5-6   ADVANTAGES OF THREE–WAY CROSSBREEDING OVER PUREBRED

Factor	Percentage Increase
Litter size	12
Survival	14
154-day pig weight	14
154-day litter weight	40
Feed efficiency	slight
Meatiness	none

Source: University of Nebraska.

purebreds and the first-crossbreds in performance. The superiority of the last two groups over the first-crossbreds indicates the importance of the crossbred sow.

There is some evidence that the benefits in crossbreeding may be greater when proven, highly efficient lines are used than when average animals are crossed. Recent tests in Nebraska indicated that a three-way crossbreeding program resulted in 12 percent larger litters, 14 percent increase in survival rate, and 40 percent heavier 154-day litter weights than when purebreds were mated to other purebreds of the same breed. The Nebraska tests are summarized in Table 5-6.

Inbreeding

The mating of boars with sows of the same breed which are closely related is called *inbreeding.* The mating of brothers with sisters and of sires with daughters are examples of inbreeding. After several years of inbreeding it is possible to obtain lines which produce uniform offspring. The crossing of two or more of the inbred lines results in hybrid vigor which is greater than that obtained when two noninbred purebred lines are crossed.

Line Breeding

Line breeding involves the mating of purebred boars of one line or family with sows of the same line that are not so closely related as those used in inbreeding. The mating of second cousins is an example of line breeding.

Crossline Breeding

When this system is used, inbred boars of one line are mated to inbred sows of another line of the same breed. The crossing of two distinct lines of a breed produces a type of hybrid vigor similar to, but in smaller amounts than, that resulting from the crossing of two breeds.

Hybrid vigor is increased from about 12 to 15 percent as a result of the crossing of two inbred lines. By crossing four lines, as is done in producing some hybrid corn, the hybrid vigor is further increased. By crossing lines of different breeds, it is possible to obtain higher levels of performance than can be obtained by crossing two lines of one breed.

MULTIPLE FARROWING

Packers and marketing specialists have encouraged farmers to follow a multiple-farrowing system in order to make economical use of the capital invested in breeding stock and equipment and to distribute the marketing of hogs throughout the year.

Nearly 52 percent of the 1973 U.S. pig crop was farrowed during the December to May period, whereas 48 percent was farrowed from May through November. In 1972, 51 percent of the nation's pigs were farrowed during the December to May period, and 49 percent were farrowed from June through November.

The percentages for Iowa were the same as for the U.S. in 1973. Fifty-one percent of the pigs farrowed in Missouri, Illinois, and Wisconsin in 1972 were farrowed during the June through November period. In Ohio, 52 percent of the pigs were farrowed from June through November.

The farrowing system that will be most profitable on your farm will depend upon the size of the hog enterprise and its relationship to the other enterprises on the farm. On most farms the advantages of the two-litter and multiple-farrowing systems outweigh the advantages of the one-litter system.

A study was conducted by the South Dakota State University to compare one- and two-litter systems. The income and net return data showed an advantage of farrowing two litters per sow. This was particularly true of the system which farrows the first litter in the fall and the second in the spring.

Two Litters per Sow Multiple-farrowing Program

It has been pointed out that most sows are bred to farrow during two periods of the year—the spring period from February to May, and the fall period from August to November. As a result there have been seasonal runs on the markets which resulted in lower prices. This problem is being solved by a large number of farmers who split the sow herds into groups and have each group farrow twice each year. If breeders would divide their herds and have pigs farrowed at four or more times during the year, the wide fluctuations in seasonal hog prices could, at least inpart, be avoided.

More Than Two Litters per Sow per Year

The use of prestarter pig rations has made it possible to wean pigs when from ten to twenty days of age. As a result it is possible to rebreed sows and raise more than two litters from each sow each year.

Multiple farrowing permits repeated use of some or all of the specialized swine farrowing and feeding equipment as well as housing. A Purdue study indicated a 22 percent reduction in annual use-costs of buildings and equipment per sow and two litters

when going from one group of sows farrowing twice a year to three groups of sows farrowing twice a year. The study showed only a 4 percent reduction when going from a three-group to six-group plan with sows farrowing each month.

Multiple farrowing provides for more even distribution of labor and reduces the uncertainty of speculation associated with a one-litter system. The average selling price of hogs farrowed monthly or each two or three months may be lower than when hogs are farrowed in selected months in order to hit the peak market.

In Illinois tests the number of pigs weaned per litter did not change when farrowings were increased to six times per year rather than two to four times. Feed per 100

FIGURE 5-7. The investment in the specialized farrowing house shown necessitates continuous use throughout the year. Six or more crops of pigs can be farrowed in these quarters each year. (*Graves Peterson photograph. Courtesy Farmstead Industries.*)

T A B L E 5-7   MULTIPLE-FARROWING PRODUCTION SCHEDULES
(Based on 4 Farrowings per Year)

	Breeding Dates		Farrowing Dates 115 Days		Weaning Dates 35 Days		Marketing Dates 150 Days	
	Set 1	Set 2	Set 1	Set 2	Set 1	Set 2	Set 1	Set 2
A	Jan. 1	Apr. 1	Apr. 25	July 24	May 30	Aug. 28	Sept. 22	Dec. 21
B	Jan. 15	Apr. 15	May 9	Aug. 7	June 13	Sept. 11	Oct. 6	Jan. 8
C	Feb. 1	May 1	May 26	Aug. 23	June 30	Sept. 27	Oct. 23	Jan. 20
D	Feb. 15	May 15	June 9	Sept. 6	July 14	Oct. 11	Nov. 6	Feb. 3
E	Mar. 1	June 1	June 23	Sept. 23	July 28	Oct. 28	Nov. 20	Feb. 20
F	Mar. 15	June 15	July 7	Oct. 7	Aug. 11	Nov. 11	Dec. 4	Mar. 6
G	Apr. 1	July 1	July 24	Oct. 23	Aug. 28	Nov. 27	Dec. 21	Mar. 22
H	Apr. 15	July 15	Aug. 7	Nov. 6	Sept. 11	Dec. 11	Jan. 8	Apr. 5
I	May 1	Aug. 1	Aug. 23	Nov. 23	Sept. 27	Dec. 28	Jan. 20	Apr. 22
J	May 15	Aug. 15	Sept. 6	Dec. 7	Oct. 11	Jan. 11	Feb. 3	May 6
K	June 1	Sept. 1	Sept. 23	Dec. 24	Oct. 28	Jan. 28	Feb. 20	May 23
L	June 15	Sept. 15	Oct. 7	Jan. 7	Nov. 11	Feb. 11	Mar. 6	June 6
M	July 1	Oct. 1	Oct. 23	Jan. 23	Nov. 27	Feb. 27	Mar. 22	June 22
N	July 15	Oct. 15	Nov. 6	Feb. 6	Dec. 11	Mar. 13	Apr. 5	July 6
O	Aug. 1	Nov. 1	Nov. 23	Feb. 23	Dec. 28	Mar. 30	Apr. 22	July 23
P	Aug. 15	Nov. 15	Dec. 7	Mar. 9	Jan. 11	Apr. 13	May 6	Aug. 6
Q	Sept. 1	Dec. 1	Dec. 24	Mar. 25	Jan. 28	Apr. 29	May 23	Aug. 22
R	Sept. 15	Dec. 15	Jan. 7	Apr. 8	Feb. 11	May 13	June 6	Sept. 5
S	Oct. 1	Jan. 1	Jan. 23	Apr. 25	Feb. 27	May 30	June 22	Sept. 22
T	Oct. 15	Jan. 15	Feb. 6	May 9	Mar. 13	June 13	July 6	Oct. 6
U	Nov. 1	Feb. 1	Feb. 23	May 26	Mar. 30	June 30	July 23	Oct. 23
V	Nov. 15	Feb. 15	Mar. 9	June 9	Apr. 13	July 14	Aug. 6	Nov. 6
W	Dec. 1	Mar. 1	Mar. 25	June 23	Apr. 29	July 28	Aug. 22	Nov. 20
X	Dec. 15	Mar. 15	Apr. 8	July 7	May 13	Aug. 11	Sept. 5	Dec. 4

Rebreeding Schedule

	Set 1	Set 2	Set 1	Set 2	Set 1	Set 2	Set 1	Set 2
A	June 3	Sept. 1	Sept. 25	Dec. 22	Nov. 20	Feb. 18	Mar. 24	June 22
B	June 17	Sept. 15	Oct. 9	Jan. 7	Dec. 4	Mar. 4	Apr. 7	July 6
C	July 4	Oct. 1	Oct. 26	Jan. 23	Dec. 21	Mar. 20	Apr. 24	July 22
D	July 18	Oct. 15	Nov. 9	Feb. 6	Jan. 4	Apr. 3	May 8	Aug. 5
E	Aug. 1	Nov. 1	Nov. 23	Feb. 23	Jan. 18	Apr. 20	May 22	Aug. 22
F	Aug. 15	Nov. 15	Dec. 7	Mar. 9	Feb. 1	May 4	June 5	Sept. 5
G	Sept. 1	Dec. 1	Dec. 24	Mar. 25	Feb. 18	May 20	June 22	Sept. 21
H	Sept. 15	Dec. 15	Jan. 7	Apr. 9	Mar. 4	June 3	July 6	Oct. 6
I	Oct. 1	Jan. 1	Jan. 23	Apr. 25	Mar. 20	June 20	July 22	Oct. 22
J	Oct. 15	Jan. 15	Feb. 6	May 9	Apr. 3	July 4	Aug. 5	Nov. 5
K	Nov. 1	Feb. 1	Feb. 23	May 26	Apr. 20	July 21	Aug. 22	Nov. 22
L	Nov. 15	Feb. 15	Mar. 9	June 9	May 4	Aug. 4	Sept. 5	Dec. 7
M	Dec. 1	Mar. 3	Mar. 25	June 25	May 20	Aug. 20	Sept. 21	Dec. 22
N	Dec. 15	Mar. 17	Apr. 9	July 9	June 3	Sept. 3	Oct. 6	Jan. 5
O	Jan. 1	Apr. 3	Apr. 25	July 26	June 20	Sept. 20	Oct. 22	Jan. 22
P	Jan. 15	Apr. 17	May 9	Aug. 9	July 4	Oct. 4	Nov. 5	Feb. 5
Q	Feb. 1	May 3	May 26	Aug. 25	July 21	Oct. 20	Nov. 22	Feb. 21
R	Feb. 15	May 17	June 9	Sept. 8	Aug. 4	Nov. 3	Dec. 7	Mar. 7
S	Mar. 3	June 3	June 25	Sept. 25	Aug. 20	Nov. 20	Dec. 22	Mar. 24
T	Mar. 17	June 17	July 9	Oct. 9	Sept. 3	Dec. 4	Jan. 5	Apr. 7
U	Apr. 3	July 4	July 26	Oct. 26	Sept. 20	Dec. 21	Jan. 22	Apr. 24
V	Apr. 17	July 18	Aug. 9	Nov. 9	Oct. 4	Jan. 4	Feb. 5	May 8
W	May 3	Aug. 1	Aug. 25	Nov. 23	Oct. 20	Jan. 18	Feb. 21	May 22
X	May 17	Aug. 15	Sept. 8	Dec. 7	Nov. 3	Feb. 1	Mar. 7	June 5

pounds of gain did increase from 419 to 433 pounds and feed cost per pound of gain increased from 9.9¢ to 10.3¢ when there were six or more farrowings each year.

The farrowing house investments per pig raised were as follows:

1 farrowing per year, $62.50 per pig
2 farrowings per year, $31.25 per pig
4 farrowings per year, $15.62 per pig
6 farrowings per year, $10.42 per pig

The finishing house investments per pig raised under the four systems of farrowing were:

1 farrowing per year, $48.00 per pig
2 farowings per year, $24.00 per pig
4 farrowings per year, $12.00 per pig
6 farrowings per year, $8.00 per pig

In Table 5-7, pages 114-5, is a multiple-farrowing production schedule based upon the use of two groups of sows, each farrowing two litters each year. The schedule is based upon the assumptions that (1) the sows will farrow 115 days after breeding, (2) the pigs will be weaned at 35 days, and (3) the pigs will be marketed at 150 days of age.

Tests conducted at the University of Wisconsin indicate that the best time to rebreed sows whose pigs have been removed is the period 8 to 16 days after farrowing. Attempts to breed sows one to four days after farrowing were unsuccessful in their tests.

In the Wisconsin tests only 6 percent of the sows were fertile during the post-partum heat period, yet 54 percent of the sows gave signs of being in heat. Approximately 75 percent of the sows were ready to breed 8 to 16 days after farrowing, and nearly 95 percent of those that were in heat were fertile.

Investment Costs Encourage Multiple Farrowing

The percentage of hogs marketed in the U.S. by large producers is increasing rapidly. It is not possible for small growers to compete with the large producers due to the cost of equipment, housing, and labor. The use of heated farrowing facilities equipped

TABLE 5-8   INVESTMENT AND ANNUAL COST FOR BUILDINGS AND EQUIPMENT IN A TWO-HOUSE CONFINEMENT BUILDING SYSTEM—SIX FARROWINGS PER YEAR, 192 LITTERS, 1973

Facility	Cost
Farrowing, 32 crates	$ 28,800.00
Nursery building	12,480.00
Finishing building	43,200.00
Sow facilities	19,200.00
Feed processing facilities	6,000.00
Lagoon	3,000.00
Manure equipment	2,500.00
Total investment	$115,180.00
Per litter	$ 600.00
Annual Costs	
Per litter	$ 108.00
Per hundredweight of pork	$ 6.35

Source: University of Illinois.

FIGURE 5-8. These gilts are being reared on pasture. They should be in excellent condition for breeding when they weigh 250 to 275 pounds. (*Courtesy Farmers Hybrid Hogs.*)

with farrowing crates, ventilating equipment, and waste removal programs involves a large investment that must be put to continuous use. Most swine growers with modern equipment farrow several times each year, thus making economical use of their investment. As a result, they are able to market their swine throughout the year.

According to data in Table 5-8, the annual cost for buildings and equipment in Illinois was estimated to be $108.00 per litter or $6.35 per hundredweight of pork when two confinement houses were used and six farrowings were made each year.

MANAGEMENT OF THE BREEDING HERD

The profit or loss from a swine breeding herd is often determined by the number of pigs weaned and marketed per sow. An average of 6 or 7 pigs marketed per sow is usually necessary to come out even.

It has been estimated that out of 100 pigs farrowed only 65 will live to be weaned, and only 55 will be marketed. Tests conducted at Iowa State University indicated that when 11 pigs were weaned per litter, the average cost of each 8-week-old pig was about $10.45. When 9 pigs were weaned the cost was $12.01 per pig. Pigs from litters of 7 weaned cost $14.52, and pigs from litters of 5 cost $17.92 at weaning time. These figures would indicate that unless we can save 7 or more pigs per sow, we should consider buying feeder pigs, or anticipate little profit from the enterprise.

The feeding and management of the herd at breeding time and during the gestation period can greatly influence the number of pigs farrowed and weaned. Good production methods will result in the following:

1. **More pigs being farrowed per sow.**
2. **Larger and healthier pigs at birth.**
3. **Fewer dead pigs, runts, and abnormal pigs per litter.**
4. **Better production of milk by the sows.**
5. **More and heavier pigs weaned per litter.**

Feeding and Care During the Breeding Season

A farmer may decide that the small litters produced are due to the lack of prolificness in his breeding stock. Usually he is wrong in making this assumption. Small litters are very often a direct result of carelessness in the feeding and management of the herd at breeding time.

Age to breed. Gilts should be bred to farrow when they are 11 to 13 months of age if they have been well grown out. The maturity of the gilt is more important than its age. Most gilts which have done well reach puberty and come in heat when they are 5 to 6 months of age. It is not a good policy to breed gilts during their first or second heat periods. Larger litters usually result from gilts bred during their third heat periods.

Gilts should be well grown out and weigh from 225 to 250 pounds at breeding time. Boars should be a little heavier if they are to receive heavy service. Some breeders prefer to use mature animals, fall boars, or early spring boars in their breeding programs. The older boars are more dependable, especially in settling mature sows.

Heat period. It is better to breed the sow during the first or second day of the heat period than during the last day. Larger litters may result from two services 24 hours apart. Sows that are not bred will usually come in heat at intervals of about three weeks.

Gestation period. The period of time between the breeding of the sow and the farrowing of the litter is known as the *gestation or pregnancy period.* The length of period varies somewhat with sows but is usually from about 112 to 115 days. Older sows usually have longer gestation periods than gilts.

Time to breed. Some farmers who have very limited facilities and rely upon the use of wooded areas and pastures, breed their sows in the fall so that they will farrow in the spring when it is warm. In the southern states they may breed in November and December. In some northern states the sows that will farrow on pasture are bred in December and January. These growers usually raise only one pig crop each year.

Most swine producers farrow two or more times each year. When two crops are farrowed, the gilts are bred to farrow in January, February, and March, and again in July, August, and September. Most multiple-farrowing programs provide for somewhat continuous use of the farrowing and nursery facilities throughout the year. Thus sows and gilts are bred at regular intervals.

Limited feeding of gilts. Flushing is impossible if the gilts approach the breeding season in a too fat condition. Gilts to be retained in the breeding herd should be removed from other hogs when they weigh 100 to 125 pounds. They should be fed a balanced growing ration but limited to about a two-thirds ration. Tests indicate that gilts full-fed up to breeding time average 7.6 pigs per litter, whereas gilts that receive only a two-thirds ration farrow 8.8 pigs per litter. Limited feeding of gilts up to flushing time can add an extra pig or two to litter size and also reduce the cost of feeding the breeding herd.

Flushing. The condition of the sow or gilt at breeding time affects the regularity of the heat period, the number of eggs or reproductive cells produced, and the con-

ception or the settling of the sow at the first service. *Flushing* is the feeding of the sows to insure their good health and a gain of from 1½ to 2 pounds per day from about 2 weeks before breeding until after they are bred. Gilts or sows that are recovering from influenza at the time of breeding seldom produce good litters. It is better to flush these sows and breed them for a later litter.

The ration to be fed during the flushing period will vary with the age and condition of the sow and with the feeds and pastures available. It is important that the ration be well balanced and fed in the proper amount. Plenty of proteins, minerals, vitamins, and forages are desirable. Gilts should usually receive 2½ to 3 pounds of feed daily per 100 pounds of live weight, whereas mature sows may need less than 2 pounds daily for each 100 pounds of live weight. Usually a ration fed each day consisting of one-half to three-quarters of a pound of a balanced protein and mineral supplement, 1 pound of oats, and enough corn (or other grains) to produce daily gains of 1 to 2 pounds will

FIGURE 5-9 (*Above*). These sows are being fed limited rations in individual feeding stalls. (*Courtesy American Cyanamid Co.*) **FIGURE 5-10** (*Below*). Although some farmers let boars run with the sows during the breeding season, most breeders of purebreds keep their boars in separate quarters. Pictured are houses and shades for Duroc boars on an Illinois farm. (*Courtesy United Duroc Swine Registry.*)

meet the needs of a sow during the flushing period. Sows and gilts should not become overfat.

A breeding experiment at the Ohio Agricultural Experiment Station showed that sows fed alfalfa before breeding produced more eggs during the critical heat period. Gilts fed a ration containing 18 percent ground alfalfa for approximately 3 weeks before breeding farrowed an average of 1.2 more live pigs per litter. Alfalfa and other legume pasture or high quality alfalfa hay should be fed to the boar and sow previous to breeding time.

In Missouri tests the daily feeding of one-half gram of antibiotic per head for 10 to 21 days at breeding time resulted in one more pig farrowed per litter and 0.8 more pigs weaned per litter. Aureomycin and Terramycin gave equal results. Similar results have been obtained at some other stations when antibiotics have been fed to sows and gilts at breeding time.

Feeding and management of the boar. Normally, if the boar is in good, thrifty condition and is well managed, he will not affect unfavorably the size of the litters which he sires. If his vitality is too low or if he is used very heavily, the sperms, or male reproductive cells, may be so weak that only part of the eggs produced by the female will be fertilized. In that case it is probable that the litters will be small.

On some farms the boar is a neglected animal. He is purchased at breeding time and placed in a pen in the hog house. He may have been used previously by other farmers. He is in poor physical condition. Quite often he is fed too little protein, mineral supplement, and vitamins. Such a boar will be low in vitality and may prove to be a poor breeder.

The boar should receive about the same kind of a ration fed to the gilts during flushing period. He should not be fattened but kept in good, thrifty condition. The ra-

tions presented in Tables 5-9 and 5-10 are recommended.

The boar should be given a lot of about one-quarter to one-half acre, which has been seeded to alfalfa, red clover, ladino clover, or some other forage crop. A movable house should be provided at one end, and he should be fed and watered at the other end of the lot. Exercise is important in caring for a boar.

Yearling and other mature boars can be mated to 50 or 60 gilts during a breeding season if properly managed. An eight-month-old boar normally should not be expected to service more than 20 to 30 sows unless he is very carefully managed.

Pen- versus lot-breeding. It is usually better to bring the boar to the sow or the sow to the boar at the time of service rather than to permit the boar to run with the sows. One good service is normally sufficient, and a record can be kept of the breeding date. If boars are available, gilts should be bred twice at 12- to 24-hour intervals. A young boar turned in with a large number of sows may be punished and become shy.

A mature boar should not be mated to more than three sows during one day, and a young boar should not serve more than two sows per day. Some farmers who practice pen-breeding bring a sow to the boar in the morning and another sow during the evening.

Ranty and inactive boars. Some boars are very active during the breeding season and pace back and forth along fences or go through the fence. They may be difficult to manage, but are excellent breeders. Some breeders place a barrow or a bred gilt with the boar with excellent results. A better practice is to keep the boar in a lot at some distance from the sows.

Other boars are inactive—quite often because of improper feeding and management—and are slow breeders. It is a good idea to try out a boar a month or so before

T A B L E 5-9   PREGESTATION, BREEDING, AND GESTATION RATIONS
(For Boars, Sows, or Gilts Being Fed 4 Pounds per Day)[1,2]

Percent Protein	Ingredient	1	2	3	4
8.9	Ground yellow corn[3]	1,740	1,645	1,750	1,675
44	Solv. soybean meal	175[4]	170[4]	90	70
17	Dehydrated alfalfa meal	—	100	—	100
50	Meat and bone meal	—	—	100	100
	Calcium carbonate (38% Ca)	10.0	10.0	5.0	—
	Dicalcium phosphate (26% Ca, 18.5% P)	45.0	45.0	25.0	25.0
	Iodized salt	12.5	12.5	12.5	12.5
	Trace mineral premix	2.5	2.5	2.5	2.5
	Vitamin premix	15.0	15.0	15.0	15.0
	Feed additives[5]	—	—	—	—
	Total	2,000	2,000	2,000	2,000

CALCULATED ANALYSIS

Protein	%	11.59	11.91	12.27	12.34
Calcium	%	0.80	0.86	0.84	0.80
Phosphorus	%	0.69	0.68	0.68	0.68
Lysine	%	0.50	0.52	0.51	0.51
Methionine	%	0.21	0.21	0.22	0.22
Cystine	%	0.20	0.21	0.18	0.19
Tryptophan	%	0.12	0.13	0.10	0.11
Metabolizable energy	kcal./lb.	1,316	1,274	1,311	1,274

Feeding Directions

[1] See Table 4-2, for recommended feeding levels for boars, gilts, and sows when hand-fed daily in drylot or confinement. These rations can be used for gilts on pasture during gestation since they require 3 to 4 pounds of feed daily. These rations can also be used for interval-fed sows or gilts if the average daily intake is approximately 4 pounds.

[2] To simplify the feeding program these rations can also be used for lactation.

[3] Ground oats can replace corn up to 20 percent of the total ration. If more than 20 percent oats is used in the ration, the level of feeding should be increased because of the low energy content of oats. Ground milo, wheat or barley can replace the corn.

[4] If 48.5 percent soybean meal is used instead of 44, use 20 pounds less soybean meal and 20 pounds more corn. If whole cooked beans are used instead of 44 percent soybean meal, use 30 pounds more beans and 30 pounds less corn.

[5] Feed additives are not generally recommended during gestation or for gilts during the developer period after selection unless specific disease problems exist. High levels of feed additives (100 to 300 gm./ton) may be beneficial 2 to 3 weeks prior to breeding and 2 to 3 weeks prior to farrowing.

Source: *Life Cycle Swine Nutrition*, Iowa State University, 1974.

T A B L E 5-10   PREGESTATION, BREEDING, AND GESTATION RATIONS
(For Boars, Sows, or Gilts Being Fed 3 Pounds per Day)[1,2]

Percent Protein	Ingredient		1	2	3	4
8.9	Ground yellow corn[3]		1,637	1,552	1,647	1,582
44	Solv. soybean meal		250[4]	250[4]	170[4]	140
17	Dehydrated alfalfa meal		—	100	—	100
50	Meat and bone meal		—	—	100	100
	Calcium carbonate (38% Ca)		15	10	5	—
	Dicalcium phosphate (26% Ca, 18.5% P)		60	60	40	40
	Iodized salt		15	15	15	15
	Trace mineral premix		3	3	3	3
	Vitamin premix		20	20	20	20
	Feed additives[5]		—	—	—	—
	Total		2,000	2,000	2,000	2,000

CALCULATED ANALYSIS

	Protein	%	12.75	13.18	13.53	13.47
	Calcium	%	1.09	1.07	1.04	1.01
	Phosphorus	%	0.83	0.83	0.83	0.83
	Lysine	%	0.60	0.63	0.61	0.60
	Methionine	%	0.23	0.23	0.23	0.23
	Crystine	%	0.21	0.22	0.20	0.20
	Tryptophan	%	0.14	0.15	0.13	0.13
	Metabolizable energy	kcal./lb.	1,291	1,252	1,289	1,252

Feeding Directions

[1] See Table 4-2, for recommended feeding levels for boars, gilts, and sows when hand-fed daily in drylot or confinement. These rations can also be used for sows on pasture since they will require only supplemental minerals or at the most 2 to 3 pounds of complete feed. If less than 3 pounds is fed per day, free-choice minerals should be available.

[2] These rations can also be used as a silage balancer. Gestating sows and gilts will consume 5 to 7 pounds of corn silage daily which should be supplemented with 2 to 3 pounds of one of the above rations.

[3] Ground oats can replace corn up to 20 percent of the total ration. If more than 20 percent oats is used in the ration, the level of feeding should be increased because of the low energy content of oats. Ground milo, wheat or barley can replace the corn.

[4] If 48.5 percent soybean meal is used instead of 44, use 30 pounds less soybean meal and 30 pounds more corn. If whole cooked beans are used instead of 44 percent soybean meal, use 40 pounds more beans and 40 pounds less corn.

[5] Feed additives are not generally recommended during gestation or for gilts during the developer period after selection unless specific disease problems exist. High levels of feed additives (100 to 300 gm./ton) may be beneficial 2 to 3 weeks prior to breeding and 2 to 3 weeks prior to farrowing.

Source: *Life Cycle Swine Nutrition*, Iowa State University, 1974.

FIGURE 5-11 (*Above*). Many specialized swine programs involve confinement quarters for breeding animals. These Berkshire boars are active and do not need to be driven for exercise. (*Courtesy American Berkshire Association.*) FIGURE 5-12 (*Below*). A swine breeding crate. (*Courtesy Breeders Supply Company.*)

the breeding season begins to make certain the boar will serve and settle a sow.

A boar may be inactive because he has become too fat or has not had sufficient exercise. Many so-called *nonbreeders* are boars of this type. Be certain your boar is in good breeding condition. It is a good policy to turn two boars together in case one is inactive. The boar should be familiar with his surroundings because quite often a boar will not service a sow in new quarters.

A breeding crate also may be helpful in getting a boar to serve a sow. This is especially true in mating mature boars with young gilts. It is possible to make a temporary crate, or a commercially made crate may be used.

A number of products are sold as remedies for the inactivity of boars, but in most cases they are not effective. A veterinarian should be called in if it is necessary to use drugs or hormones.

Breeding records. A record of the breeding dates for the sows in the herd is very helpful in checking upon the ability of the boar to settle sows and in selecting sows to be kept in the herd. It is also helpful in selling bred sows. The buyer will need to know when to expect the sows to farrow.

ARTIFICIAL INSEMINATION

Artificial insemination in swine is currently in about the stage of development that artificial insemination in dairy cattle was in 1955. During the past 25 years, techniques for the collection of semen, short-term storage, and insemination have been developed. Problems in storing frozen semen and in easy detection of estrus have not as yet been completely solved.

Advantages of Artificial Insemination in Swine

The following are some of the benefits to be derived from the use of artificial insemination in swine:

1. It will facilitate the breeding of outstanding sires to a larger number of females.
2. It will make it possible for owners of small herds to use production tested or otherwise superior boars.
3. It will aid in stopping the spread of some swine diseases.
4. It will reduce the investment in boars and boar handling equipment by individual swine producers.
5. It will speed up improvement and uniformity in carcass quality of comercial hogs.
6. It will overcome the difficulties encountered when large differences exist in the size of boars and gilts.
7. More accurate breeding records will be possible.

8. The problems of the sterile, slow, or nonbreeding boars will be overcome.
9. A large number of gilts can be bred to one sire in a shorter period of time. This benefit is important in multiple-farrowing programs.

Problems in Artificial Insemination

1. Heat detection is a serious problem in making effective use of artificial insemination. Many swine growers have left heat detection to the boar. Lot breeding practice has been followed by a large percentage of producers. Many growers have not detected the stage of the heat period, and poor conception has resulted. Those breeders who have hand-mated previously have had less trouble in determining the best time to inseminate.
2. The average commercial producer is not used to devoting as much time to management of the herd at breeding time as is necessary when artificial insemination is used. At present, artificial insemination is being done by specialized swine producers, many of whom are seed stock growers.
3. Extra labor is required since each female must be bred and handled individually.
4. Only a small percentage of swine growers understand and can do artificial insemination. The techniques are not difficult to learn. This problem can easily be overcome.
5. For economical use of the technician's time, several animals need to be bred at the same time. Progress is being made in the synchronization of ovulation.
6. Since 1972 we have successfully frozen and stored semen. We have not solved all problems related to the freezing and storage of semen, but sufficient progress has been made to justify the establishment of boar studs by many commercial organizations and to make frozen semen available at fairly reasonable prices.

Keys to Successful Use of Artificial Insemination

1. Boar spermatozoa are sensitive to sudden changes in temperature. Care must be taken

to avoid sudden temperature changes at all stages of semen collection, dilution, storage, and insemination.

2. Fresh semen can be taken from a boar several times a week. Boars must be trained to mount a dummy and to ejaculate. The ejaculate is collected and diluted with an extender. Several commercial extenders are available. The following are examples of commonly used extenders:

   a. Whole homogenized milk   1,000 ml.
      Dihydro-Streptomycin sulfate   1 gram
      Penicillin   one million I.U.
      One ejaculate added to the above combination. This quantity is sufficient for 20 inseminations.

   b. Glucose   15 grams
      Sodium bicarbonate   0.75 grams
      Egg yolk   150 milliliters
      Distilled water   350 milliliters
      Penicillin   5,000 I.U.
      Streptomycin   500 mg.
      One ejaculate added to the above is sufficient to inseminate 10 to 15 females.

3. Fresh semen should be used within 24 hours. Conception rate decreases rapidly after 36 to 48 hours.

4. The best time to inseminate is near the middle of the estrous period, which is usually 12 to 24 hours after first signs of heat in gilts, or 24 to 36 hours with mature sows.

5. Heat can be detected by placing weight on the back of the gilt. If she stands, she is usually ready for insemination. It is a good policy to have a boar available to see if the gilt will stand for natural breeding.

6. Adequate semen must be used whether fresh or frozen. Equal conception is possible, usually 70 percent or higher. From 50 to 70 ml. of the extended semen is used for each insemination.

7. The directions provided by the supplier of semen should be followed carefully. Frozen semen is being supplied by some technicians at a cost of $5.00 to $8.00 per sow, depending on volume.

8. Care should be taken to make certain that semen used was obtained from the boar best suited to mate with a specific gilt or sow.

The conformation, pedigree and production performance of the ancestry represented by the boar should be studied. Carcass data and Sonoray or EMME information should be evaluated.

9. It is essential that a microscope be available for the testing of the semen for mobility of the sperm cells. From 75 to 85 percent should be active. There should be 2 to 4 billion live sperm for each insemination.

10. Insemination should be done in quarters familiar to the gilt and at the time when she is in standing heat. Insemination kits are available. Cattle insemination equipment may be used with minor alterations. The insemination process involves the insertion of the pipette or rod in the vagina and working it to 6 to 8 inches. It will stop at this point and the technician will need to twist it slightly to penetrate the cervix. At this point the syringe containing the semen solution is attached to the pipette and the semen deposited. A second insemination 12 to 24 hours later is recommended.

Ovulation Control

In May, 1968, a heat synchronization drug was submitted to the Food and Drug Administration for approval. The drug called Aimex, Match, or ICI was developed in England and was approved there for use by farmers for a three-year period. A supplier of artificial insemination service in Wisconsin used the drug experimentally in its operations for two years.

The product is to be used as a feed additive. Gilts are fed five pounds daily of a ration containing the product for a 20-day period.

Dosage per gilt is 125 mg. fed over the 20-day period. The drug inhibits estrus (heat period). When the drug is withdrawn, the gilts or sows tend to come in heat within 3 or 4 days. Tests conducted by the U.S.D.A. involving more than 1,000 sows indicate that the drug has practical application in hog production. In Montana tests 166 of 171 treated gilts came in heat with 151 of them showing estrus between the fourth and seventh day. Seventy percent of those treated were bred in 2 days and 151, or about 90 percent, were ready to breed in 4 days. Ninety-seven percent of those bred settled and 91 percent conceived at first service. Litter size averaged 9.7 pigs born alive.

Estrous control programs in the future may combine the use of Aimex or ICI-like compounds with injections of two hormones. The treated feed containing Aimex or ICI-like compounds will be fed to gilts or sows for 20 days. They will then be given an injection of PMS (Pregnant Mare Serum). Four days later they will be given an injection of HCG (Human Chorionic Gonadotrophin). All animals should ovulate and be bred within 24 hours. This procedure could greatly enhance the use of artificial insemination in swine production.

The pharmaceuticals described above have not been approved by the Federal Food and Drug Administration for use in livestock feeding. Other drugs are being used experimentally, and it is anticipated approved materials will be available in the near future. At present the best method of bringing a large number of gilts in heat at the same time is to move them to new quarters adjacent to boar quarters. Sows whose pigs are weaned at the same time will usually come in heat 4 to 8 days after weaning of pigs. It is hoped that a new hormone-like substance called prostaglandin, used successfully with cattle, may have application to ovulation control in swine.

SUMMARY

The reproductive cell of the female is called an ovum. The male cell is called a sperm. Each parent contributes equally in the reproduction process.

The ovum is produced in a follicle of the ovary and is dropped into the oviduct. It is fertilized by a sperm in the upper end of the oviduct. From there it moves to the uterus, where it becomes attached to the wall and develops.

The sperm is produced in the testicle and moves through the sperm ducts to the urethra and the penis. The seminal vesicles, prostate, and Cowper's glands add secretions to the seminal fluid.

The heat period occurs about every 21 days and ovulation usually takes place the second day. Breeding should be timed so that the sperm is in the oviduct when ovulation occurs.

The chromosomes in the reproductive cells carry genes which determine the characteristics to be transmitted to the progeny. Chromosomes occur in pairs, one of each pair in the male cell uniting with one of each pair in the female cell to produce the new offspring. The reproductive cell in swine has 19 pairs.

Some factors of inheritance are dominant, whereas others are recessive. A hybrid condition exists when the germ cell has one dominant and one recessive gene.

The crossing of two superior animals of different breeds usually results in increased growth rate, increased feed efficiency, improvement in body conformation, and increased fertility. These increases are usually referred to as hybrid vigor or heterosis.

The systems used in swine breeding are upgrading, purebred breeding, crossbreeding, crisscrossing, rotation breeding, inbreeding, line breeding, and crossline breeding.

Crossbreeding and linecrossing are recommended in commercial pork production. A rotation system involving three or more breeds of purebred, inbred, or hybrid hogs is recommended.

It is usually more profitable to raise two litters from each sow each year. By early weaning it is possible to obtain five litters from a sow in two years.

FIGURE 5-13. These Poland China boars have plenty of pasture and range for exercise. (*Courtesy Land O Lakes, Felco Division.*)

Gilts should be bred to farrow when they are from 11 to 13 months of age. It is usually best to breed gilts at the end of the first day of the third heat period. A second mating 12 hours after the first mating is also recommended. Gestation takes about 115 days.

Boars and sows should be in gaining condition but should not be fat at breeding time. Rations high in protein, minerals, and vitamins are important during the flushing or prebreeding period. Legume pasture or high quality legume hay should be provided.

Pen breeding is preferred, and breeding records should be kept. Unless artificial insemination is used, not more than one or two sows should be mated to a young boar in one day. Mature boars may be mated to three or four sows in one day if matings are properly spaced. A boar needs exercise to be in good breeding condition.

Artificial insemination in swine is now possible. One ejaculate may be used in breeding 15 to 25 females. Ovulation control or estrus-synchronization drugs supplemented with injections of two hormones will result in 80 to 90 percent of females coming in heat in a two- to four-day period. Effective drugs have not as yet been approved by the Food and Drug Administration.

QUESTIONS

1   What is meant by ovulation, and when does it take place in relation to the heat period?

2   What is the difference between a gene and a chromosome?

3   What contributes more to inheritance, the ovum or the sperm? Explain.

4   Explain hybrid vigor in terms of the gene composition of the chromosomes.

5   Explain the difference between inbreeding and line breeding.

6   How do upgrading, crossbreeding, and rotation breeding differ? Explain.

7   Which method of breeding is best adapted to your swine enterprise? Why?

8   Of what advantage is crossbreeding in commercial hog production?

9   How often will gilts come in heat?

10   When during the heat period is the best time to breed gilts?

11   What breeding and management practices would have to be followed in order to get three litters from each sow on your farm in one year? Explain.

12   Which is better on your farm, a one-litter per year or multiple-litter breeding program? Why?

13   When should sows and gilts be bred to farrow litters in February and March?

14   What care and management should be given the boar during the breeding season?

15   ·Plan a good ration for growing out and flushing a group of 225-pound gilts.

16   What is the future of artificial insemination in commercial hog production?

17   Of what value is estrous control in swine production?

18   Outline a program of breeding practices which will insure more profitable returns from the swine enterprise on your farm.

REFERENCES

Holden, Palmer, Vaughan Speer, and E. J. Stevermer, *Life Cycle Swine Nutrition,* Pm. 489, Ames, Ia.: Iowa State University, 1974.

Krider, J. L., and W. E. Carroll, *Swine Production,* 4th ed. New York, N.Y.: McGraw-Hill Book Company, 1971.

Leman, A. D., *Estrous Control and Artificial Insemination of Swine,* Cir. 1029. Urbana, Ill.: University of Illinois, 1970.

National Pork Producers Council, *Productive Efficiency of Swine.* 3101 Ingersoll, Des Moines, Ia.: 1969.

Sherritt, G. W., and D. E. Younkin, *Improving Swine Through Genetics,* Sp. Cir. 172. University Park, Pa.: Pennsylvania State University, 1973.

Stevermer, Emmett, Palmer Holden, and John Berthelsen, *Feeding and Managing the Swine Breeding Herd,* Pm. 583. Ames, Ia.: Iowa State University, 1974.

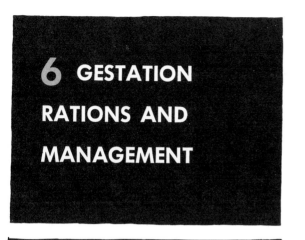

6 GESTATION RATIONS AND MANAGEMENT

The vigor of the pigs and the number produced in the litter at farrowing time are determined by the number of eggs fertilized by the boar at the time of breeding and by the methods used in feeding and managing the sows during pregnancy. Not all eggs fertilized at breeding time produce live healthy pigs at farrowing time.

The ration fed the sow has much to do with the type of litter which she will farrow. The methods used in feeding, exercising, and housing the sow also influence the number of healthy pigs which she will farrow. Tests have shown that a three and one-half pound pig at birth has a six times better chance to reach weaning than a one and one-half pound runt.

NUTRITIONAL NEEDS DURING GESTATION

Bred sows use feed to maintain their bodies and also to produce litters of pigs. Young gilts also need feed to grow and become mature sows. If the rations are inadequate, the sows will be ineffective in maintaining themselves and in producing strong litters. Weak pigs may result. Young sows may remain small or stunted in growth and be poor milkers. The feeding of a balanced ration in adequate amounts during pregnancy is a must in profitable pork production.

The age and condition of the sow determine the amount and kind of a ration which she should receive. Thin sows require more feed than do sows in good flesh. Large sows need more feed than do small sows, and young sows must have more feed per hundred pounds of live weight than must mature sows. The stage of pregnancy also must be considered.

Nutritional Allowances

As shown in Chapter 4, pregnant sows, fed 4 pounds of feed daily should receive rations which contain 12 percent protein,

1 Growing Breeding Stock Self-Feeding	2 Pregestation and Gestation Hand-Feeding		
600	1200	1600	1
600	400	—	5
600	—	—	—
200	400	400	500

0.8 percent calcium, 0.55 percent phosphorus, 2,250 units of vitamin A and 300 units of vitamin D per pound, and the following milligrams per pound of the B vitamins and antibiotics: riboflavin, 1.5; niacin, 11; pantothenic acid, 6; and choline, 400 milligrams. In addition the ration should contain about 0.5 percent salt and 7.5 micrograms of vitamin B_{12} per pound. These nutritional needs may be supplied by feeding grains, mixed concentrates, milk products, hays, and silages, and by providing adequate pasture.

Gain Desired

The amount of gain desired during the gestation period varies with the age and condition of the animal. A gain of 70 to 90 pounds in gilts during pregnancy will allow for the growth of the gilt and her litter. Mature sows should gain from 60 to 80 pounds during gestation. These figures can be reduced somewhat if the litter is to be weaned when from one to three weeks of age. Sows and gilts should not be permitted to become fat. An overweight condition reduces the number and health of the pigs farrowed.

Amount of Feed Required

The amount of feed required by sows during the pregnancy period varies with the age, condition, and stage of pregnancy. Up to about 1953, it was assumed that a thin mature sow would require from 1 to 1¼ pounds of feed daily, and a young gilt would need from 1½ to 2 pounds daily, for each 100 pounds of live weight. Recent tests, however, indicate that we can greatly reduce the amounts of concentrates. This is especially true in feeding mature sows whose litters will be weaned at an early age.

Tests conducted in Iowa, Illinois, Ohio, and South Dakota indicate that about 4 pounds of feed per day will result in maximum birth and weaning weights of pigs. Ohio tests indicated that sows fed 5 pounds per day became fat. Those fed 3 pounds per day became skinny and weak. Limited feeding is recommended especially when environmental temperature is at 50 degrees or above. In South Dakota tests gilts fed 3 pounds of feed daily gained 80 pounds during their first gestation and 89 pounds during the second farrowing. Gilts fed 5 pounds per day gained 148 and 124 pounds during the two gestation periods. The amount of feed necessary will vary with quality of ration, temperature, and condition of the sow.

Critical Periods During Gestation

There are two critical periods during pregnancy, the early part of gestation and the latter third of pregnancy. Proteins and vitamins are very important during the embryo formation period. At 30 days, each embryo will weigh about 0.06 oz.; at 60 days, 3.3 oz.; at 80 days, 11.75 oz.; at 90 days, 24 oz.; and at 106 days, 40 oz.

Rations

The ration best suited to your farm must be based upon the kinds of feeds and forages available and the prices of these feeds in your community. The key to feed efficiency is the use of high quality, cheap farm-grown feeds, properly supplemented with proteins, minerals, vitamins, and antibiotics.

The rations presented in Table 5-9 may be fed to gilts and sows during gestation at the rate of 4 pounds per day. The rations in Table 5-10 may be fed at the rate of 3 pounds per day. In Table 6-1 are rations recommended for feeding gestating gilts and sows when fed 5 or more pounds per day.

Grain sorghum may be substituted for corn, oats, or barley in the rations, but it should be ground. Rye may be fed in small quantities, but it cuts down the palatability.

T A B L E 6-1   PREGESTATION, BREEDING, AND GESTATION RATIONS
(For Sows or Gilts Being Fed 5 or More Pounds per Day)[1]

Percent Protein	Ingredient	1	2	3	4
8.9	Ground yellow corn[2]	1,838	1,753	1,853	1,803
44	Solvent-process soybean meal	100[3]	85	—	50
17	Dehydrated alfalfa meal	—	100	—	50
50	Meat and bone meal	—	—	115	50
	Calcium carbonate (38% Ca)	15	10	5	5
	Dicalcium phosphate (26% Ca, 18.5% P)	25	30	5	20
	Iodized salt	10	10	10	10
	Trace mineral premix	2	2	2	2
	Vitamin premix	10	10	10	10
	Feed additives[4]	—	—	—	—
	Total	2,000	2,000	2,000	2,000

CALCULATED ANALYSIS

			1	2	3	4
	Protein	%	10.38	10.52	11.12	10.80
	Calcium	%	0.62	0.66	0.63	0.60
	Phosphorus	%	0.49	0.53	0.51	0.53
	Lysine	%	0.41	0.41	0.41	0.41
	Methionine	%	0.20	0.20	0.20	0.20
	Cystine	%	0.18	0.20	0.16	0.18
	Tryptophan	%	0.10	0.11	0.08	0.10
	Metabolizable energy	kcal./lb.	1,338	1,298	1,334	1,317

Feeding Directions

[1] See Table 4-2, for recommended feeding levels for gilts and sows when hand-fed daily in drylot or confinement. These rations can also be used for interval-fed sows or gilts if the average daily intake is 5 pounds or more per day.

[2] Ground oats can replace corn up to 20 percent of the total ration. If more than 20 percent oats is used in the ration, the level of feeding should be increased because of the low energy content of oats. Ground milo, wheat or barley can replace the corn.

[3] If 48.5 percent soybean meal is used instead of 44, use 15 pounds less soybean meal and 15 pounds more corn. If whole cooked beans are used instead of 44 percent soybean meal, use 20 pounds more beans and 20 pounds less corn.

[4] Feed additives are not generally recommended during gestation or for gilts during the developer period after selection unless specific disease problems exist. High levels of feed additives (100 to 300 gm./ton) may be beneficial 2 to 3 weeks prior to breeding and 2 to 3 weeks prior to farrowing.

Source: *Life Cycle Swine Nutrition*, Iowa State University, 1974.

Hand-feeding Versus Self-feeding

Swine producers must control the rate of gain of bred sows and gilts. They can be fed by hand the right amounts of feed daily or self-fed a bulky ration. Either method may be used successfully. The self-feeding method requires less labor, while hand-feeding usually takes less feed.

The ration may be mixed with water and fed as a slop or as a paste. Sows will usually consume more feed and more water when slop fed. The additional labor, however, is an important factor when many sows are involved.

In self-feeding, the proportion of bulky feeds in the ration must be governed by the condition of the sows. Concentrates are added if the sows do not gain well. When the sows gain too rapidly, the amount of corn or grain is reduced, and ground alfalfa is added.

FIGURE 6-1. Fourteen pigs from this Chester White litter were raised. Two pens from the litter made excellent performance records in Iowa and Wisconsin boar testing stations. (*Courtesy Dave Huinker.*)

In Wisconsin tests it cost $6.00 more per litter to feed a self-fed bulky ration to sows during gestation than to hand-feed. Each sow received 5 pounds of feed a day.

Bred sows and gilts should receive about ½ to ¾ pounds of crude protein each day. It may be hand-fed in supplement or supplied in the self-fed ration.

Feeding Grain and Supplement Separately

Some swine growers prefer to hand-feed grain and supplement separately to gilts and sows during gestation. This system encourages the feeder to give more careful attention to the condition of the sows and the feeding of rations best suited to their physical condition. Presented in Table 6-4 are rations recommended for self-feeding during gestation.

In Table 6-5 are suggested diets for meeting the daily amino acid requirements for sows and gilts during pregnancy and lactation under various feeding and management situations.

FEEDING METHODS

Feeding Stalls

The use of sow feeding stalls permits feeding each sow the amount of feed that she needs. Thin sows can be fed heavier rations and fat sows can be fed limited rations. The use of stalls also decreases the fighting of sows competing for feed.

The stalls may be constructed of wood, steel, or concrete. A 24-stall setup constructed of concrete will cost from $400 to $500. In South Dakota tests, stalls have been constructed of wood for $6 to $10 per stall.

Tie Stalls

The Ontario Agricultural College has recommended the use of tie stalls for sows

T A B L E 6-2   35 PERCENT SOW SUPPLEMENT

(For Growing-breeding Stock, Gilts, and Sows During Pregestation, Gestation, and Lactation)

Percent Protein	Ingredients	1	2	3
16	Wheat middlings	230	260	—
50	Meat and bone scrap	300	400	—
44	Soybean meal	1050	1000	1520
17	Dehydrated alfalfa meal	200	200	200
32	Fish solubles	50	—	—
	Calcium carbonate (38% Ca)	20	15	50
	Dicalcium phosphate (26% Ca, 18% P)	60	35	140
	Iodized salt	40	40	40
	Trace mineral premix*	10	10	10
	Vitamin premix*	40	40	40
	Totals	2000	2000	2000

CALCULATED ANALYSIS

		1	2	3
Protein	%	34.90	35.78	35.14
Fat	%	1.92	2.17	0.58
Fiber	%	7.80	7.90	7.92
Calcium	%	2.95	3.02	3.10
Phosphorus	%	1.74	1.75	1.74
Vitamin A, from:				
Carotene	I.U./lb.	10,000	10,000	10,000
True A	I.U./lb.	6,000	6,000	6,000
Vitamin D	I.U./lb.	1,600	1,600	1,600
Riboflavin	mg./lb.	9.8	9.7	9.6
Pantothenic acid	mg./lb.	22.1	21.8	21.9
Niacin	mg./lb.	53.8	53.3	46.5
Choline	mg./lb.	868.0	850.0	952.0
Vitamin B_{12}	mcg./lb.	46.2	45.0	40.0
Metabolizable energy	kcal./lb.	825	859	825

*Table 7-2.

Feeding Directions

The 35 percent Sow Supplement may be hand-fed to sows and gilts at the rate of 0.5-0.75 pound per head daily during pregestation and gestation. Add shelled corn or equivalent in amounts to total the feeding levels that are shown in Tables 5-9, 5-10, and 6-1 for the complete rations.

Source: *Life Cycle Swine Nutrition (Revised)*, Iowa State University, 1968.

TABLE 6-3 SUGGESTED BROOD SOW RATIONS

Ration Number	1	2	3	4	5	6
INGREDIENT				Pounds		
Milo, ground (8.5% Protein)[1]	1432			742	742	
Wheat, ground (11.0% protein)[1]		1542		742		772
Barley, ground (11.0% protein)[1]			1542		742	772
Soybean meal (44.0% protein)	412	304	304	360	360	302
Alfalfa meal, dehydrated (17.0% protein)	100	100	100	100	100	100
Dicalcium phosphate	30	26	26	28	28	26
Calcium carbonate	16	18	18	18	18	18
Salt	10	10	10	10	10	10
Vitamin-trace mineral mix[2]	+	+	+	+	+	+
Total	2000	2000	2000	2000	2000	2000

[1] Rations were formulated on the basis of cereal grain protein content as indicated. Adjustments should be made when using cereal grains with different protein content.

[2] Vitamin-trace mineral mix should be added in sufficient amounts so that the total ration contains the level suggested in Tables 5-9, 5-10 and 6-1.

Source: Oklahoma State University.

TABLE 6-4 BULKY RATIONS FOR SELF-FEEDING SOWS AND GILTS DURING GESTATION

Ingredient	Ration 1 (Pounds)	Ration 2 (Pounds)
Ground shelled corn	600	600
Ground oats	600	600
Ground alfalfa hay	600	600
40% protein supplement[1]	200	—
Soybean meal (44%)	—	120
Meat and bone meal	—	60
Salt	—	10
Vitamin-trace mineral mix	—	10
	2,000	2,000

[1] It is assumed that the 40% protein supplement will be fortified with salt, vitamins, and minerals.

Source: Michigan State University.

T A B L E 6-5   SUGGESTED DIETS FOR MEETING THE DAILY ESSENTIAL AMINO ACID
REQUIREMENTS FOR PREGNANCY AND LACTATION

Daily Feed Intake (pounds)	Pregnancy[1]				Lactation[2]
	3.5	4.0	4.4	5.0	12.0
INGREDIENT	PERCENT OF DIET				
Ground yellow corn	87.0	89.5	90.75	92.5	85.0
Solvent soybean meal (44%)	10.0	7.5	6.25	4.5	12.0
Mineral and vitamins	3.0	3.0	3.00	3.0	3.0
	100.0	100.0	100.00	100.0	100.0
ANALYSIS (CALCULATED)					
Crude protein (%)	12.1	11.2	10.7	10.1	12.8
Most limiting essential amino acid	Tryptophan	Tryptophan	Tryptophan	Tryptophan	Lysine Tryptophan Methionine?

[1] Several feeding levels suggested for different environmental situations as follows:
3.5 lbs. — Complete confinement environmental control, tethered.
4.0 lbs. — Spring, summer, fall conditions, drylot or concrete, open-front shed.
4.4 lbs. — N.R.C. recommended feeding level.
5.0 lbs. — Winter conditions, drylot or concrete, open-front shed.

[2] Recommendation based on 3-week lactation for sows nursing nine pigs. For each pig above or below nine add or subtract 1 pound daily feed.

Source: Iowa State University.

during the gestation period. The tethering of sows has been popular in England and in some European countries.

The stalls are approximately 30 x 72 inches and include a feed trough. The size varies with breed and age of animal. The floors slope slightly to the gutter. Some use slatted floors. About 25 percent more sows can be housed in a given area by using tie stalls.

Metal collars rather than leather straps are recommended. Two bars joined at the top and bottom will lock behind the sow's jaws without being uncomfortable to the animal. The use of tie stalls permits individual feeding and the keeping of breeding records. Sows adjust readily to the new quarters and become gentle. Injuries to the sows are less frequent than when other methods are used.

Tests by swine producers in the United States have not shown the use of tie stalls to be as advantageous as Canadian tests have indicated.

In Purdue tests tethered sows required more intensive care and observation than other methods. The rations, temperature control, and ventilation within confinement

FIGURE 6-2. These sows were encouraged to feed in good weather in a large field. (*Courtesy Kent Feeds.*)

buildings had to be watched closely. It was found that sows required a minimum of 12 square feet of floor space, but 14 to 16 square feet were preferred.

Skip-a-day Feeding

Tests in Nebraska indicate that sows do well if fed each third day. In Nebraska tests the sows were permitted to eat at a self-feeder for 1½ to 2 hours each 72-hour interval. The number of pigs at birth, birth weights, number of weaned, and weaning weight of pigs were similar for the daily- and interval-fed sows. The sows fed daily received 4 pounds of feed per day in stalls. The interval-fed sows ate slightly more feed during the test period than the others did.

Feeding Silage to Bred Sows and Gilts

The Illinois, Indiana, Minnesota, and Iowa experiment stations have conducted research which indicates that good corn or grass silage, properly supplemented, may constitute the major part of the ration for sows and gilts during gestation. The quantitative requirements of bred sows and gilts are not high, but the qualitative requirements are quite exacting.

In Minnesota tests grass-legume silage replaced 42 percent of the daily mixed ration fed to bred sows. The sows gained 25 percent in weight, and there was a saving in feed cost amounting to about $3.60 for each 100 pounds gained.

The feed costs during gestation on a per pig weaned basis in Illinois tests were reduced 28 percent by feeding either grass silage or corn silage with the proper supplementation.

Kinds of silage to feed. Corn alfalfa, and brome-alfalfa silages have been fed successfully. Only high quality silage should be fed. Silage made from immature or too dry corn is not satisfactory for brood sows. Sows require much the same quality of silage required by dairy cattle. Moldy silage should not be fed.

Amounts to feed. Tests conducted at Minnesota, Iowa, and Illinois indicate that gilts may be fed from 5 to 7 pounds of corn or grass silage per day. Sows can consume as much as 8 to 12 pounds of corn or grass silage.

Preservatives in silage. Corn silage usually requires no preservative. Grass silage for sow feeding should be preserved with either corn or molasses to increase the palatability.

Supplementation. Corn silage is high in carotene (pro-vitamin A), but is low in protein, minerals, and other vitamins. Grass silage is higher than corn in calcium and in protein as well as in all vitamins except vitamin B_{12}, but is low in phosphorus and may need additional protein to provide a desirable amino acid balance.

Sufficient amounts of grains and protein, mineral and vitamin supplements should be provided to maintain the sows in medium condition. Shown in Tables 6-6 and 6-7 are

T A B L E 6-6   16 PERCENT CORN SILAGE BALANCER

(Fed with Corn Silage to Sows and Gilts During Pregestation and Gestation)

Percent Protein	Ingredients	1	2	3
8.8	Ground yellow corn*	1312	1212	1227
12.0	Ground oats	—	200	—
50.0	Meat and bone scrap	100	100	—
44.0	Soybean meal	300	300	450
32.0	Fish solubles	50	—	—
17.0	Dehydrated alfalfa meal	150	100	—
13.0	Alfalfa meal (sun cured)	—	—	200
	Calcium carbonate (38% Ca)	10	10	20
	Dicalcium phosphate (26% Ca, 18% P)	40	40	60
	Iodized salt	15	15	20
	Trace mineral premix**	3	3	3
	Vitamin premix**	20	20	20
	Total	2000	2000	2000

CALCULATED ANALYSIS

Protein	%	16.95	16.48	16.60
Fat	%	3.12	3.18	2.59
Fiber	%	4.78	5.19	6.41
Calcium	%	1.36	1.33	1.34
Phosphorus	%	0.89	0.89	0.85
Vitamin A, from:				
Carotene	I.U./lb.	8156	5606	1613
True A	I.U./lb.	3000	3000	3000
Vitamin D	I.U./lb.	800	800	800
Riboflavin	mg./lb.	5.2	5.0	5.0
Pantothenic acid	mg./lb.	11.8	11.5	11.7
Niacin	mg./lb.	30.0	28.1	27.4
Choline	mg./lb.	411	401	423
Vitamin B_{12}	mcg./lb.	23.8	21.2	20.0
Metabolizable energy	kcal./lb.	1234	1244	1185

*Molasses may be added to aid in pelleting by substituting 100 pounds molasses for 100 pounds corn.

**Table 7-2.

Feeding Directions

1. Corn silage should be fed at the rate of approximately 6 pounds (wet basis) per head daily.
2. The amounts of 16 percent Corn Silage Balancer to be fed in addition to corn silage are shown in Table 6-7.

Source: *Life Cycle Swine Nutrition*, Iowa State University, 1968.

TABLE 6-7   FEEDING LEVELS

The Levels of Complete Gestation Ration or Corn Silage Balancer to Feed During Different Stages of Pregnancy (For Rations in Table 6-6)

	Pregestation	Flushing*	Gestation	Last 6 Wks. of Gestation
Complete gestation ration (pounds per day)	4	6	4	6
Corn silage balancer (pounds per day)	2	4	2	4

*2 to 3 weeks prior to breeding.

Source: *Life Cycle Swine Nutrition*, Iowa State University, 1968.

recommendations for supplementation of silage in feeding sows and gilts during pregnancy.

Hay for Pregnant Sows

Ground alfalfa or clover hay should make up one-third of the ration for brood

FIGURE 6-3. These sows are on good pasture and have ample shade available. (*Courtesy American Yorkshire Club.*)

sows and gilts. In a Minnesota test a group of gilts were fed a ration containing 45 percent ground alfalfa hay. The gilts produced litters of pigs averaging three pounds in weight.

Ground Corncobs for Bred Sows

Illinois tests showed that a ration consisting of 700 pounds of finely ground cobs, 600 pounds ground shelled corn, 200 pounds alfalfa meal, and 500 pounds of protein supplement proved adequate in feeding both bred sows and bred gilts. Ground cobs contain less nutrients than either oats or alfalfa meal, so adjustments must be made in the amounts of grains and supplements when ground cobs are included in the ration.

Pasture

A brood sow during gestation can make very effective use of good pasture. Sows that have access to alfalfa, clover, rape, or other good pasture need little or no grain during the first half of the pregnancy period. They will need only about one-half as much grain during the latter part of the pregnancy period as they would have needed had they been in dry lot.

FIGURE 6-4 (*Above*). These sows fed in confinement have room for exercise in outside pens. They are fed limited rations. (*Courtesy Starcraft Livestock Equipment.*) FIGURE 6-5 (*Below*). Bred gilts can be forced to exercise during the winter by hand-feeding them away from their sleeping quarters. (*Courtesy Kent Feeds.*)

Ladino clover and alfalfa have proved to be our best pastures for brood sows in most of the hog production areas.

Tests conducted in Illinois indicate that bred sows on lush Ladino clover pasture may need no other feed than a balanced mineral mixture. They found, however, that bred gilts needed at least 1½ to 2 pounds of grain per day, in addition to the minerals, to make them gain the desired three-quarters of a pound per day.

Antibiotics for Bred Sows

Numerous tests have been conducted to determine the advisability of feeding antibiotics to bred sows and gilts. Tests conducted at Purdue University proved that Aureomycin was valuable in rations for bred gilts. Pigs farrowed by gilts which had received 20 to 30 milligrams of Aureomycin per ton of feed were 0.2 to 0.3 pounds heavier at birth, and more of them were stronger than pigs farrowed by gilts which received only 10 milligrams or no antibiotics.

The number of pigs farrowed and the number of pigs weaned were not affected by the feeding of the antibiotic. The pigs, however, from the gilts which had been fed 30 milligrams of Aureomycin per ton of feed were stronger and more uniform in appearance at weaning time.

The addition of antibiotics to the ration of healthy sows during gestation has not provided enough extra pounds of pigs at weaning time to pay the extra cost of the antibiotics. Sows that are unthrifty, however, usually respond satisfactorily to the feeding of antibiotics.

Water

Plenty of water should be fed to sows during pregnancy either in the form of slop feeds, in troughs twice daily, or in automatic waterers. Water is especially impor-

FIGURE 6-6 (*Above*). These Hampshire gilts are in excellent condition. (*Courtesy Land O Lakes, Felco Division.*)   FIGURE 6-7 (*Below*). These gestating sows are fed in confinement. Note the feed spouts over each pen. A complete pelleted ration is fed. (*Courtesy Big Dutchman.*)

tant during the summer when the temperature is high. Fresh water should be available to the sows at all times.

GESTATION MANAGEMENT

Exercise

During the early stages of gestation it is desirable to have the sows forage in the cornstalk fields or in the pasture. Brood sows need exercise, and if they do not rustle out into the fields themselves, they should be fed at a distance of from 10 to 15 rods from their housing quarters. Inactive, fat sows rarely produce and raise good litters.

The sows shown in Figure 6-7 are in excellent condition and should produce healthy litters.

Separation of Gilts and Sows

It has been pointed out that breeding animals should be separated from fattening animals. It is a good practice, also, to separate bred gilts from mature sows. Gilts require more feed than mature sows; they require more proteins and minerals and can make effective use of less silage and hay.

Sows should not be kept in the same lots or buildings with other types of livestock. Sows heavy with pig may be injured by cattle or horses, or may injure themselves by crowding or by going up steep inclines, under creeps, or in low doors. Sows should not be in the cattle feed lots.

Shelter

During the summer months pregnant sows need only a wooded area, open sheds, or shades to protect them from the sun and rain. Portable houses with open doors or straw sheds are satisfactory during the winter months in many areas. In the northern Corn Belt warmer houses may be necessary. Several sows may sleep in one house, but

FIGURE 6-8. Gestating sows and boars gleaning a cornfield will need protein, mineral and vitamin feeds to supplement the corn. (*Courtesy Land O Lakes, Felco Division.*)

they should not be crowded. Bred gilts should have 11 to 14 square feet of shelter per head in cold weather and 15 to 22 square feet of shade per head in warm weather. Mature sows needs 16 to 20 square feet of shelter per head in cold weather and 20 to 30 square feet of shade per head in warm weather.

The hog house does not need to be warm for bred sows and gilts, but it should be dry and free from drafts and dust.

Health Care During Gestation

Adequate rations, proper housing, desirable exercise, and a good water supply can do much to prevent disease problems at and after farrowing. Additional preventative measures are also necessary.

Assuming that the gilts and sows were tested and found to be free from brucellosis, and were wormed before breeding, it is often advisable to feed antibiotics during the first three weeks of the gestation period. This is especially necessary if there is a high disease level on the premises. The antibiotics should also be included in the ration during the last two or three weeks before the sows and gilts are expected to farrow.

Nasal swabs should be taken by a veterinarian at about four-week intervals to aid in culling out animals infected with Bordetella bronchiseptica, which is associated with atrophic rhinitis. Gestating females should be vaccinated for clostridial enteritis twice, six weeks and two weeks prior to farrowing. They should also be vaccinated for transmissible gastorenteritis (TGE) six weeks and two weeks before farrowing.

In most areas it is recommended that sows and gilts be vaccinated for erisipelas both before and during gestation. The latter should be administered from two to four weeks before farrowing.

About a week before farrowing, the gestating females should be wormed with a broad spectrum wormer, treated for mange and lice, and started on the lactation ration.

SUMMARY

Pregnant sows should receive rations which contain 10.4 to 12.8 percent protein, 0.7 percent calcium, 0.55 percent phosphorus, 2,250 units of vitamin A and 300 units of vitamin D per pound depending on whether they receive three, four, or five pounds of feed per day. Each ration should also include the following milligrams per pound of B vitamins and antibiotics: riboflavin 1.5; niacin, 11; and pantothenic acid 6. In addition each ration should contain about 0.5 percent salt and 7.5 micrograms of vitamin B_{12} per pound. These needs may be supplied in the form of grains, mixed concentrates, milk products, hays, silage, or pasture.

Gilts should gain from 70 to 90 pounds and sows 60 to 80 pounds during pregnancy

FIGURE 6-9. Gilts housed in an open-front, breeding-gestation building in North Carolina. (*Courtesy J. Ray Woodard, North Carolina State University.*)

if they are to suckle litters for six to eight weeks. If the pigs are to be weaned during the first three weeks after farrowing, less gain on the sows and gilts is needed.

With limited forage sows need from 1 to 1¼ pounds of feed and gilts 1½ to 2 pounds daily. Sows and gilts on good pasture need little or no feed during the first part of pregnancy. Sows and gilts should not be permitted to get fat.

Sows and gilts may be hand-fed the grain and supplement needed to keep them in good condition. Alfalfa hay may be self-fed.

Pregnant sows and gilts may be self-fed a bulky ration. It should contain from 25 to 35 percent alfalfa.

Gilts may be fed from 5 to 7 and sows 8 to 12 pounds of corn or grass-legume silage. The silage must be supplemented with protein, mineral, and vitamin feeds, and additional grain when necessary, to keep them in a satisfactory gaining condition.

Alfalfa and ladino clover pastures will provide for most of the nutritional needs of bred sows and much of the needs of bred gilts.

The feeding of antibiotics may increase pig size when fed in adequate amounts. It does not affect the weight of pigs at weaning time unless the pigs receive antibiotics in the creep ration. Usually it does not pay to feed antibiotics to healthy sows.

It is good practice to separate bred sows and gilts from other hogs and other livestock. Usually gilts should be separated from the mature sows while feeding.

Sows and gilts should be fed at some distance from sleeping quarters or be given the opportunity to glean a cornstalk field or good pasture.

From 12 to 20 square feet of dry, draft-free housing should be provided for each sow during winter. From 16 to 30 square feet of shade per animal should be provided during the summer.

In order to maintain the health of females and their litters, the gestating gilts and sows should be swabbed by a veterinarian for Bordetella bronchiseptica at four-week intervals, vaccinated for TGE six weeks and two weeks before farrowing, and vaccinated for clostridial enteritis twice, six weeks and two weeks before farrowing. A week before farrowing the sows and gilts should be wormed with a broad spectrum wormer, treated for mange and lice, and started on a lactation ration.

QUESTIONS

1   What effect does the feed and management given a bred sow during pregnancy have upon the size and health of the litter which she will farrow? Explain.

2   What nutritional allowances are recommended for bred sows and gilts during gestation?

3   How much should sows and gilts gain during pregnancy?

4   Should sows which are to suckle litters six to eight weeks be fed differently from those from which the pigs will be weaned within three weeks after farrowing? Explain.

5   When are the most critical periods in feeding bred gilts or sows during gestation? Why?

6   Formulate a ration which when self-fed would meet the nutritional needs of the sows and gilts on your home farm during gestation.

7   Describe the best methods of making effective use of silage in feeding pregnant sows and gilts.

8   How much alfalfa hay should be included in a self-fed ration for pregnant sows?

9   What are the grain and supplement needs of sows and gilts on alfalfa and ladino pasture during gestation? Explain.

10   Does it pay to feed brood sows antibiotics? Explain.

11   Describe the methods that you would use in managing the sow herd on your home farm to provide adequate rations, exercise, housing, and protection from injury.

12   Outline the parasite and disease control practices that should be followed in managing gestating sows and gilts.

REFERENCES

Becker, D. E., A. H. Jensen, and B. G. Harmon, *Balancing Swine Rations,* Circular 866. Urbana, Ill.: University of Illinois, 1966.

Ensminger, M. E., *Swine Science,* 4th ed. Danville, Ill.: Interstate Printers and Publishers, 1970.

Holden, Palmer, Vaughan Speer, and E. J. Stevermer, *Life Cycle Swine Nutrition,* Pm. 489. Ames, Ia.: Iowa State University, 1974.

Krider, J. L., and W. E. Carroll, *Swine Production,* 4th ed. New York, N.Y.: McGraw-Hill Book Company, 1971.

Luce, William, and Charles Maxwell, *Management and Nutrition of the Bred Gilt and Sow,* No. 3653. Stillwater, Okla.: Oklahoma State University, 1971.

Miller, E. C., J. A. Hoefer, and D. E. Ullrey, *Nutrition: Swine Feeds and Feeding,* Ext. Bul. 537. East Landing, Mich.: Michigan State University, 1967.

Moyer, Wendell, and B. A. Koch, *Swine Nutrition,* C-333. Manhattan, Kan.: Kansas State University, 1967.

National Research Council, *Nutrient Requirements of Swine,* 6th ed. Washington, D.C.: National Academy of Sciences, 1968.

Stevermer, Emmett, Palmer Holden, and John Berthelsen, *Feeding and Managing the Swine Breeding Herd,* Pm. 583. Ames, Ia.: Iowa State University, 1974.

7 NURSERY MANAGEMENT

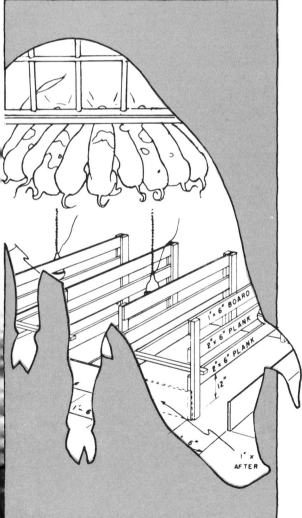

Critical periods in hog production occur at farrowing time, during the suckling period, at weaning time, and during the period from weaning until the pigs have been marketed. The successful producer plans carefully the practices which will be followed during each of these periods. If the hogs are not cared for properly during any one of the periods, serious losses may result.

CARE AND MANAGEMENT AT FARROWING TIME

Some veteran hog producers maintain that the care of the sow during gestation and the care of the pigs through the first few days of life account for more than one-half the job of raising hogs.

Assuming that the breeding herd has been properly managed up to farrowing time and that large litters are farrowed, there is still a chance that the pigs will not be saved. It is estimated that from 25 to 30 percent of all pigs farrowed never reach weaning age, and 80 or 90 percent of the death losses occur within three or four days after farrowing. Each pig that is dead at birth or dies shortly after birth costs the producer about $8.00 invested in feeding the sow from breeding to farrowing, and perhaps a potential profit, if grown to market, of $12.00 to $16.00.

Big Litters Lower Pig Costs

It was estimated in 1970 by the University of Maryland that the labor income per pig sold in the Maryland Feeder Pig Sales from litters of 7 pigs weaned was $6.71; those from litters of 8 pigs, $7.70; those from litters of 9 pigs, $8.59; and those sold from litters of 10 pigs weaned netted a labor income of $9.23 per pig. Average prices from 1965 to 1969 were used, and it was assumed that each sow farrowed two litters per year. The total production costs per pig from litters of seven pigs weaned was $14.19;

FIGURE 7-1. Large litters must be saved to keep production costs down. A crossbred Berkshire X Yorkshire X Hampshire X Duroc litter. (*Courtesy American Berkshire Association.*)

those from litters of 8 pigs, $13.13; those from litters of 9 pigs weaned, $12.31; and the cost of those from litters of 10 pigs weaned was estimated to be only $11.67.

In a 1974 study completed in Iowa, it was found that in a herd averaging 1.7 litters of 7 pigs each weaned per year from each sow, the total cost per 40-pound pig produced was $25.53. The total cost per sow was $258.00. Each additional pig farrowed and marketed at 40 pounds cost only $7.10, thus reducing rapidly the average cost per pig. The average cost when 8 pigs were produced was $23.22. With 9 pigs produced in each litter, the cost was $21.43. Much of the profit from a swine enterprise is determined by the number of pigs weaned per litter.

Time of Farrowing

Some farmers have been successful in producing only one litter of pigs per sow per year by having the sows farrow in late May or June in timber areas or on legume pasture with temporary shelters or houses. They give the sows little or no attention at farrowing time but plan on saving fewer pigs per litter and keep a few more sows than they would keep if they were to give them more attention. Whether this practice is economically sound depends upon a number of factors: the price of hogs, the capital and facilities available, the labor supply, and the feed situation. Many farmers now plan on two or more litters per year, which usually necessitates having sows farrow during the cold months of January, February, and March, and again during the hot months of July, August, and September. With the latter system special care of the sows and litters at farrowing time is necessary.

Housing

Types of hog houses are described in Chapter 10. Either portable or central-type

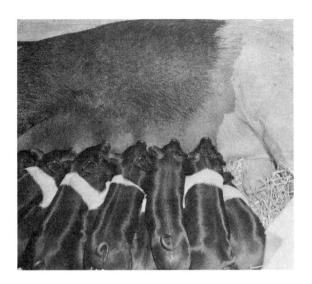

houses may be used to provide quarters at farrowing time. Portable hog houses, pens, and farrowing stalls in the permanent, central-type house should be cleaned and ready for use several days before the sows are due to farrow. The sow should be penned up or placed in farrowing stall a few days before farrowing so that she will be familiar with the surroundings and less likely to be nervous. The size of pen needed will vary with the age and size of the sow and whether or not guard rails, pig brooders, or farrowing stalls are used. Pens 7 or 8 feet wide and from 8 to 10 feet long are preferred when guard rails are used. Some producers prefer using free stalls 6 feet wide and 12 feet long. The sow stall is 7 feet long with guard rails on each side. The sows are fed in a 3- by 6-foot area at one end, and the pigs have a 2-foot creep area in the front of the 6- by 12-foot pen. In central-type houses the short

FIGURE 7-2 (*Above*). Comfortable, clean, warm, and roomy farrowing quarters are essential. (*Courtesy Hampshire Swine Registry.*) FIGURE 7-3 (*Below*). Note the feeder and waterer for the sow and the heat lamp and brooder area for the pigs. (*Courtesy Confinement Builders, Inc.*)

FIGURE 7-4. Farrowing quarters should be scrubbed with hot soap and lye water. (*Courtesy John Simpson.*)

side of the pen should be next to the wall or alley.

The house should be clean, dry, relatively warm, and well ventilated. The floors of movable houses should not be drafty; floor cracks should be fixed. It is sometimes a good policy to bank the sides and ends of movable houses with baled straw or with earth.

Some farmers raise the front side of the movable hog house off the ground from about 6 to 8 inches, so that the sow will lie in the pen with her back toward the door. Some breeders have tilted floors in permanent hog houses. A floor slope of one inch to the foot is recommended.

Sanitation

Losses due to swine diseases and worms can be avoided in part by properly cleaning the farrowing pen before it is used. All dust, dirt, and litter should be removed, and the floors and walls from 1½ to 2 feet from the floor should be scrubbed with boiling lye water or disinfected by use of a steam cleaner. One pound of lye should be added to each 25 gallons of water. After the pen has been cleaned, it should be sprayed with a good disinfectant.

Saponated solution of cresol is very effective. Soluble compounds of cresol should be mixed with soap. A 2 to 3 percent solution is recommended.

Sodium orthophenylphenate is also very effective and leaves no objectionable odor. It is not poisonous. It should be mixed with hot water and used as a 1 percent solution.

The lye water loosens the dirt and kills the worm eggs, while the distinfectant kills the germs of diseases. Equipment used in the pen should also be cleaned and sprayed.

Purdue and North Carolina tests indicate that dichlorvos (Atgard) pig wormer given to sows two to three weeks before farrowing resulted in fewer dead pigs and an average of nearly one more pig per litter at four weeks of age. Piperazine and other wormers may also be used.

The sides, underline, feet, and legs of the sow should be brushed and washed with soap and warm water before she is placed in the farrowing pen. Any dirt left on the sow may contain worm eggs which, if eaten by the pigs, could produce worm infection and runty pigs. In moving the sows or the pigs from one lot to another, it is best to haul them so that worm eggs or disease germs will not be picked up en route. Traffic into the farrowing house should be restricted.

Bedding

Coarse-ground corncobs, wood shavings, fine straw, or sawdust may be used for bedding. Too much bedding or coarse hay and straw bedding may result in the loss of pigs. A bushel basket or two of bedding in a pen is much better than a 4-inch layer covering the entire pen. The bedding should be kept clean, dry, and well distributed. Removing the sow from the pen or stall each morning and evening, for brief exercise, helps to keep the pen clean and dry.

Small pigs in damp or wet quarters are more subject to scours and pneumonia than pigs in dry quarters. Loss of tails may also be a problem. Medicated salves may not save the tails of pigs subjected to continued damp quarters.

Guard Rails

A number of devices have been developed to keep sows from lying on the pigs. A simple device is the guard rail which is placed on the three back sides of the pen, about 8 to 10 inches from the wall, and 8 to 10 inches from the floor. These may be made of metal pipe, native poles, or two-by-fours. They should be installed several days before the sow is due to farrow, so that she will be used to them.

Heat Lamps and Heated Floors

The temperature in the farrowing house should range from about 50 to 60 degrees. When the temperature drops below 50 degrees, the little pigs chill and may catch cold. The use of heat lamps and/or heated floors is recommended. The heat lamp or heated floor not only provides heat but also attracts the pigs away from the sow where they will be less likely to be stepped on or lain upon.

The infrared heat lamp unit usually consists of a drop cord with a mechanical support, a socket, a heat bulb, and a protective reflector with a screen over the end of the reflector. The bulb should be a 250-watt lamp. The hard Pyrex heat-resisting bulb is preferred. In using heat lamps, care should be taken to avoid fire. Many hog houses burn each year because of carelessness in the use of heat lamps. The cord should be heavy and rubber-covered, with a block-type con-

nector. The socket should be porcelain and keyless, and the lamp should be supported by a chain, wire, or bracket. The electric cord should never be used to support the heating unit.

The unit should be placed about 20 inches above the top surface of the bedding, and care should be taken that water cannot drip down on the heat lamp. To prevent the sow from getting into contact with the heat unit, a barrier should be built with only room enough for the pigs to get under the lamp for heat. The barrier should be pen height or 36 inches.

Tube-type quartz lamps are available that provide the highest intensity infrared heat. They are usually $\frac{3}{8}$ inch in diameter and from 8 to 48 inches long. They are placed in a horizontal position above the creep section of free stalls, or shorter lamps are placed above the creep sections of farrowing stalls.

Metal sheath heaters consisting of a nichrome heating element and a supporting metal sheath are available in many sizes and may be adapted to situations demanding rugged equipment and large amounts of heat.

Heat lamps should be plugged into permanent wiring circuits only, and care should be taken that adequate size of wire is used. No more than seven 250-watt lamps, or equivalent wattage, should be used on one No. 12 wire circuit.

In wiring a building for pig brooders and heat lamps, provide a permanent and separate circuit with a maximum of 1,600 to 1,800 watts on each circuit, protected by a 15-ampere fuse. The wire should be No. 12 size or heavier. The extension cords to the brooders should not be longer than six feet, and each pen should have a separate outlet. The pig brooder, like the guard rails, should be installed several days before the sows are expected to farrow.

Catalytic heaters using LP gas are available and may be less costly than electricity.

They are made in various sizes and may be used over the creep area of both farrowing stalls and free stalls. They are flameless and provide a radiant heat.

Electrically heated concrete slabs under the creep area of farrowing stalls and free stalls have proved very successful in providing heat for the young pigs. Electric heating cable is stapled to an asphalt-fiber board base, and 2 to 2½ inches of a dry, but workable, concrete mix is placed above the cable. About 20 to 25 watts of heat per square foot of heated area is recommended. When 2.75 watts per foot cable is used, it should be spaced about 1½ inches apart.

FIGURE 7-5 (*Above*). What not to do. Note unprotected drop cord and heat lamp, and absence of guard rails. FIGURE 7-6 (*Below*). An excellent farrowing stall arrangement. The solid creep dividers help prevent drafts and spread of disease. (*Courtesy Farmstead Industries and Kemlite Corp.*)

Farrowing Stalls

The farrowing stall in Figure 7-7 shows the use of both guard rails and heat lamps,

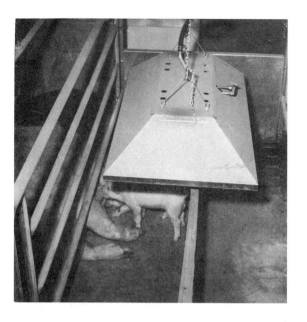

which is the most effective means of providing heat and protection for the pigs. One large producer who used farrowing stalls saved all but two pigs farrowed by a herd of 22 brood sows. It is estimated that farmers can save at least one more pig per litter by using these stalls.

Tests conducted by the Kentucky Agricultural Experiment Station involving 22 sows and litters confined to farrowing crates and an equal number kept in conventional pens equipped with guard rails and modified hovers proved the superiority of farrowing stalls. Only 3.2 percent of the pigs farrowed by sows in crates were crushed, as compared to 8.1 percent crushed by sows in conventional pens.

FIGURE 7-7 (*Above*). Gas or electric pig brooders provide zone heating in a confinement house. The pigs are warm, yet the sows are comfortable. (*Courtesy A. R. Wood Mfg. Co.*) FIGURE 7-8 (*Below*). A farrowing crate with slatted floors under the waterer and at the rear of the crate. (*Graves Peterson photograph. Courtesy Farmstead Industries.*)

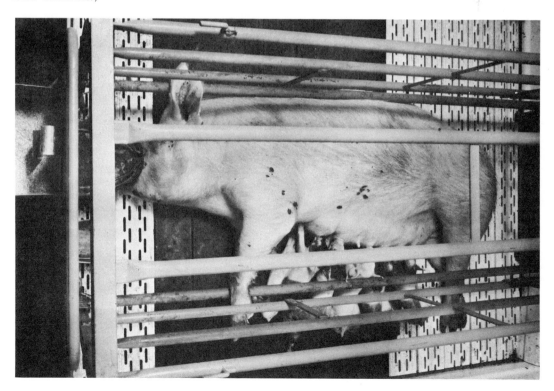

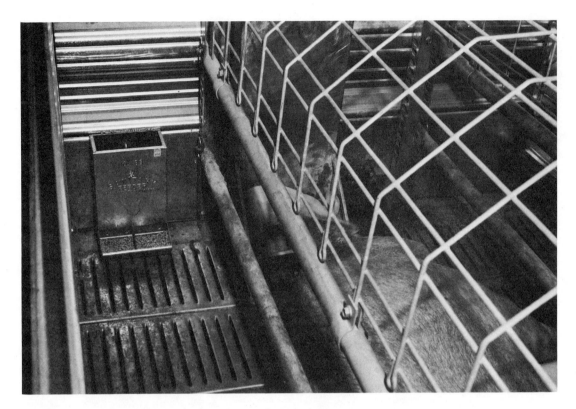

FIGURE 7-9 (*Above*). A crate with slatted floor for both pigs and sow. Note the feeder in the pig brooder area. (*Courtesy Behlen Mfg. Co.*)   FIGURE 7-10 (*Below*). Plans for a farrowing stall. (*Courtesy University of Illinois.*)

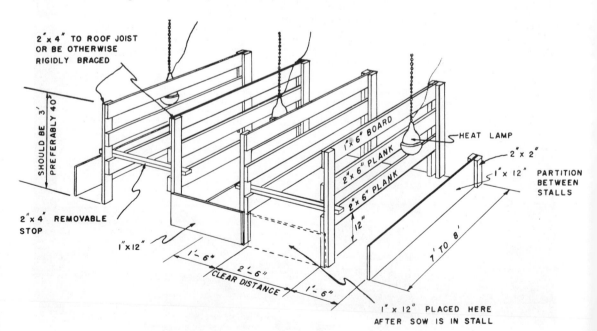

Shown in Figure 7-10 is a drawing of a recommended farrowing stall. The sow is confined in the stall (which is 24 inches wide) except when she is let out for feed, water, and exercise. The little pigs can run around the sow and under the heat lamps on either side. Each heat lamp may be used by two litters at the same time, since a divider is provided to keep the litters separated from each other.

The stall should be at least 6 feet long, but a length of 8 feet is preferred. A two-by-four across the back of the stall may be used to regulate length. The partition panels should be from 10 to 12 inches above the floor, depending upon the size of the sow.

Sows may be left in the farrowing stall for just a few days or for several weeks. By moving the sows to regular pens after two or three days it is possible to use the same stall for several litters. The sow may be removed from the stall each morning and evening for exercise, feed, and water.

Farrowing units are being used by large scale producers. These units may have 10 to 50 farrowing crates or stalls. The floors may be completely or partially slatted. Concrete slats 3 to 4 inches wide with ⅜-inch spacings are common. The spacings in the two-foot area behind the sow are widened to one inch to permit ease in cleaning. A grate is used over the wide spacing during farrowing and for a few days following. Some crates have solid floors under the sow and under the heated pig brooder area with slats used only in the area back of the sow.

Slatted floors can reduce 8 to 10 minutes per sow the time necessary to clean farrowing quarters. Illinois studies show feeding and cleaning took 2 minutes per day for each

FIGURE 7-11. A confinement farrowing setup using free stalls and pig brooder areas. (*Courtesy Moorman Mfg. Co.*)

sow kept and fed on slats and 10 to 12 minutes per sow on solid floors.

Most commercially constructed farrowing crates have doors that will swing open from either side, and many provide for a feeder and waterer in one of the doors. All have metal rails lengthwise the topside of the crate to keep the sows from getting out.

Free Stalls

In Figure 7-11 on page 153 is one free-stall arrangement. There are many types. The free stall provides more room for the sow, and the animal adjusts to it more readily than to a crate. Since there is room for the sow to back out of the stall area for feeding and dunging, she has greater opportunity for exercise. The stall requires more floor space than does a crate, but the pigs can remain in the free stall until weaned.

Feeding and Management

Rations at farrowing time. The sow should be fed a bulky, laxative ration in moderate amounts just before farrowing. She should receive all the water that she can drink, but cold water should be avoided. It is usually best to reduce the ration just before farrowing and to give her no feed for 12 hours after farrowing, unless she is nervous and appears hungry.

The rations presented in Tables 7-1 and 7-2 are recommended for sows at farrowing time.

The sow may be fed the same ration the three days after farrowing as the three days before. It is usually best to feed about a half-ration the first day and to increase the ration gradually until she is on full feed. Heavy milking sows should be hand-fed for the first week after farrowing. Other sows may be self-fed.

Assistance at Farrowing Time

Although most sows do not need assistance at farrowing time, it is a good policy to be on hand. Sows become nervous as they approach farrowing time and may pace the pen or crate and scrape up bedding materials. Most sows farrow within 24 hours after milk develops in the nipples.

If the building is cold at farrowing time, you may need to dry the pigs and place them under the heat lamp as they are farrowed. The pigs should be returned to the sow to suckle as soon as possible. Pigs normally will

T A B L E 7-1   BULKY RATIONS TO BE FED AT FARROWING TIME

Ingredient	Pounds	
	RATION 1	RATION 2
Ground corn		300
Ground sorghum or corn	300	
Ground oats	300	250
Linseed oilmeal		50
Alfalfa meal	100	150
Commercial vitamin-mineral fortified protein balancer	100	50
Wheat bran	200	200
	2,000	2,000

T A B L E 7-2 COMPOSITION OF VITAMIN AND TRACE MINERAL PREMIXES
For Balanced Swine Ration Formulas

Vitamin Premix[1]	
ESSENTIAL	
Vitamin A, million I.U.	3.0
Vitamin D, million I.U.	0.4
Riboflavin, grams	2.0
Pantothenic acid, grams	8.0
Niacin, grams	15.0
Vitamin B_{12}, milligrams	10.0
OPTIONAL	
Vitamin E, thousand I.U.[2]	10.0
Menadione (source of vitamin K), grams[3]	2.0
Carrier	?
	10 lbs.

[1] A feed additive if desired may also be placed in the vitamin premix.

[2] Supplemental E is needed only in certain areas of the United States where the selenium content of feed ingredients is extremely low because of the low selenium level in the soil. If deficient areas exist in Iowa, they have not been differentiated at the present time.

[3] The vitamin K requirement is normally met by the level present in natural feedstuffs and by intestinal synthesis. A hemorrhagic or bleeding syndrome has been diagnosed which is apparently due to a vitamin K antimetabolite which interferes with the use of vitamin K. The antimetabolite is thought to be produced by a mold occurring in one or more of the ration ingredients. When this has occurred, adding menadione has been helpful in preventing or overcoming the problem.

Trace Mineral Premix[1]	
The swine trace mineral premix should contain approximately the following percent of trace elements:	
Iron (Fe)	7.0%
Copper (Cu)	0.45%
Manganese (Mn)	5.5%
Zinc (Zn)	8.0%
Selenium (Se) Optional	0.005%

Trace Mineral Premix (Example)	
Ferrous sulfate	35.0
Copper sulfate	2.0
Manganese sulfate	24.0
Zinc sulfate	36.0
Potassium iodide	0.05
Sodium selenite	0.011
Carrier	2.939
	100.00 lbs.

[1] If iodized salt is not used at the rate of 0.25 to 0.50 percent, then 0.03 percent iodine (I) should be included in the swine trace mineral premix.

Source: *Life Cycle Swine Nutrition*, Iowa State University, 1974.

suckle each two or three hours. If pigs are removed from the sow, they should be returned each two or three hours for nourishment. Some pigs may not show interest in nursing even though they are a day or two old. These pigs may respond to a hand-fed solution of half sirup and half water.

Hogs whose handlers have worked with them are usually not nervous if the handlers are present at farrowing time. If the hogs are nervous, it is best to leave them alone or to work very quietly.

Generally speaking, when sows have difficulty farrowing, it is best to have a veterinarian render assistance. A veterinarian with a small hand may straighten out a pig which a sow has been unable to deliver. However, under no conditions should one attempt to do this without the use of a rubber glove which has been coated with vaseline. Brucellosis in hogs in readily transferable to humans and undulant fever may result.

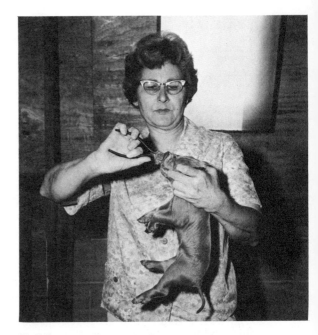

FIGURE 7-12. *Above:* The upper and lower incisor teeth are being clipped to avoid injury to the sow or the other baby pigs. (*Courtesy* Hog Farm Management.) *Below:* Drying, warming, and weighing newborn pigs. (*Courtesy American Cyanamid Company.*)

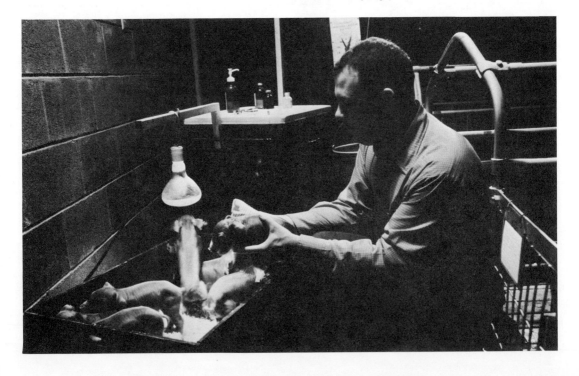

Navel Cord

Sometimes the navel cord is long and it is impossible for the pig to move about freely. The cord should be cut, and the navel daubed with a tincture of iodine solution. The latter should be used on all pigs regardless of the length of the cord. It may be necessary to tie the cord if bleeding continues.

Needle Teeth

The clipping of needle teeth is a controversial issue. Some hog raisers do not clip the teeth because they believe that the mouth and gums may be injured, allowing infectious organisms to enter the body. However, the practice, if done properly, will not cause disease infections. It may even prevent them, because, when fighting, pigs will not be able to inflict wounds about each other's faces that often permit necrotic and rhinitis organisms to enter the body.

Clippings should be done with a pair of cutters in such a way that there is a clean, smooth break with no injury to the gums and no jagged edges.

Ear-notching

Breeders of purebred hogs and most commercial hog producers make a practice of notching the pigs' ears at farrowing time. It is the most practical method of identifying the pigs of a litter so that the productiveness of a sow can be determined and considered in the selection of breeding stock. All sow-testing programs begin with the ear-notching of the pigs.

Most producers number the litters in the order in which they are farrowed. The first litter farrowed is No. 1, the second litter farrowed is No. 2, and so forth. The notches

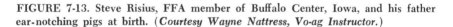

FIGURE 7-13. Steve Risius, FFA member of Buffalo Center, Iowa, and his father ear-notching pigs at birth. (*Courtesy Wayne Nattress, Vo-ag Instructor.*)

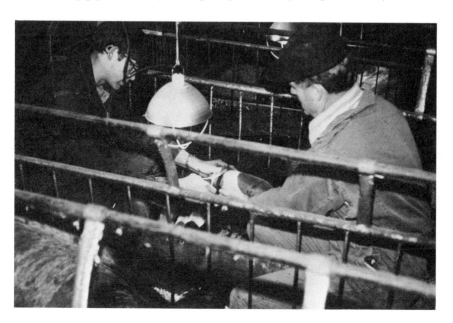

may be made by using a notching instrument or scissors.

Although there are numerous marking systems, the following is most commonly used:

STANDARD EAR-NOTCHING SYSTEM
One notch in the lower right ear is 1
One notch in the lower left ear is 3
One notch in the upper right ear is 10
One notch in the upper left ear is 30

EXAMPLES:
Litter No. 2—Two notches in the lower right ear.
Litter No. 14—One notch in the upper right ear, one notch in the lower left ear, and one notch in the lower right ear.

Swine producers and research specialists usually mark each pig so that it can be identified by litter number and by number within the litter. This is accomplished by using a dual system of ear-notching. One ear is used to identify the litter; the other is used to identify the various pigs of the litter.

The Universal Ear-Notching System is approved by all breed associations and is used in registration of breeding animals.

The right ear is used in the universal system to indicate litter number, and the left ear to indicate the pig number in the litter. All notches are made in the one-third of the ear nearest the body, and in the one-third nearest the tip of the ear. No notches are made in the middle of either the upper or the lower ear.

One notch near the body in the lower ear represents 1, one notch in the lower ear near the tip represents 3, and one notch in the upper ear near the tip represents 9 in both the right and the left ears. One notch in the upper right ear near the body indicates 27, and a notch in the tip of the ear represents 81. This ear notching system is illustrated in Figure 7-15.

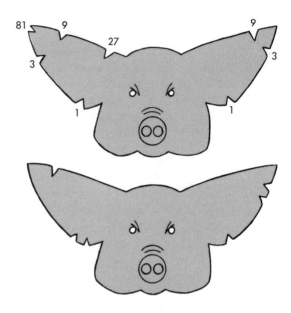

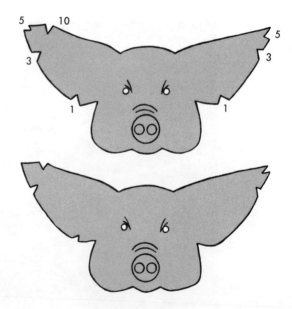

FIGURE 7-14 (*Above*). Standard ear-notching system. Notches in right ear indicate litter number; notches in left ear indicate pig number. The lower drawing shows pig No. 7 of litter No. 11. FIGURE 7-15 (*Below*). Alternate ear-notching system. Notches in right ear indicate litter number; notches in left ear indicate pig number. Pig No. 8 of litter No. 14 is illustrated in lower drawing. (*Figures 7-14 and 7-15 are original drawings by Linda and Jerry Geisler.*)

Treating Tails

The tails of young pigs may become wet and chapped. Mange or other infections may enter the breaks in the skin and cause the loss of the tails. Although hogs without tails or with short tails bring the same price at the packing plant as do pigs with tails, buyers of purebreds and exhibitors of hogs prefer animals with tails. The loss of tails can be avoided (1) by keeping the bedding and quarters dry, warm, and free from drafts, and (2) by coating the tails with Vaseline, a medicated salve, or iodine.

Docking of Tails

Many feeder-pig producers and commercial swine growers dock the tails at time of birth when the needle teeth are clipped, the navel is treated, and the ears are notched. Usually the tails are clipped near the body and a disinfectant is used. This is considered a necessary procedure when the pigs are to be grown out in confinement.

MANAGEMENT DURING THE SUCKLING PERIOD

It has been estimated that from 25 to 30 percent of pigs that are farrowed fail to reach a weaning age of five to six weeks. Although many of the pigs die because they are lain upon, injured, diseased, or chilled, many of them also die because they do not receive sufficient milk and other feeds. The ration given the sow, the ration provided for the pigs, and the management given the sow and litter determine whether or not a high percentage of the pigs in the litter will survive and attain a healthy heavy weight at weaning time.

Feeding

Feeding the sow. A sow, when fed a good ration, may produce from 6 to 8 pounds of milk per day. When the size of the litter is small, 5 or 6 pounds of milk may be sufficient. Sows, however, which are nursing from 10 to 12 pigs must produce the maximum of milk to satisfactorily meet the needs of the pigs. As we increase the size of the litters produced by our sows, we must also improve the rations fed to both the sows and their litters. A sow's milk contains about 81 percent water, nearly 6 percent fat, slightly more than 6 percent protein, about 6 percent lactose sugar, and about one percent ash or mineral. The sow must receive feeds containing these nutrients in sufficient amounts to produce the milk required by the litter.

The bulky, laxative ration recommended for the sow at farrowing time is usually continued for several days after farrowing. The amount is increased daily. Some producers are having success feeding a 12 or 13 percent protein, bulky ration from two weeks before farrowing to about 10 days following farrowing. Sometime during the second week after farrowing the sow should be on full feed. By that time the pigs require large amounts of milk, and changes must be made in the ration. Most sows by this time can, and should be, self-fed.

The condition of the sows, the number of pigs being nursed, and the type of pasture available influence the amount and kind of ration which should be fed. The rations presented in Tables 7-3 and 7-4 are recommended for sows during the suckling period.

Thyroprotein. When thyroprotein, frequently called iodinated casein, is included in the sow's ration, pig gains have increased 20 to 30 percent during the first week of lactation. Thyroprotein is a product manufactured from milk protein and iodine. It contains about 1 percent thyroxine activity. Thyroxine is the hormone produced by the thyroid gland. Thyroprotein is fed to sows to stimulate more milk production.

Creep-feeding young pigs. Pigs will begin to nibble at feeds when a few days old

T A B L E 7-3 LACTATION RATIONS[1]

Percent Protein	Ingredient		1	2	3	4
8.9	Ground yellow corn[2]		1,677	1,577	1,577	1,522
44.0	Solvent-process soybean meal[3]		250	250	250	130
8.0	Dried beet pulp		—	100	—	—
14.5	Wheat bran		—	—	100	100
17.0	Dehydrated alfalfa meal		—	—	—	100
50.0	Meat and bone meal		—	—	—	100
	Calcium carbonate (38% Ca)		15	15	15	—
	Dicalcium phosphate (26% Ca, 18.5% P)		25	25	25	5
	Iodized salt		10	10	10	10
	Trace mineral premix		3	3	3	3
	Vitamin premix		20	20	20	20
	Feed additives[4]		—	—	—	—
	Total		2,000	2,000	2,000	2,000

CALCULATED ANALYSIS

	Protein	%	12.96	12.92	13.24	13.73
	Calcium	%	0.64	0.67	0.65	0.62
	Phosphorus	%	0.52	0.51	0.56	0.55
	Lysine	%	0.61	0.63	0.62	0.61
	Methionine	%	0.23	0.22	0.23	0.23
	Cystine	%	0.22	0.21	0.23	0.21
	Tryptophan	%	0.14	0.15	0.15	0.14
	Metabolizable energy	kcal./lb.	1,318	1,300	1,294	1,249

Feeding Directions

[1] These rations may be full-fed from a few days after farrowing until the baby pigs are weaned.

[2] Ground oats can replace corn up to 15 percent of the total ration. Ground milo, wheat, or barley can replace the corn.

[3] If 48.5 percent soybean meal is used instead of 44 percent soybean meal, use 30 pounds less soybean meal and 30 pounds more corn. If whole cooked beans are used in place of 44 percent soybean meal, use 30 pounds more soybeans and 30 pounds less corn.

[4] A high level of feed additive (100 to 300 gm./ton) may be beneficial 2 to 3 weeks before farrowing until 7 to 10 days after farrowing, but will be of little benefit thereafter.

Source: *Life Cycle Swine Nutrition*, Iowa State University, 1974.

and can consume a considerable amount of feeds by the time they are two or three weeks old. The milk production of the sow usually declines after the third week, and it is important that the little pigs be fed a palatable ration in ample amounts. This is especially true when sows are suckling large litters, and when the sows are in poor condition.

T A B L E 7-4   LACTATION RATIONS FOR SOWS AND GILTS[1]

Ingredient		1	2	3
Ground yellow corn		800	1,400	1,400
Ground oats		690	—	—
Beet pulp		—	100	—
Ground wheat		—	—	100
Solvent-process soybean meal (44%)[2]		230	404	404
Dehydrated alfalfa meal (17%)		200	—	—
Meat and bone meal (50%)		70	—	—
Calcium carbonate (38% Ca)		—	18	18
Dicalcium phosphate (26% Ca, 18.5% P)		30	45	45
Iodized salt		10	10	10
Trace mineral premix		3	3	3
Vitamin premix		20	20	20
Feed additives[3]		—	—	—
CALCULATED ANALYSIS:				
Protein	%	16.30	15.50	15.76
Calcium	%	0.95	0.99	0.97
Phosphorus	%	0.76	0.72	0.77
Lysine	%	0.67	0.83	0.83
Methionine	%	0.32	0.25	0.26
Cystine	%	0.24	0.25	0.26
Tryptophan	%	0.20	0.18	0.19

Source: South Dakota State University

[1] These rations may be full-fed a few days after farrowing until pigs are weaned.

[2] If 48.5 percent soybean meal is used instead of 44 percent, use 25 to 50 pounds less soybean meal and 25 to 50 more pounds of corn. If whole cooked beans are used instead of 44% soybean meal, use 100 pounds more beans and 100 pounds less corn.

[3] High level feed additives (100 to 300 gm./ton) may be beneficial 2 to 3 weeks prior to farrowing until 7 to 10 days after farrowing, but will be of little benefit thereafter.

The little pigs need a more palatable ration, higher in protein and in antibiotics and lower in fiber content, than do their mothers. As a result, the feeding of the pig starters or creep rations is recommended. These rations should be fed in a trough or feeder in a pen separate from that of the mother. The gates may be adjusted so that the little pigs can get in but the sows cannot. The use of prestarters and creep rations is very important if the pigs are to be weaned from the sows at an early age.

PRESTARTERS AND CREEP RATIONS

There have been a number of important developments in the creep feeding of pigs during the past few years. A common practice among hog producers 30 years ago was to feed suckling pigs rations of rolled or hulled oats, cracked wheat, or cracked corn. These feeds when fed in a creep aided materially in supplementing the sow's milk. More young pigs were saved, they were heavier at weaning time, and there was less

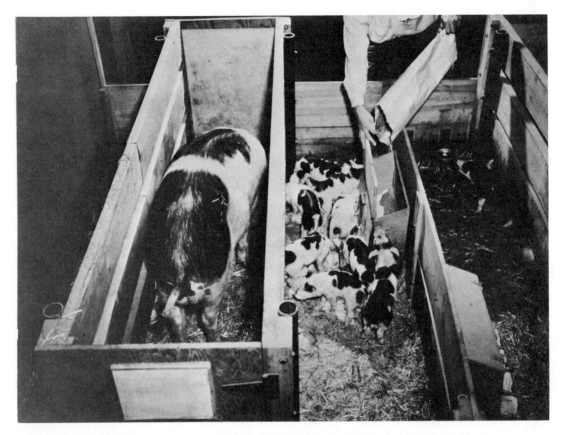

FIGURE 7-16. These pigs have a prestarter ration available in a creep. (*Courtesy Eli Lilly and Company.*)

loss of weight on the sow during the suckling period than when creep rations were not fed to the pigs.

A few years ago much research was done in developing synthetic milk rations in the hope that pigs could be weaned from the sows when only a few days old. While excellent feeds were developed, they required more exacting procedures than the average farmer was willing or had the time to use, and the cost of the feeding program was high.

Swine nutritionists next concentrated their efforts toward building dry pig starters which were more palatable and better fortified with vitamins, minerals, and antibiotics than creep rations previously recommended.

A number of excellent rations have been developed.

Pig Creep-feeding Program

Shown in Table 7-5 is the Iowa State University pig creep-feeding program which involves the feeding of three creep rations.

Prestarter period. The milk replacer rations are self-fed for a period of one to three weeks beginning when the pigs are seven days old. They usually are not weaned from the sow. Each pig will consume about 3 to 5 pounds of the milk replacer ration during the period and will weigh at the close of the period about 10 to 12 pounds.

TABLE 7-5   PIG CREEP-FEEDING PROGRAM

Period	Days	Weight	Feed % Protein	Amount of Feed Needed	Feed per Pound Gain
Prestarter	7 to 21	10 to 12 lbs. (minimum weaning weight)	21 to 23	3 to 5 lbs. total	1.2 to 1.25
Starter	21 to 42 or weaning	up to 40 lbs.	18 to 19	36 to 44 total	1.2 to 1.6

Source: Adapted from *Life Cycle Swine Nutrition*, Iowa State University, 1974.

Shown in Table 7-6 are the formulas of the rations. No more than ten pigs should be penned together, and the pigs should be of the same size. Approximately 4 to 6 square feet of floor space per pig should be provided, and floor drafts should be avoided. Solid pen walls should be used. Heat lamps or electrically heated floors should be provided, and the bedding should be kept clean and dry. Young pigs will need plenty of fresh water. Automatic fountains are preferred, and they should be clean and sanitary at all times.

Starter period. As the pigs reach 10 to 12 pounds in weight the ration is shifted from the milk replacer to a highly fortified starter feed which has been pelleted. See Table 7-7 for formulas of starter rations for weaned and unweaned pigs. Weaned pigs should be fed 18 to 19 percent protein rations.

The starter period usually begins when the pigs are about three weeks old. It ends when the pigs weigh an average of about 40 pounds. They are then about 6 or 7 weeks old.

Pigs like pellets better than meals or crumbled feeds. Dusty or finely ground feeds are not eaten readily. Most pig starters on the market are in pellet form, and it has been found that pigs will consume nearly twice as much feed in pellet form as in the form of meal.

In feeding the starter ration, the pigs should not be crowded. At least four inches of feeder space should be provided for each two pigs. An adequate supply of fresh water should be available at all times. Each pig will require 36 to 44 pounds of the starter ration.

The pig grower ration is substituted for the starter ration as the pigs reach an average weight of 40 pounds. The protein requirements of pigs of this weight are considerably less than for younger and lighter pigs. The milk replacer ration contains 21 to 23 percent protein, the starter 18 to 19 percent, and the grower 14 to 16 percent protein.

The grower ration may be fed to weaned or unweaned pigs. Weaning is a normal process if the pigs are provided adequate amounts of the creep-fed rations.

PREWEANING MANAGEMENT

Anemia Control

Small pigs which have had little exercise and no opportunity to pick up iron and copper from the soil may develop anemia.

TABLE 7-6   MILK REPLACERS[1]
(For Baby Pigs Before 3 Weeks of Age)

Percent Protein	Ingredient	1	2	3	4
8.9	Ground yellow corn	881	652	551	530
48.5	Solv. soybean meal	455	500	600	500
15.0	Rolled oat groats	—	—	—	200
33.0	Dried skim milk	200	400	200	200
12.0	Dried whey	200	200	400	400
31.0	Fish solubles	—	50	50	—
	Sugar	200	100	100	100
	Stabilized animal fat	—	50	50	20
	Calcium carbonate (38% Ca)	14	10	10	10
	Dicalcium phosphate (26% Ca, 18.5% P)	22	10	10	10
	Salt	5	5	5	5
	Trace mineral premix	3	3	3	3
	Vitamin premix	20	20	20	20
	DL-methionine	—	—	1	2
	Feed additives[2] (gm/ton)	100-300	100-300	100-300	100-300
	Total	2,000	2,000	2,000	2,000

CALCULATED ANALYSIS

Protein	%	19.45	23.61	23.48	21.69
Calcium	%	0.98	0.72	0.69	0.69
Phosphorus	%	0.81	0.62	0.61	0.61
Lysine	%	1.23	1.58	1.56	1.40
Methionine	%	0.33	0.43	0.44	0.45
Cystine	%	0.37	0.35	0.36	0.34
Tryptophan	%	0.24	0.30	0.30	0.28
Metabolizable energy	kcal./lb.	1,380	1,429	1,418	1,405

Feeding Directions

[1]The milk replacer ration is normally fed in only limited amounts. It should be used for pigs weaned prior to 3 weeks of age until they reach approximately 12 pounds. Then they can be switched to a starter ration. It is a good ration to feed orphan pigs when the sow dies, extreme disease outbreak (TGE) occurs, or the sow is agalactic.

[2]The feed additive may be part of the vitamin premix, or if a separate premix, it should replace an equal amount of corn.

Source: *Life Cycle Swine Nutrition*, Iowa State University, 1974.

TABLE 7-7   PIG STARTER RATIONS[1]

Percent Protein	Ingredient	1	2	3	4	5
8.9	Ground yellow corn	1,091	1,087	1,216	966	887
48.5	Solv. soybean meal	440	—	450	450	—
44.0	Solv. soybean meal	—	550	—	—	—
37.0	Soybeans, whole cooked	—	—	—	—	750
33.0	Dried skim milk	—	—	50	50	—
12.0	Dried whey	400	300	200	300	300
31.0	Fish solubles	—	—	—	50	—
	Sugar	—	—	—	100	—
	Stabilized animal fat	—	—	20	20	—
	Calcium carbonate (38% Ca)	15	10	10	10	10
	Dicalcium phosphate (26% Ca, 18.5% P)	20	25	25	25	25
	Iodized salt	5	5	5	5	5
	Trace mineral premix	2	3	3	3	3
	Vitamin premix	25	20	20	20	20
	DL-methionine[2]	2	—	1	1	—
	Feed additives[3] (gm/ton)	100-300	100-300	100-300	100-300	100-300
	Total	2,000	2,000	2,000	2,000	2,000

CALCULATED ANALYSIS

Protein	%	18.1	18.62	18.28	18.66	19.51
Calcium	%	0.7	0.72	0.69	0.73	0.75
Phosphorus	%	0.6	0.63	0.62	0.64	0.66
Lysine	%	1.06	1.11	1.06	1.12	1.16
Methionine	%	0.38	0.30	0.35	0.36	0.30
Cystine	%	0.30	0.31	0.30	0.30	0.31
Tryptophan	%	0.22	0.23	0.22	0.22	0.26
Metabolizable energy	kcal./lb.	1,350	1,309	1,375	1,359	1,422

Feeding Directions

[1]The pig starter ration should be used as a creep ration before weaning and fed after weaning until the pigs reach approximately 40 pounds. Then the pigs can be switched to a grower-finisher ration.

[2]Rations 2 and 5 are borderline in methionine content. The addition of 1 pound of DL-methionine to these rations may improve performance and feed efficiency slightly, although pigs will perform very satisfactorily on these rations without the DL-methionine addition.

[3]The feed additive may be part of the vitamin premix, or if a separate premix, it should replace an equal amount of corn.

Source: *Life Cycle Swine Nutrition*, Iowa State University, 1974.

FIGURE 7-17 (*Above*). Pigs like crumbled or pelleted feeds. (*Courtesy Eli Lilly and Company.*)   FIGURE 7-18 (*Below*). The heat source is placed over the creep area on each side of the stall. These pigs are all crowded in on one side of the sow. (*Agricultural Associates photograph. Courtesy Starcraft Livestock Equipment.*)

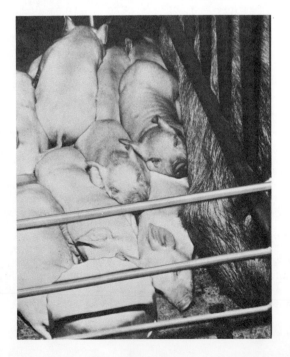

Anemia in pigs is a blood condition usually caused by a lack of copper and iron.

The anemic pigs usually show a loss of appetite, become weak, have trouble in breathing, and sometimes have a swollen condition around the head and shoulder. Sometimes they "thump."

There are at least four ways of preventing anemia. First, pigs which are raised in the late spring or summer and have access to earth and sod do not get anemia. The little pigs get enough iron and copper from the sod to satisfy their needs. Second, if the pigs are confined and cannot get sod, bring it to them. Get the sod from a disease-free field or from along the roadside. Keep sod before them until they can get out on clean ground pasture.

The third method of controlling anemia is to use drugs obtained from the drugstore or veterinarian. Iron and copper pills can be given weekly to a pig, starting when it is about five days old. The udder of the sow may be brushed daily with a solution of copperas (ferrous sulfate), which can be purchased at any drugstore.

Although anemia can be avoided by swabbing the sow's udder or providing fresh soil if practiced faithfully, it has been found that the injection of 2 cc. of iron-dextrans or iron-dextrins at three days and at three weeks is the most economical control.

Tests conducted at the University of Tennessee indicated no significant difference in weaning weights of pigs when pigs were provided iron in (1) clean sod three times each week for four weeks, (2) commercial iron supplements in pellet form three times

each week for four weeks, and (3) injectable iron shots at two to four days containing 100 mg. of elemental iron, and a second injection of 100 mg. of iron at 10 to 12 days.

A large number of preparations are being marketed for anemia control. The material used (1) should be rapidly absorbed, and the total iron should be available for the production of hemoglobin, (2) should not produce pain or inflammation at the site of injection, (3) should be stable in storage, (4) should possess a low toxicity, and (5) should have a sufficient concentration of iron to keep injection volume at a minimum.

Most anemia control products contain 50 mg. of iron per cubic centimeter. The

FIGURE 7-19 (*Below*). Sod is provided to prevent anemia. (*Courtesy Rath Packing Company.*) FIGURE 7-20 (*Right*). A young pig being treated for scours on an Illinois farm. (*Courtesy* National Hog Farmer.)

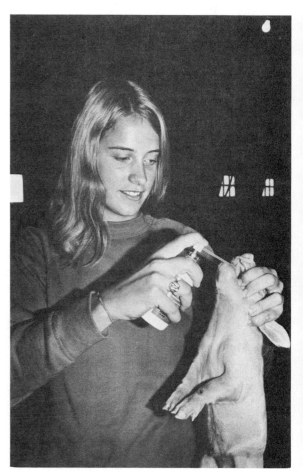

iron-dextrans and iron-dextrins have in the main proven superior to other injectable materials. It is important that the first injection be made in the neck muscle of each pig when it is three days old. The second shot of 1 to 2 cc. may be given when the pigs are about three weeks old, if they are not eating freely a creep ration, or if they are unthrifty.

The pigs will need supplemental iron until they start eating solid food in the form of creep rations adequately fortified with iron. The sow's milk provides only about one-sixth of the pigs iron needs.

In using injectable iron products, it is important that the needle be sufficiently large to get the solution readily into the syringe. Eighteen-, 19-, or 20-gauge needles, ¾ to 1 inch long, are recommended. Larger needles may leak and smaller needles may make it difficult to fill the syringe.

There is some question concerning the need for a second injection in the control of anemia. Pigs that are healthy and consume normal amounts of a well-fortified milk replacer creep ration may not need the second injection. Many producers give the second injection as an insurance measure. It is likely that some pigs in any group of pigs may need the shot although it would appear to be unnecessary for the group as a whole.

Castration

Pigs should be castrated when they are one to two weeks of age. The later they are castrated, the greater the setback. Under no conditions should the pigs be castrated,

FIGURE 7-21. The "Pig Grip" is a handy piece of equipment for castrating and vaccinating pigs. (*Courtesy Caswell Manufacturing Co.*)

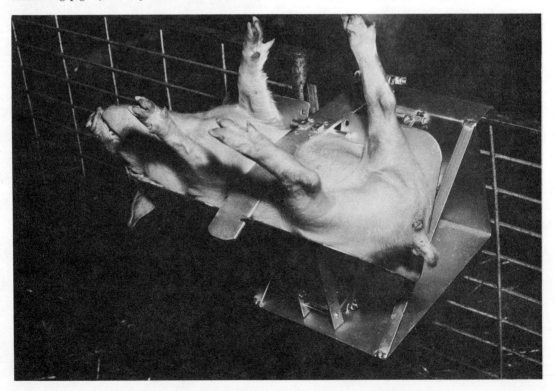

wormed, vaccinated, or weaned at the same time. These operations should be spaced at two-week intervals.

Unless heat is provided by the use of heated floors, brooders, or heat lamps, pigs should be castrated when the weather is warm. The hog house and pen should be clean, dry, and well bedded.

Unless a mechanical holder is used, two persons are usually needed when castrating, one to hold the pig while the other operates. The pig may be held by a front and a hind leg on opposite sides with its back on the floor, or by the hind legs with its head and shoulders between the assistant's knees. The scrotum should be washed with a mild antiseptic solution or with soap and water. The sharp knife used in castrating and the operator's hands or rubber gloves should be carefully cleaned and disinfected.

The incision may be made over each testicle, or in between the testicles, parallel to the middle line of the body. The incision should pass through the skin near the top of the testicle and through the testicle covering. The testicle is pulled slowly through the incision, and the attachments are separated with little bleeding. The second testicle is removed in the same manner.

Care should be taken to make the incision fairly long and low so that it will drain well. As much of the cord should be removed as possible. Usually there is no need to apply disinfectant or healing oil after the operation.

Sometimes pigs are ridglings or "originals." They have but one testicle in the scrotum. The other testicle is in the body of the pig. It is often best to turn over these pigs, as well as those that are ruptured, to a veterinarian.

Vaccination

Vaccination of pigs for cholera has been a common and necessary practice for years.

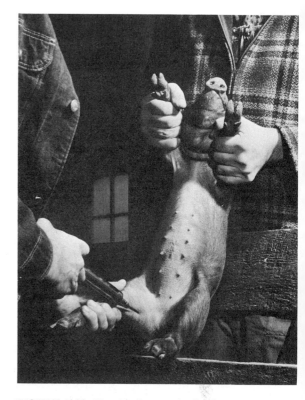

FIGURE 7-22. Vaccinating a pig for leptospirosis or for erysipelas at about 8 weeks of age. Vaccinating in neck is preferred. (*Hufnagle photograph. Reprinted from* Successful Farming.)

As a result of a vigorous four-phase cholera eradication program, it is now not only unnecessary but also illegal to vaccinate for cholera.

In some communities erysipelas has become a problem, and it is necessary to have the veterinarian vaccinate the pigs for this disease when they are from eight to ten weeks of age. This normally should be done about ten days or two weeks after weaning.

Antibiotics

Young pigs with a high disease-level environment will show greatest response to

FIGURE 7-23. Effect of feeding vitamin B_{12} and antibiotics to pigs. *Top left:* Pigs received no vitamin B_{12} or antibiotics. *Top right:* Pigs received vitamin B_{12}. *Bottom:* Pigs received vitamin B_{12} and Aureomycin in addition to basal ration. (*Courtesy U.S.D.A. Bureau of Animal Industry.*)

antibiotics. It is recommended that 200 grams of antibiotics be added to each ton of the milk replacer ration used during the pre-starter period, 100 to 200 grams to each ton of starter ration fed to pigs weighing 12 to 25 pounds, and 50 to 100 grams per ton of feed fed to pigs 25 to 75 pounds in weight (a pound equals 453.6 grams, and a gram equals 1,000 milligrams). Experimental work indicates that Aureomycin, Terramycin, tylosin, and penicillin are about equal in increasing rate of gain and in protecting pigs

from disease. Mixtures of these and other antibiotics, however, may be more effective than a single antibiotic.

Antibiotics vary in their effectiveness against different kinds of disease organisms. As a result, there is an advantage in feeding combinations of antibiotics or combinations of antibiotics and other drugs. It is desirable also to change the antibiotics used from time to time. Combinations of (1) Aureomycin, sulfamethazine, and penicillin, (2) tylosin and sulfamethazine, (3) penicillin and streptomycin, and (4) Terramycin, sulfamethazine, and penicillin are recommended. The manufacturer's recommendations should be followed.

Tests conducted at Illinois and Minnesota stations indicate that the implanting of antibiotic pellets in baby pigs did not increase the weaning weights or the survival of pigs to weaning age.

Arsenicals

Arsenicals such as sodium arsanilate or arsenilic acid may be used alone or with antibiotics in feeds. The University of Illinois recommends the addition of 90 grams of arsenilic acid or 22 grams of 3-nitro-4-hydroxy-phenylarsonic acid per ton of pig starter. Arsenicals have much the same effect upon pigs as antibiotics. The higher the disease level the greater the effect. Feeds containing 50 to 200 grams of antibiotics per ton are used with arsenicals in preventing or controlling infectious enteritis in swine.

Presented in Table 7-8 are recommended levels of feeding antibiotics and arsenicals to hogs.

Copper Sulfate

Copper sulfate increases rate of gain and feed efficiency when added to the ration of pigs under 100 pounds in weight. Research shows that about two pounds of copper sulfate or copper oxide per ton of hog ration gives the best gain and conversion. By destroying streptococcus and other organisms, copper sulfate or oxide acts in a manner similar to some antibiotics.

Pasture

Many hog producers like to move the sows and litters to clean legume pasture as soon as possible after the pigs are a week or two old. Pigs can be raised efficiently while confined, but care must be taken to provide sanitary quarters, and a ration which contains the minerals, vitamins, and proteins, which are provided by pasture crops.

Care should be taken in moving pigs to pasture so that no infections are picked up en route. It is better to haul the sow and litter than to drive them. On some farms concrete runways have been provided for this purpose.

WEANING

Advantages of Early Weaning

With the use of fortified milk replacer, starter, and grower rations, it is possible to wean the pigs in advance of the usual age of eight weeks. Swine specialists are convinced that early weaning of pigs will reduce loss by death of pigs, prevent large losses in weights of the sows, and permit

T A B L E 7-8   RECOMMENDED ANTIBIOTIC AND ARSENICAL LEVELS

| | | Grams Per Ton of Complete Feed | |
Ration		Antibiotic	Arsenilic Acid
Milk replacer	(up to 10 to 12 pounds)	200	45 to 90
Starter	(12 to 25 pounds)	100 to 200	45 to 90
Grower	(25 to 75 pounds)	50 to 100	45 to 90
Developer	(60 to 125 pounds)	20 to 50	45 to 90
Finisher	(125 pounds to market)	20 to 50	45 to 90

earlier rebreeding of the sows. They also think that early weaning will reduce the number of runts, prevent the spread of disease from the sow to the pigs, economize on feed for the sows, save labor, and make more economical use of farrowing house and equipment. Less floor space is required when pigs are weaned early.

Disadvantages of Early Weaning

Early weaning of pigs requires special rations which must be carefully formulated. These rations are expensive. Not all hog producers are willing to buy or mix these highly fortified feeds. When they try to wean early and feed ordinary rations, the results are shocking.

Good management is essential to successful early weaning of pigs. The hog farmer must assume full responsibility when the sow is removed from the litter.

Early weaning requires better than average equipment. The sow may be able to do a fair job of raising the litter even with inadequate equipment.

Weaning Pointers

1. Pigs less than 21 days of age and weighing less than 10 to 12 pounds should not be weaned from the sow unless adequate equipment, heat, labor, and highly fortified rations are available.
2. The tendency among swine producers is to wean pigs when they are from three to five weeks of age.
3. It is a good idea to wean pigs by weight rather than by age. Wean the larger pigs from the litter first. Some producers wean pigs as they reach 20 pounds in weight. The pigs left with the sows make rapid gains.
4. Avoid making any changes in the ration at weaning time. It is good practice to have the pigs eating the weanling ration for several days before weaning time.
5. The sows should be moved from the pigs rather than the pigs from the sow at weaning time. A new location plus the loss of the sows will make the pigs uneasy. The sows should be completely isolated from the pigs.
6. Care should be taken to make certain that feed and water are fresh and plentiful.
7. Warm, dry, well-ventilated quarters are necessary during winter months. Shade, cool, and good ventilation are important in hot weather.
8. It is best to separate pigs by weight and keep not more than about 20 pigs of the same size in a pen at weaning time.
9. In Minnesota tests, pigs that were weaned at 3 weeks of age and received rations containing 40 grams of Aureomycin and 210 grams of arsenilic acid per ton of feed, averaged from 3 to 7½ pounds heavier at 8 weeks of age than those that received no antibiotics or arsenicals.

Weaning Day-old Pigs

Primary SPF pigs are obtained by hysterectomy on the 112th day of gestation, and are reared in isolation in a pathogen-free environment. They are fed a sterile modified cow's milk or infant formula diet. Several commercial diets are available. These pigs do not have an opportunity to consume the colostrum milk produced by their mothers, yet a high percentage of them are reared to maturity.

It has been found that pigs that have been farrowed naturally and remain with their mothers 6 to 12 hours and have consumed the colostrum, can be successfully reared in nonisolated yet rigidly controlled environments. There has been much interest in weaning one- or two-day-old pigs.

Early weaning of day-old pigs can eliminate losses due to overlay by the sows, can reduce fighting of pigs for food, and can reduce runts. The sows will be able to produce more litters in a year, with a resulting smaller feed cost per pig.

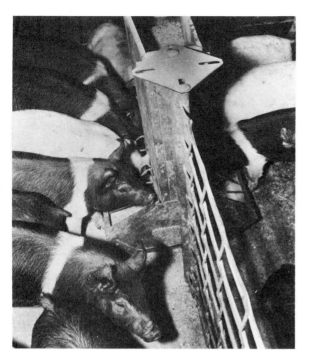

FIGURE 7-24. These pigs are weaned and doing well on a starter ration in the pig nursery. (*Courtesy Big Dutchman.*)

Several commercial devices are being marketed for use in feeding and rearing early weaned pigs. Many producers have constructed mechanical devices that have met their needs.

Presented in Table 7-9 is a ration used by Michigan State University in feeding pigs weaned at two or three days of age. It is fed as a liquid and care is taken to maintain sanitary, warm, and dry pen environments.

In Table 7-10 is a dry ration for pigs weaned at two to three days of age. Fresh water is available at all times. In Michigan tests the early weaned pigs were not as heavy at two weeks of age as those reared on the sow, but had caught up with the sow-reared pigs by the time they were five weeks old. The pigs fed the dry ration did not do as well the first day or two and were a few pounds lighter than those fed a liquid diet when five weeks old.

Recommendations for feeding and managing pigs weaned at different ages are presented in Table 7-11.

SUMMARY

From 25 to 30 percent of the pigs farrowed never reach weaning age, and 80 to 90 percent of death losses occur shortly after farrowing time. The cost of producing a pig doubles when one-half of the litter is lost.

Sows should be penned up two or three days before farrowing time. The farrowing pen or stall should be thoroughly cleaned and disinfected. Farrowing stalls are recommended over farrowing pens and guard rails. The use of free stalls is increasing. They represent a combination of the advantages of pen and farrowing crate facilities. Pens 6 by 10 or 12 feet with a creep at one end for pigs, and a feeding and dunging area for sows at the other end are recommended. Heat lamps or heated floors should be provided on each side of the sow. Farrowing quarters should be bedded lightly with dust-free dry material.

A bulky, 12 to 13 percent protein, laxative ration in moderate amounts should be fed just prior to farrowing. Fresh chill-free water should be available. It is good practice to be on hand at farrowing time. The navel cord should be clipped and treated with tincture of iodine. Pigs should be ear-marked at birth to facilitate record keeping.

Needle teeth should be clipped in case pigs are inclined to fight. The tails of pigs to be reared in confinement should be docked at birth.

Sows should be fed lightly after farrowing but brought to full feed on a 12 to 13 percent protein ration in about a week or ten days.

Pigs should be fed a 21 to 23 percent protein, highly fortified prestarter or milk replacer feed until they weigh 10 to 12 pounds. They then should be fed a starter ration containing 18 to 19 percent protein.

T A B L E 7-9   DRY SEMI-PURIFIED RATION FOR PIGS WEANED AT 2 OR 3 DAYS OF AGE

Ingredient	Percent
Purified milk or soybean protein + DL-Methionine	30
Lard	5
Glucose and/or lactose	53
Cellulose	3
Mineral Mixture supplying 14 mineral elements	6
Corn oil with 4 fat soluble vitamins	1
Water with 12 water soluble vitamins	2
	100%

Protein		Mineral	Vitamins	
Total protein	25%	Ca — 0.8-1.0%	Thiamine	3 ppm
Arginine	1.0-2.0	P — 0.6-0.8	Riboflavin	6
Histidine	0.7	K — 0.4	Niacin	40
Isoleucine	1.8	Na — 0.4	Calcium pantothenate	30
Leucine	2.5	Cl — 0.4	Pyridoxine	2
Lysine	2.0	Mg — 400 ppm	Para aminobenzoic acid	13
Methionine	0.7	Fe — 100	Ascorbic acid	80
Phenylalanine	1.5	Zn — 80	α-Tocopheryl acetate	10
Threonine	1.2	MN — 20	Inositol	130
Tryptophan	0.3	Cu — 10	Choline Chloride	1300
Valine	1.7	Co — 1	Folic acid	260 ppb
		I — 1	Biotin	50
Energy		Se — 0.1	B_{12}	100
		SO_4	Vitamin K_3	40
Gross energy = 1850 kcal./lb.		HCO_3	Vitamin A palmitate	1500
		Co_3	Vitamin D_2	10

Source: Michigan State University.

At 40 pounds they should be fed a grower ration containing 14 or 15 percent protein.

The milk replacer ration should contain 200 grams of antibiotics per ton; the starter ration, 100 to 200; and the grower ration should contain 50 to 100 grams per ton.

Most pigs can be weaned when three to five weeks of age if adequate rations and quarters are provided. Pigs should be weaned by weight, rather than by age.

Pigs like and do better on pellet feeds than on meals. Creep feeds that are highly fortified with vitamins, minerals, antibiotics, and arsenicals help to increase swine production.

Anemia may be controlled by providing clean sod, by use of iron-dextrin injections or by use of copper pills that should be given each pig once each week. Boars should be castrated by the time they are one to two weeks old.

Vaccination for erysipelas should be done when pigs are eight to ten weeks old if needed. Vaccination, castration, and weaning should not be done at the same time.

Early weaning will save feed and loss of weight on the sow and permit early rebreeding. Early weaning also aids in disease control and reduces death loss. The key to success in early weaning rests with the adequacy of ration, the quarters available, and the attention given the pigs by the producer.

T A B L E 7-10   DRY NATURAL INGREDIENT RATION FOR PIGS WEANED AT 2 OR 3 DAYS OF AGE

Ingredient	%
Dried Skim milk (33% C.P.)	10
Dried whey (16% C.P.)	20
Soybean meal (50% C.P.)	40
DL-menthionine	0.3
Dicalcium phosphate	2
Vitamin-trace mineral premix[1]	1
Salt	0.5
Corn oil	2
Corn dextrose	24.2

[1]One pound supplies the following: vitamin A, 400,000 IU; vitamin D_2, 60,000 IU; vitamin E, 2,000 IU; riboflavin, 0.3gm.; calcium pantothenate, 1.2gm.; niacin, 1.6gm.; choline chloride, 10gm.; B_{12}, 1.8 mg.; zinc, 6.8gm.; iron, 5.4gm.; manganese, 3.4gm. and iodine, 0.2gm.

Source: Michigan State Universtiy.

T A B L E 7-11   RECOMMENDATIONS FOR FEEDING AND MANAGING PIGS WEANED AT DIFFERENT AGES

Weaning Age	Birth	2 Days	10 Days	3 Weeks	5 Weeks
Frequency of feeding	90 min. intervals	3 times daily	Ad libitum	Ad libitum	Ad libitum
Type or ration	Liquid	Liquid or dry	Dry	Dry	Dry
Type or protein	Colostrum then milk protein	Milk and soy proteins	Soy, milk and grain proteins	Soy, grain and milk proteins	Soy and grain proteins
Level of protein, %	30	25	22	19	16
Type of carbohydrate	Lactose or glucose	Lactose or glucose	Lactose, glucose or starch	Starch, sucrose, lactose or glucose	Starch, sucrose, lactose or glucose
Type of pen	Individual	Individual	Group	Group	Group
Floor area, ft^2/pig	1.5	1.5	2	2.5	3
Ambient temperature, °F	90+	85-90	80	75	70

Source: Michigan State University.

QUESTIONS

1 Outline the plan which you should use on your home farm in caring for brood sows at farrowing time.

2 Which method is best in saving newborn pigs: using pens with guard rails and heated floors or heat lamps? farrowing stalls? or using free-stall facilities? Why?

3 Make a diagram of an economical and serviceable farrowing stall.

4 What attention should be given pigs at farrowing time?

5 Outline a system that you can use on your farm in ear-marking the pigs.

6 Describe the differences in milk replacer, starter, and grower rations.

7 How many pounds of each of these feeds will be consumed by an average pig before it reaches 40 pounds in weight?

8 Which antibiotics and arsenicals should be fed to young pigs and in what amounts?

9 At what age should pigs be vaccinated for erysipelas? castrated?

10 At what age should the pigs be weaned on your farm?

11 How can anemia be controlled in young pigs?

12 What are the advantages and disadvantages of early weaning?

13 Outline a program of feeding and management of the sows and litters on your farm through the weaning period.

REFERENCES

Aanderud, Wallace G., *Practical Hog Production and Marketing for S. D. Farmers*, EC 649. Brooking, S.D.: South Dakota State University, 1973.

Ensminger, M. E., *Swine Science*, 4th ed. Danville, Ill.: Interstate Printers and Publishers, 1970.

Holden, Palmer, Vaughan Speer, and E. J. Stevermer, *Life Cycle Swine Nutrition*, Pm. 489. Ames, Ia.: Iowa State University, 1974.

Krider, J. L., and W. E. Carroll, *Swine Production*, 4th ed. New York, N.Y.: McGraw-Hill Book Company, 1971.

Luce, William G., *Managing the Sow and Litter*, No. 3650. Stillwater, Okla.: Oklahoma State University, 1971.

Moyer, W. A., L. T. Wendling, and B. A. Koch, *Successful Swine Management*. Manhattan, Kan.: Kansas State University, 1971.

8 RATIONS AND MANAGEMENT FOR GROWING AND FINISHING PIGS

Approximately one-third of the total cost of producing a 220-pound market hog is involved in raising the pig to weaning age. It takes the remaining two-thirds of the total cost to grow out the pig from weaning time until it is ready for market. From the production standpoint, the period from weaning to market is an important one since more than twice as much capital is involved in growing out pigs as is involved in producing weanling pigs.

The efficient hog raiser carefully plans the production program. Adequate rations are to be fed in proper amounts. The feeding method chosen is the one best suited to the farm situation. Pasture of good quality and in the desired amount is also provided in the plan. Efficiency in housing the hogs and in controlling diseases and parasites is another consideration. The profit from the enterprise is determined by the cost of production and by the selling price. Usually the hog raiser can influence the production cost more easily than the selling price.

Worming

Most farms are contaminated with round and other worms. These pests may be controlled in part by the use of rotated pastures, but we are never certain that the mature animals in the herd are free from the parasites. Furthermore, more and more hogs are being raised in confinement. The only safe way is to treat the pigs for worms.

With confinement feeding, the worm problem may become acute. The only safe way is to worm the sows 2 to 3 weeks before farrowing and worm the pigs 2 to 3 weeks after weaning.

There are at least nine commercial wormers available:

Levamisole HCl (Tramisol) is the most recently developed wormer. It was approved in 1971 for control of lung-worms. It is a broad spectrum wormer and will control also round and nodular worms. It is used in drinking water as a one-dose treatment.

Dichlorvos (Atgard) is another recently developed wormer and rates high. It is more powerful than previous wormers and can be used with younger pigs than other wormers. It expels adult and young adult ascarid round worms, nodular worms, and whipworms. It is a single treatment wormer. It is mixed with feed and must be hand-fed.

Thiabendazole is a drug recently approved as an aid in the prevention of infestations of large round worms. It is fed at the rate of 1 pound per 1,000 pounds of feed for a 14-day period. In Florida tests it was very effective in eliminating the eggs of *Strongyloides ransomi* worms from the feces of weanling pigs. This worm is the smallest type of round worm that infests pigs. It is called intestinal threadworm and is quite prevalent in southern states.

Hygromycin is an antibiotic and should be fed continuously from the time the pigs start eating until they are marketed. It should be mixed with the feed according to directions of the manufacturer. Continued feeding may impair the hearing of hogs. Hygromycin B is sold in a premix called Hygromix. Five pounds of the premix mixed in a ton of feed yields the desired level of drug concentration. It controls three major hog worms—round worms, whipworms, and nodular worms.

Piperazine compounds should be fed at the rate of one-fifth of an ounce of anhydrous piperazine per 100 pounds of animal weight. The pigs should be wormed at weaning time and at 50-day intervals until marketed. Piperazine wormers are administered in water or in feed after the pigs have been fasted for a 12- or 18-hour period. The drug should be mixed thoroughly with either the feed or the water that will be consumed in a 24-hour period. This wormer, like hygromycin B, is not toxic and does not upset the digestion of the animal. It is also effective against nodular worms.

Other swine wormers are discussed in Chapter 11.

PASTURE VERSUS CONFINEMENT FEEDING

Some hog producers make effective use of pastures. Legume pastures supply needed

FIGURE 8-1. These Hampshires are being grown out on pasture. The pigs on the right have been weaned. (*Courtesy Hampshire Swine Registry.*)

proteins, vitamins, and minerals and reduce the quantities of feeds needed in growing out a group of pigs. The use of pasture may also aid in the prevention of disease and parasite losses.

It is usually more difficult to maintain sanitary quarters for confined pigs than for pigs on pasture. In some areas nearly 60 percent of the pigs are reared in confinement. The advantages and disadvantages of pasture and confinement feeding are presented in Chapter 9.

Rations

Nutritionists have formulated rations which will provide the nutritive requirements of pigs in dry lot. The rations are more highly fortified with proteins, vitamins, minerals, and antibiotics than are the rations recommended for use in feeding pigs on pasture. Tests indicate that pigs in confinement can make rapid and economical gains.

FIGURE 8-2. A typical confinement finishing facility. Note the mechanical feeding equipment and the open sides. (*Courtesy Starcraft Swine Equipment.*)

Carcass Quality

Many swine feeders have assumed that pasture feeding will result in the production of better carcasses than when dry-lot feeding is practiced. The assumption is probably true if both groups of hogs are fed the same ration. Tests indicate, however, that carcasses superior to pigs grown on pasture may be produced by pigs fed in confinement, providing satisfactory rations are fed and sanitary quarters are maintained.

NUTRITIONAL NEEDS OF GROWING-FINISHING PIGS

The nutritional needs of pigs from weaning to market weight have been described in Chapter 4. We present in this chapter the methods which may be used in meeting these needs.

Farm Grains

Farm grains should make up the main part of the ration. Corn, when available, produces the most rapid and economical gains. Barley, wheat, oats, and grain sorghums may be fed, but they should be ground. A common practice in the Corn Belt is to feed rations made up quite largely of corn supplemented with soybean oilmeal, or with a mixed protein feed. In areas where corn is not obtainable, it is possible to build good rations by using ground barley or ground grain sorghums.

Gibberella-infected corn can not be fed to swine breeding stock, but can be fed to growing-finishing pigs if less than five percent of the corn is infected. Infected corn should be mixed with good corn (one to three ratio), if fed, and increased amounts of vitamins A and K should be added to the ration.

Pigs prefer high moisture corn to dry corn, but storage costs are higher. Corn can be harvested earlier, however, and there is less field loss. Growth, feed efficiency, and performance are about equal when both types of corn are fed.

Since grains and pasture cannot provide all the proteins, vitamins, and minerals pigs need, protein and mineral supplements fortified with vitamins, antibiotics, and arsenicals must be added to the ration.

Proteins

It was pointed out in Chapter 4 that proteins are made up of amino acids, and that ten of these amino acids are essential for animal health. Corn is low in two of these essential amino acids: lysine and tryptophan. Therefore, it is necessary to supplement corn with protein feeds high in these two amino acids. Most proteins of plant origin are deficient in lysine. The exception is soybean oilmeal, which is a good source of lysine. Animal proteins, such as tankage, meat scraps, and fish meal, are also good sources of this amino acid. A good protein supplement should contain soybean oilmeal or a meat origin protein, or both.

Plant proteins, in general, are good sources of tryptophan, whereas meat and bone scraps, if fed alone with corn, will not provide an adequate amount of this amino acid for maximum growth. Most mixed protein and mineral supplements contain several protein feeds of plant origin, which provide an adequate supply of tryptophan.

Value of protein supplements. In general, protein feeds are more important in feeding pigs weighing from 40 to 75 pounds than in feeding those weighing 100 pounds and over. Light pigs fed corn alone will gain less than one-third of a pound per day. The same pigs fed a protein supplement will gain more than a pound a day. Heavier pigs fed corn alone will require about 200 pounds more feed to produce 100 pounds of gain than will pigs fed corn and a protein supplement. Tests show that 100 pounds of soybean meal may save from 375 to 450 pounds of corn.

Protein feeds are more important in feeding pigs in confinement than in feeding those on pasture. The use of good legume pastures can reduce the amount of protein concentrates by nearly 50 percent.

Until the discovery of vitamin B_{12}, it was necessary to provide rations containing from 18 to 20 percent protein for weanling pigs. By adding B_{12}, we can reduce this percentage to 14 to 16 percent, providing the ration is well balanced. Pigs weighing 40 to 120 pounds require 14 to 16 percent protein rations; heavier hogs, from 120 to 240 pounds, require 12 to 14 percent rations if the rations are balanced and vitamin B_{12} has been added. The addition of B_{12} in the ration can reduce the amount of protein supplement to the extent of about 80 to 110 pounds of 40 percent supplement saved per 240-pound market hog. Be certain that you feed a protein supplement containing vitamin B_{12}.

Following are examples of recommended protein supplements for pigs on pasture and on dry lot:

Ration 1
(Purdue Supplement 5—for Pigs in Confinement)

	Percent
Meat and bone scraps	20
Menhaden fish meal	20
Soybean oilmeal	40
Cottonseed meal	10
Alfalfa leaf meal	10
Total	100

Ration 2
(Purdue Supplement C—for Pigs on Pasture)

	Percent
Meat and bone scraps	20
Fish meal	20
Soybean oilmeal	40
Cottonseed meal	10
Linseed meal	10
Total	100

Source: Purdue University.

Antibiotics, arsenicals, and minerals are discussed later on in this chapter. It is suggested that these materials be mixed with the protein supplement and be self-fed. For pigs weighing from 40 to 120 pounds, 160 pounds of corn and 41 pounds of 40 percent protein supplement will make the recommended 15 percent protein ration. For pigs weighing more than 120 pounds, 172 pounds of corn and 29 pounds of 40 percent supplement will make a 13 percent ration. The amounts of protein supplements and grains needed to make up rations of different levels of protein are shown in Tables 8-1 and 8-2 (pages 182 and 184). We are indebted to nutritionists of Iowa State University for this information.

Though soybean meal is the most commonly used protein supplement during the growing-finishing period, the use of roasted soybeans has increased. In Michigan tests, roasted soybeans adequately replaced soybean meal in corn-soybean meal rations fed to growing-finishing pigs. Similar results have been obtained in other states. Since roasted whole soybeans contain about 18 percent fat (oil), it is necessary to reduce the amount of corn in the ration when the roasted beans are fed.

The feeding of ground, unprocessed soybeans is not recommended. The heating process changes the oil in the soybean making it more desirable as a feed supplement.

Recent tests indicate that both single-cell protein and whey can be used as supplements in swine feeding.

With the high price of soybeans, it is necessary to consider alternative protein sources. Listed below are the number of pounds of various substitutes that equal the

FIGURE 8-3. Weaned pigs in a partially slatted floor facility. Note automatic waterer over slatted floor. (*Courtesy Big Dutchman.*)

T A B L E 8-1   AMOUNT OF SUPPLEMENT OF DIFFERENT PERCENT PROTEIN NEEDED WITH CORN TO FORMULATE RATIONS OF DIFFERENT LEVELS OF PROTEIN

(Shelled Corn Only Figured at 8.5% Protein)

% Protein in Supplement		Percent of Protein in Total Ration										
		10	11	12	13	14	15	16	17	18	19	20
30	Grain (lb.)	1860	1767	1674	1581	1488	1395	1302	1209	1116	1023	1070
	Suppl. (lb.)	140	233	326	419	512	605	698	791	884	977	930
	Lb. grain per 1 lb. suppl.	13.3	7.6	5.1	3.8	2.9	2.3	1.9	1.5	1.3	1.0	1.2
31	Grain	1867	1778	1689	1600	1511	1422	1333	1244	1156	1067	1022
	Suppl.	133	222	311	400	489	578	667	756	844	933	978
	Lb. grain per 1 lb. suppl.	14.0	8.0	5.4	4.0	3.1	2.5	2.0	1.6	1.4	1.1	1.0
32	Grain	1872	1787	1702	1617	1532	1447	1362	1277	1191	1106	1021
	Suppl.	128	213	298	383	468	553	638	723	809	894	979
	Lb. grain per 1 lb. suppl.	14.6	8.4	5.7	4.2	3.3	2.6	2.1	1.8	1.5	1.2	1.0
33	Grain	1877	1796	1714	1633	1551	1469	1388	1306	1224	1143	1061
	Suppl.	123	204	286	367	449	531	612	694	776	857	939
	Lb. grain per 1 lb. suppl.	15.3	8.8	6.0	4.4	3.5	2.8	2.3	1.9	1.6	1.3	1.1
34	Grain	1882	1804	1725	1647	1569	1490	1412	1333	1255	1176	1098
	Suppl.	118	196	275	353	431	510	588	667	745	824	902
	Lb. grain per 1 lb. suppl.	15.9	9.2	6.3	4.7	3.6	2.9	2.4	2.0	1.7	1.4	1.2
35	Grain	1887	1811	1736	1660	1585	1509	1434	1358	1283	1208	1132
	Suppl.	113	189	264	340	415	491	566	642	717	792	868
	Lb. grain per 1 lb. suppl.	16.7	9.6	6.6	4.9	3.8	3.1	2.5	2.1	1.8	1.5	1.3
36	Grain	1891	1818	1745	1673	1600	1527	1455	1382	1309	1236	1164
	Suppl.	109	182	255	327	400	473	545	618	691	764	836
	Lb. grain per 1 lb. suppl.	17.3	10.0	6.8	5.1	4.0	3.2	2.7	2.2	1.9	1.6	1.4
37	Grain	1895	1825	1754	1684	1614	1544	1474	1404	1333	1263	1193
	Suppl.	105	175	246	316	386	456	526	596	667	737	807
	Lb. grain per 1 lb. suppl.	18.0	10.4	7.1	5.3	4.2	3.4	2.8	2.4	2.0	1.7	1.5

38	Grain	1220	1288	1356	1424	1492	1559	1627	1695	1763	1831	1898
	Suppl.	780	712	644	576	508	441	373	305	237	169	102
	Lb. grain per 1 lb. suppl.	1.6	1.8	2.1	2.5	2.9	3.5	4.4	5.6	7.4	10.8	18.6
39	Grain	1246	1311	1377	1443	1508	1574	1639	1705	1770	1836	1902
	Suppl.	754	689	623	557	492	426	361	295	230	164	98
	Lb. grain per 1 lb. suppl.	1.7	1.9	2.2	2.6	3.1	3.7	4.5	5.8	7.7	11.2	19.4
40	Grain	1270	1333	1397	1460	1524	1587	1651	1714	1778	1841	1905
	Suppl.	730	667	603	540	476	413	349	286	222	159	95
	Lb. grain per 1 lb. suppl.	1.7	2.0	2.3	2.7	3.2	3.8	4.7	6.0	8.0	11.6	20.1
41	Grain	1292	1354	1415	1477	1538	1600	1662	1723	1785	1846	1908
	Suppl.	708	646	585	523	462	400	338	277	215	154	92
	Lb. grain per 1 lb. suppl.	1.8	2.1	2.4	2.8	3.3	4.0	4.9	6.2	8.3	12.0	20.7
42	Grain	1313	1373	1433	1493	1552	1612	1672	1731	1791	1851	1910
	Suppl.	687	627	567	507	448	388	328	269	209	149	90
	Lb. grain per 1 lb. suppl.	1.9	2.2	2.5	2.9	3.5	4.2	5.1	6.4	8.6	12.4	21.2
43	Grain	1333	1391	1449	1507	1565	1623	1681	1739	1797	1855	1913
	Suppl.	667	609	551	493	435	377	319	261	203	145	87
	Lb. grain per 1 lb. suppl.	2.0	2.3	2.6	3.1	3.6	4.3	5.3	6.7	8.9	12.8	22.0
44	Grain	1352	1408	1465	1521	1578	1634	1690	1746	1803	1859	1916
	Suppl.	648	592	535	479	422	366	310	254	197	141	84
	Lb. grain per 1 lb. suppl.	2.1	2.4	2.7	3.2	3.7	4.5	5.5	6.9	9.2	13.2	22.8
45	Grain	1370	1425	1479	1534	1589	1644	1699	1753	1808	1863	1918
	Suppl.	630	575	521	466	411	356	301	247	192	137	82
	Lb. grain per 1 lb. suppl.	2.2	2.5	2.8	3.3	3.9	4.6	5.6	7.1	9.4	13.6	23.4

Source: Iowa State University, 1968.

T A B L E 8-2 AMOUNT OF SUPPLEMENT OF DIFFERENT PERCENT PROTEIN NEEDED WITH 3/4 CORN AND 1/4 OATS TO FORMULATE RATIONS OF DIFFERENT LEVELS OF PROTEIN

(3/4 Shelled Corn at 8.5% Protein and 1/4 Oats at 12% Protein)

% Protein in Supplement		Percent of Protein in Total Ration										
		10	11	12	13	14	15	16	17	18	19	20
30	Grain (lb.)	1939	1842	1745	1648	1552	1455	1358	1261	1164	1067	1030
	Suppl. (lb.)	61	158	255	352	448	545	642	739	836	933	970
	Lb. grain per 1 lb. suppl.	31.8	11.7	6.8	4.7	3.5	2.7	2.1	1.7	1.4	1.2	1.1
31	Grain	1942	1850	1757	1665	1572	1480	1387	1295	1202	1110	1017
	Suppl.	58	150	243	335	428	520	613	705	798	890	983
	Lb. grain per 1 lb. suppl.	33.5	12.3	7.2	5.0	3.7	2.8	2.3	1.8	1.5	1.2	1.0
32	Grain	1945	1856	1768	1680	1591	1503	1414	1326	1238	1149	1061
	Suppl.	55	144	232	320	409	497	586	674	762	851	939
	Lb. grain per 1 lb. suppl.	35.4	12.9	7.6	5.2	3.9	3.0	2.4	2.0	1.6	1.4	1.1
33	Grain	1947	1862	1778	1693	1608	1524	1439	1355	1270	1185	1100
	Suppl.	53	138	222	307	392	476	561	645	730	815	900
	Lb. grain per 1 lb. suppl.	36.7	13.5	8.0	5.5	4.1	3.2	2.6	2.1	1.7	1.5	1.2
34	Grain	1949	1868	1787	1706	1624	1543	1462	1381	1300	1218	1137
	Suppl.	51	132	213	294	376	457	538	619	700	782	863
	Lb. grain per 1 lb. suppl.	38.2	14.2	8.4	5.8	4.3	3.4	2.7	2.2	1.9	1.6	1.3
35	Grain	1951	1873	1795	1717	1639	1561	1483	1405	1327	1249	1171
	Suppl.	49	127	205	283	361	439	517	595	673	751	829
	Lb. grain per 1 lb. suppl.	39.8	14.7	8.8	6.1	4.5	3.6	2.9	2.4	2.0	1.7	1.4
36	Grain	1953	1878	1803	1728	1653	1578	1502	1427	1352	1277	1202
	Suppl.	47	122	197	272	347	422	498	573	648	723	798
	Lb. grain per 1 lb. suppl.	41.6	15.4	9.2	6.4	4.8	3.7	3.0	2.5	2.1	1.8	1.5
37	Grain	1955	1882	1810	1738	1665	1593	1520	1448	1376	1303	1231
	Suppl.	45	118	190	262	335	407	480	552	624	697	769
	Lb. grain per 1 lb. suppl.	43.4	15.9	9.5	6.6	5.0	3.9	3.2	2.6	2.2	1.9	1.6

38	Grain	1258	1328	1397	1467	1537	1607	1677	1747	1817	1886	1956
	Suppl.	742	672	603	533	463	393	323	253	183	114	44
	Lb. grain per 1 lb. suppl.	1.7	2.0	2.3	2.8	3.3	4.1	5.2	6.9	9.9	16.5	44.5
39	Grain	1283	1350	1418	1485	1553	1620	1688	1755	1823	1890	1958
	Suppl.	717	650	582	515	447	380	312	245	177	110	42
	Lb. grain per 1 lb. suppl.	1.8	2.1	2.4	2.9	3.5	4.3	5.4	7.2	10.3	17.2	46.6
40	Grain	1306	1372	1437	1502	1567	1633	1698	1763	1829	1894	1959
	Suppl.	694	628	563	498	433	367	302	237	171	106	41
	Lb. grain per 1 lb. suppl.	1.9	2.2	2.6	3.0	3.6	4.4	5.6	7.4	10.7	17.9	47.8
41	Grain	1328	1391	1455	1518	1581	1644	1708	1771	1834	1897	1960
	Suppl.	672	609	545	482	419	356	292	229	166	103	40
	Lb. grain per 1 lb. suppl.	2.0	2.3	2.7	3.1	3.8	4.6	5.8	7.7	11.0	18.4	49.0
42	Grain	1349	1410	1471	1533	1594	1655	1716	1778	1839	1900	1962
	Suppl.	651	590	529	467	406	345	284	222	161	100	38
	Lb. grain per 1 lb. suppl.	2.1	2.4	2.9	3.3	3.9	4.8	6.0	8.0	11.4	19.0	51.6
43	Grain	1368	1428	1487	1546	1606	1665	1725	1784	1844	1903	1963
	Suppl.	632	572	513	454	394	335	275	216	156	97	37
	Lb. grain per 1 lb. suppl.	2.2	2.5	2.9	3.4	4.1	5.0	6.3	8.3	11.8	19.6	53.1
44	Grain	1386	1444	1502	1560	1617	1675	1733	1791	1848	1906	1964
	Suppl.	614	556	498	440	383	325	267	209	152	94	36
	Lb. grain per 1 lb. suppl.	2.3	2.6	3.0	3.5	4.2	5.2	6.5	8.6	12.2	20.3	54.6
45	Grain	1404	1460	1516	1572	1628	1684	1740	1796	1853	1909	1965
	Suppl.	596	540	484	428	372	316	260	204	147	91	35
	Lb. grain per 1 lb. suppl.	2.4	2.7	3.1	3.7	4.4	5.3	6.7	8.8	12.6	21.0	56.1

Source: Iowa State University, 1968.

protein equivalent of 100 pounds of 44 percent soybean meal:

Cottonseed meal	116 pounds
Linseed meal	123 pounds
Fish meal	85 pounds
Meat and bone scrap	99 pounds
Tankage	91 pounds
Blood meal	81 pounds

Effect of Protein Level on Performance of Pigs

The high feed costs during 1973 and 1974 stimulated much experimentation to determine the minimum protein requirements of pigs during the growing and finishing periods yet producing adequate growth and feed efficiency. Presented in Table 8-3 is a summary of tests conducted in Nebraska.

The Nebraska tests indicated that most rapid gains made by pigs weighing 40 to 100 pounds resulted from rations containing 18 percent protein. Pigs fed 16 percent rations made most effective use of feed. These data indicate that pigs of this weight should be fed rations containing 16 to 18 percent protein.

Pigs weighing 100 to 170 pounds made most rapid gains when fed rations containing 16 percent protein, but best feed efficiency resulted from rations containing 14 percent protein. The improvement in feed efficiency appears to justify a 14 percent protein ration.

Though pigs weighing 170 to 250 pounds gained most rapidly on rations containing 12 percent protein, they were more efficient in use of feed when fed 14 percent protein rations.

Antibiotics

Antibiotics authorized by the Food and Drug Administration have been fed to hogs for the past 20 years, and they continue to increase rate of gain, increase feed efficiency, and help control certain types of nonspecific scours. The improvement resulting from the feeding of antibiotics varies greatly. There is greatest response when the antibiotics are fed to pigs weighing from 15 to 40 pounds. Pigs in unsanitary environments show greater

TABLE 8-3   EFFECT OF PROTEIN LEVEL ON GROWTH RATE AND FEED EFFICIENCY

Weight of Pigs	Percent Protein	Daily Gain	Feed Efficiency
40 to 100 pounds	12	1.27	3.16
	14	1.38	2.58
	16	1.46	2.48
	18	1.51	2.53
100 to 170 pounds	10	1.26	5.08
	12	1.65	3.56
	14	1.70	3.30
	16	1.73	3.39
170 to 250 pounds	10	1.71	4.37
	12	1.80	3.92
	14	1.76	3.80

Source: University of Nebraska.

FIGURE 8-4. Note that these pigs have had their tails docked to prevent tail biting. Self-feeders with mechanical feed handling equipment are being used. (*Courtesy Hawkeye Steel Products Inc.*)

response than do pigs reared under low disease-level conditions.

Under university experiment station conditions, pigs 15 to 45 pounds in weight had 20 to 40 percent increase in rate of gain and 7 to 15 percent increase in feed efficiency when fed antibiotics. Pigs 40 to 120 pounds showed a 9 percent increase in gain and less than 2 percent increase in feed efficiency. Pigs 100 pounds in weight fed antibiotics until marketed gained 8 percent faster on 4 percent less feed.

In 61 experiment station comparisons, pigs fed antibiotics from weaning until marketing gained 10.7 percent faster on 5.1 percent less feed than pigs that received no antibiotics. In Purdue tests antibiotic-fed pigs gained 6.7 percent faster and used about 2 percent less feed.

Some studies indicate that combinations of selected antibiotics are more effective than others, especially in controlling scours. Specialists are not able to recommend specific antibiotics for general use under all conditions. Swine producers should check with their veterinarians and feed service consultants in selecting the antibiotic, or combination of antibiotics to use in their ration formulation.

There is some evidence that under certain conditions it is desirable to rotate the use of antibiotics during the growing and finishing periods.

Though some swine producers may wish to omit antibiotics from rations of pigs between weaning and marketing, it is recommended that the average producer will profit by feeding rations containing 50 to 100 grams to pigs weighing 25 to 75 pounds, and rations containing 20 to 50 grams per ton of feed to pigs weighing more than 50 to 75 pounds.

The following antibiotics and antibiotic combinations have been approved for swine feeding:

1. **Bacitracin (Baciferm, Bio-Best, Noptranic, Fortracin, Kenitracin)**
2. **Chlortetracycline (Aureomycin)**
3. **Erythromycin**
4. **Flavomycin**
5. **Neomycin (Neomix)**
6. **Oleandomycin**
7. **Oxytetracycline (Terramycin)**
8. **Penicillin (Pro-Pen, Micro-Pen, Nopcaine)**
9. **Streptomycin**
10. **Tylosin (Tylan)**
11. **Virginiamycin (Staphylomycin)**
12. **Chlortetracycline, sulfamethazine, and penicillin (ASP-250)**
13. **Neomycin and oxytetracycline (Triple acting Neo-Terramycin)**
14. **Penicillin and streptomycin (Pro-Strep, Strepcillin, Streptocaine)**
15. **Tylosin and oxytetracycline (Tylan-Sulfa)**
16. **Chlortetracycline, sulfathiazole and penicillin (CSP-250)**

The feeding instructions of the manufacturer and of the veterinarian or feeding consultant should be carefully followed.

There has been some thought that the feeding of antibiotic materials to pigs throughout the growing and finishing period might increase the thickness of backfat. Tests, however, do not prove this to be true. Antibiotics fed to young pigs on high disease-level farms will produce maximum results. Tests also show that antibiotics can be effective only when fed with a balanced ration.

Antibiotics may be purchased in the form of vitamin-antibiotic premixes in 10-, 25-, and 50-pound packages. It is possible to add them to the other feeds included in the mixed ration. The directions of the manufacturer, or distributor, should be followed closely, and care should be taken to mix the materials uniformly with the other feeds.

A popular and easy way to obtain feeds containing antibiotics is to buy a ready-mixed feed. An analysis of the feed tag and the price should be made to determine the best feed and the best buy.

Several of the antibiotics and other feed additives must be removed from the ration several days or weeks before the animals are marketed. Swine producers marketing animals having traces of these additives in the carcasses at time of slaughter may be subjected to legal action. The directions of the manufacturers must be carefully followed. See Table 11-2 to determine the time specific feed additives should be withdrawn from swine rations.

Carbadox

Carbadox has been recently approved by the Food and Drug Administration as a synthetic antibacterial agent in swine feeding. It is sold under the name of Mecadox. It may be fed to pigs under 75 pounds. Carbadox must be withdrawn from the ration 10 weeks before the pigs are marketed. It

FIGURE 8-5. These pigs are being finished in a steel building equipped with slatted floors, self feeders, mechanical waterers, automatic feed conveyors, and a controlled ventilation system. (*Courtesy Butler Manufacturing Co.*)

has been proven effective in control of dysentery during the early life of the pig and there is some carryover effect after its removal from the ration.

Vitamins

Pigs require thiamine, riboflavin, niacin, pantothenic acid, pyrdoxine, biotin, folic acid, vitamin B_{12}, and choline for normal health. These are water-soluble vitamins.

In addition to the water-soluble vitamins, pigs require vitamins A, D, and E, which are fat-soluble. Yellow corn and green hays and pastures supply adequate amounts of vitamin A; sunlight, vitamin D; cereal grains and green hays or pastures, vitamin E.

Some of the water-soluble vitamins are available to the pigs in farm grains and forages. Thiamine or vitamin B_1, pyridoxine, biotin, and folic acid are available in cereal grains or in forage crops, and need not be provided from other sources.

Riboflavin should be provided at the rate of 1 milligram per pound of ration for pigs weighing from 40 to 240 pounds. Niacin, or nicotinic acid, is required at the rate of 20 milligrams for young pigs under 50 pounds in weight, and 75 milligrams per pound of ration for heavier pigs.

Pantothenic acid is required by growing pigs at the rate of 4 milligrams per pound of ration. Much of the choline may be supplied by feeding corn which contains 200 milligrams per pound, and oats, wheat, and barley, which contain about 450 milligrams per pound. Meat scraps, fish meal, and soybean meal are excellent sources of this vitamin.

Vitamin B_{12} was discovered to be an important part of APF (animal protein factor) and is a by-product in the manufacture of antibiotics. About 5 micrograms (1,000 micrograms equals one milligram) of vitamin B_{12} are needed per pound of dry ration for the most rapid gains of young pigs.

FIGURE 8-6. A pig showing riboflavin deficiency. (*Courtesy U.S.D.A. Bureau of Animal Industry.*)

The best means of providing the vitamins which cannot be supplied by farm grains, protein supplements, and forages is (1) to give them in the form of a vitamin-mineral premix, which can be mixed with the protein supplement or with the complete ration, (2) to buy a ready-mixed supplement which contains the needed vitamins and minerals, or (3) buy separate vitamin and mineral premixes and mix them with other ingredients in formulating a complete ration.

The composition of a recommended vitamin premix is presented in Table 8-4. Complete vitamin-mineral mixes for corn-soybean meal rations for use in feeding growing-finishing pigs are presented in Table 8-5.

Minerals

Pigs on good legume pasture that receive meat and bone meal usually need little additional mineral feeds. Pigs that are fed plant proteins on poor pasture need additional minerals. Cereal grains are very low in mineral content, and it is a good idea to self-feed a mineral mixture or a protein supplement to which minerals have been added. Calcium, phosphorus, and salt are the most important mineral materials in feeding hogs, and they are easily obtained. Ground limestone is an excellent source of calcium, and steamed bonemeal supplies both calcium and phosphorus. A low-fluorine calcium phosphate may also be used, but an excess of 0.4 percent of fluorine is toxic to the pigs. Suggested mineral mixtures are in Table 8-5.

The National Research Council recommends that rations fed growing-finishing pigs contain 0.4 percent phosphorus. The ratio of phosphorus to calcium in the ration should be 1 to 1. Illinois recommends 0.5 percent phosphorus and 0.5 percent calcium. Iowa recommends 0.6 percent calcium and 0.5 percent phosphorus during the growing period, and 0.5 percent calcium and 0.4 percent phosphorus during the finishing period.

Salt. Swine rations should contain about one-half pound of salt per 100 pounds. In tests conducted at Purdue University a penny's worth of salt saved 287 pounds of feed. Salt may be fed free choice in a feeder, it may be mixed in with the mineral, or it may be mixed in with a complete ration.

Trace minerals. Small quantities of so-called *trace minerals* should also be supplied in the mineral mixture. Iodine, iron, copper, zinc, and manganese are considered most important. They are most easily provided by purchasing trace mineral premixes, or ready-mixed mineral feeds, or ready-mixed protein and mineral supplements. A trace mineral premix is found in Table 7-2. Two pounds of the mix should be added to each ton of feed fed to growing-finishing pigs.

Arsenicals

Minnesota tests showed that pigs fed sodium arsanilate at levels of 0.005 percent or 0.01 percent gained weight at a rate approximately 11 percent faster than a control group of pigs that received no arsenicals.

Later tests conducted at the same station indicated that pigs fed arsenilic acid had a growth rate of approximately 5 percent over that of a control group. There were no differences in feed requirements per pound of gain.

Tests conducted in Texas show that the feeding of one ounce of arsenilic acid per ton of balanced ration resulted in an im-

TABLE 8-4 COMPOSITION OF VITAMIN PREMIX[1]

Essential	
Vitamin A, million I.U.	3.0
Vitamin D, million I.U.	0.4
Riboflavin, grams	2.0
Pantothenic acid, grams	8.0
Niacin, grams	15.0
Vitamin B_{12}, milligrams	10.0
Optional	
Vitamin E, thousand I.U.[2]	10.0
Menadione (source of vitamin K), grams[3]	2.0
Carrier	?
	10 lbs.

[1] A feed additive if desired may also be placed in the vitamin premix.

[2] Supplemental E is needed only in certain areas of the United States where the selenium content of feed ingredients is extremely low because of the low selenium level in the soil. If deficient areas exist in Iowa, they have not been differentiated at the present time.

[3] The vitamin K requirement is normally met by the level present in natural feedstuffs and by intestinal synthesis. A hemorrhage or bleeding syndrome has been diagnosed which is apparently due to a vitamin K antimetabolite which interferes with the use of vitamin K. The antimetabolite is thought to be produced by a mold occurring in one or more of the ration ingredients. When this has occurred, adding menadione has been helpful in preventing or overcoming the problem.

Source: *Life Cycle Swine Nutrition*, Iowa State University, 1974.

TABLE 8-5   COMPLETE VITAMIN-MINERAL MIXES FOR CORN-SOYBEAN MEAL
RATIONS[1,2]

Ingredients	1	2
Calcium carbonate (38% Ca)	500	400
Dicalcium phosphate (26% Ca, 18.5% P)	1,000	—
Defluorinated treble phosphate (32% Ca, 18% P)	—	1,100
Iodized salt[3]	400	400
Trace mineral premix[4]	75	75
Vitamins[5]	25	25

120 million I.U. Vitamin A
16 million I.U. Vitamin D
80 grams riboflavin
320 grams pantothenic acid
600 grams niacin
400 milligrams B_{12}

Total	2,000	2,000

CALCULATED ANALYSIS (to be adjusted for guaranteed analysis)

Elemental calcium	%	22.50	25.20
Elemental phosphorus	%	9.25	9.90
Salt	%	20.00	20.00

[1]For growing-finishing pigs, 50 pounds of the vitamin-mineral mix can be mixed with 250 to 450 pounds of 44 percent soybean meal and 1,500 to 1,700 pounds of ground corn to make 1 ton of a complete, well-balanced ration. For a gestation ration to be fed at the level of 5 pounds per day, 50 to 60 pounds of the mix should be mixed with 200 pounds of 44 percent soybean meal and 1,740 to 1,750 pounds of corn. Also, 80 pounds of the mix can be mixed with 300 pounds of 44 percent soybean meal and 1,620 pounds of ground corn for a gestation ratio to be fed at the level of 4 pounds per day or for a lactation ration.

[2]Do not add or feed free choice additional vitamins or minerals with any of the suggested balanced swine ration formulas since they contain sufficient vitamins and minerals. These mixes can be fed free choice to sows on pasture or in other instances where free-choice vitamins and minerals are needed.

[3]If plain salt is used, iodine should be added to give a level of 0.001 percent in the vitamin-mineral mix.

[4]Trace mineral premix is presented in Table 7-2.

[5]High potency vitamin sources should be added rather than a vitamin premix, since the large amount of carrier used in most premixes will take up critical volume in the vitamin-mineral mix.

Source: *Life Cycle Swine Nutrition*, Iowa State University, 1974.

provement in feed efficiency, although not as great as that received when pigs were fed 2 grams of penicillin per ton of feed.

In Purdue tests arsenilic acid increased gains in growing-finishing swine 4 percent and resulted in a 3.5 percent improvement in feed efficiency. Similar results have been obtained in other states. Michigan nutritionists recommend 90.0 grams of arsenilic acid per 10 pounds of vitamin-antibiotic-trace mineral premix (VATH). Ten pounds of the premix is used in each ton of complete growing-finishing ration. The University of Illinois also recommends 90 grams of arsenilic acid per ton of grower-finisher ration. The Iowa State University recommends 0.0025 percent of 3-nitro in the grower-finishing rations.

Arsenicals play an important part in control of infectious swine enteritis. They should not be fed for a period of seven to ten days before the hogs are marketed.

GROWING-FINISHING RATIONS

Rapid and economical gains may be brought about by full-feeding in confinement pigs that have access to good legume pasture. When feeds are high in price or scarce in quantity, it is sometimes desirable to feed a limited ration and make more effective use of pasture. The rate of gain is decreased, and it takes the pigs a longer time to reach a marketable weight. The quality of the carcasses produced, however, may in some cases be improved. High-priced land, however, limits greatly the economical use of pasture in swine production during the growing and finishing periods.

Systems of Feeding

There are two common systems of feeding pigs from weaning to market: (1) self-feeding corn or other grains and a balanced supplement, free-choice; and (2) self-feeding a complete mixed ration.

Grain and supplement self-fed, free-choice. This system is easy to follow and requires less work than other systems, but the efficiency of the system varies with the palatability of the grain, supplement, and pasture. Undereating or overeating of supplement is common.

Complete ration, self-fed. It is possible to formulate a complete ration which will provide the food nutrient needs of the pigs

FIGURE 8-7. High moisture shelled corn moves automatically from the oxygen-free silo to the self-feeders. (*Courtesy Edward Lodwick, Cincinnati Gas and Electric Co.*)

FIGURE 8-8. *Top:* Interior of a large open-front, confinement finishing house. (*Courtesy Anderson Box Co.*) *Bottom:* Exterior of the finishing house shown above. Note provisions for closing side openings. (*Courtesy Anderson Box Co.*)

by mixing ground corn, or other grains, and a balanced supplement in proper proportion. According to several swine specialists, this is the recommended system of feeding pigs whether on pasture or in confinement. The system requires the grinding of the grains and the mixing of the feeds. Overeating or undereating of supplement ceases to be a problem.

Nutritionists at Iowa State University recommend the feeding of a complete ration to growing-finishing pigs. Presented in Table 8-6 are 13.5 to 15.5 percent complete rations for pigs weighing 40 to 120 pounds. Rations containing 12–14 percent protein for pigs

weighing 120 to 240 pounds are also shown in Table 8-6. These rations are designed as "least time" rations. They are to be used when the market justifies rapid growth of pigs.

Some producers purchase or have mixed 35 to 40 percent protein and mineral supplements and then mix the supplement with ground yellow corn to formulate complete growing-finishing rations. Presented in Table 8-8 is a 40 percent protein and mineral supplement. In Table 8-7 is a 35 percent protein supplement. In Table 8-9 are complete rations formulated by using 35 and 40 percent protein supplements and ground corn to make up rations containing 13 to 16 percent protein.

T A B L E 8-6   GROWING-FINISHING RATIONS[1]

Percent Protein	Ingredients	For Pigs 40-120 Pounds			For Pigs From 120-240 Pounds[2]		
		1	2	3	4	5	6
8.9	Corn[3,4]	1660-1590	1550-1450	1666-1601	1733-1658	1648-1548	1747-1667
44.0	Solv. Soybean meal[5]	280-350	—	200-265	205-280	—	120-200
37.0	Soybeans, whole cooked[6]	—	390-490	—	—	290-390	—
50.0	Meat and Bone Meal	—	—	100	—	—	100
	Calcium carbonate (38% Ca)	15	15	6	15	15	6
	Dicalcium Phosphate (26% Ca, 18.5% P)	23	23	4	25	25	5
	Salt	10	10	10	10	10	10
	Trace minerals	2	2	2	2	2	2
	Vitamins	10	10	10	10	10	10
	Feed additives (gm./ton)[7]	0-100	0-100	0-100	0-100	0-100	0-100
	Total	2000	2000	2000	2000	2000	2000
CALCULATED ANALYSIS							
Protein	%	13.54-14.77	14.10-15.51	14.32-15.46	12.22-13.54	12.70-14.10	12.92-14.32
Calcium	%	0.62-0.63	0.63-0.64	0.60-0.61	0.64-0.64	0.64-0.66	0.60-0.61
Phosphorus	%	0.50-0.52	0.52-0.53	0.51-0.52	0.51-0.52	0.52-0.54	0.51-0.52
Lysine	%	0.65-0.75	0.68-0.79	0.66-0.75	0.55-0.65	0.58-0.68	0.55-0.66
Methionine	%	0.24-0.25	0.24-0.26	0.25-0.26	0.22-0.24	0.22-0.24	0.23-0.25
Cystine	%	0.23-0.24	0.23-0.25	0.21-0.23	0.21-0.23	0.21-0.23	0.19-0.21
Tryptophan	%	0.14-0.17	0.16-0.18	0.13-0.15	0.12-0.14	0.14-0.16	0.11-0.13
Metabolizable energy	kcal./lb.	1324-1318	1382-1391	1321-1316	1330-1324	1373-1382	1328-1321

[1] Start with the higher level of soybean meal (lower level of corn) with lighter pigs in each group, and decrease the soybean meal (increase the corn) in 50- to 100-pound increments until you reach the lower level. If you prefer, one level of protein can be fed from 40 pounds to 240 pounds with similar results as with the varying levels. In this case use a level of soybean meal and corn which is approximately the high point of the listed range for growing pigs (for example, in Ration No. 1 you might use 1,590 pounds of corn and 350 pounds of soybean meal). If barrows and gilts are separated, use the higher end of the range for soybean meal for the gilts and the lower end for the barrows.

[2] For pigs going to market the level of dicalcium phosphate may be reduced by 10 pounds which provides a level of about 0.51% calcium and 0.41% phosphorus. Replacement gilts and boars should be fed the levels provided in the above rations.

[3] Ground milo, wheat or barley can replace the ground corn. Ground oats can replace corn up to 20 percent of the total ration.

[4] If the ration is to be pelleted, 25 to 50 pounds of molasses or binder can replace 25 to 50 pounds of corn.

[5] Three pounds of L-lysine, 1 pound of DL-methionine, and 96 pounds of corn can be substituted for 100 pounds of soybean meal.

[6] Since the high fat content of whole cooked soybeans increases the energy content of the ration, the protein level in a ration utilizing whole cooked soybeans should be approximately 1 percent higher than a similar ration with soybean meal in order to maintain the same energy-to-protein ratio.

[7] The feed additive may be part of the vitamin premix, or if it is a separate premix, it should replace an equal amount of corn.

Source: *Life Cycle Swine Nutrition*, Iowa State University, 1974.

TABLE 8-7   35 PERCENT SUPPLEMENTS[1,2]

Percent Protein	Ingredient	1	2	3	4
44.0	Solvent-process soybean meal	1,670	1,575	1,100	1,050
17.0	Dehydrated alfalfa meal	—	100	—	200
50.0	Meat and bone meal[3]	—	—	400	400
8.9	Corn[4]	—	—	290	145
	Calcium carbonate (38% Ca)	80	75	50	45
	Dicalcium phosphate (26% Ca, 18.5% P)	140	140	60	60
	Iodized salt	50	50	40	40
	Trace mineral premix	10	10	10	10
	Vitamin premix	50	50	50	50
	Feed additives[5]	—	—	—	—
	Total	2,000	2,000	2,000	2,000

CALCULATED ANALYSIS					
Protein	%	36.74	35.50	35.49	35.45
Calcium	%	3.55	3.51	3.49	3.52
Phosphorus	%	1.80	1.78	1.74	1.74
Salt added	%	2.50	2.50	2.00	2.00
Lysine	%	2.51	2.40	2.21	2.20
Methionine	%	0.53	0.51	0.50	0.49
Cystine	%	0.56	0.54	0.44	0.45
Tryptophan	%	0.53	0.52	0.41	0.43
Metabolizable energy	kcal./lb.	1,022	991	1,035	958

Feeding Directions

[1] These supplements can be used to make growing-finishing, gestation, or lactation rations.

[2] Supplements 3 and 4 may be self-fed free choice with shelled corn.

[3] The meat and bone meal was considered to have 8.10 percent calcium and 4.10 percent phosphorus. If meat and bone meal with a higher concentration of calcium and phosphorus is used, the amount of dicalcium phosphate should be reduced accordingly.

[4] The corn can be replaced by wheat midds, corn distiller grains with solubles, or other grain by products.

[5] The level of feed additives will depend on the type of ration in which the supplement is going to be used, but should be 3 to 5 times higher than desired in the complete ration.

Source: *Life Cycle Swine Feeding*, Iowa State University, 1972.

Least-Time Rations

The price of hogs and anticipated changes in price may make it desirable to produce rapid gains with less attention given to the cost of gain. "Least-time" rations may be formulated that are palatable and contain the balance of nutrients necessary to produce rapid gains. Shown in Table 8-6 are "least-time" rations recommended by Iowa State University swine nutritionists.

Least-Cost Rations

"Least-cost" rations must be developed around the most economical sources of proteins, carbohydrates, and fats. Most research has been based upon corn-soybean oilmeal rations fortified with minerals, vitamins, and antibiotics. Corn and soybean oilmeal alone will not make a balanced ration.

Grain sorghum, barley, wheat or wheat mids, or molasses may be cheaper feeds than corn at times and can be substituted for it. Other protein feeds may also be substituted for soybean oilmeal when price permits.

It may be more convenient and economical to use a balanced supplement containing proteins, minerals, vitamins, and antibiotics than to use soybean oilmeal and mineral-

TABLE 8-8   40 PERCENT SUPPLEMENTS[1,2]

Percent Protein	Ingredients	1	2	3	4
48.5	Solvent-process soybean meal	1,628	—	1,238	1,458
44.0	Solvent-process soybean meal	—	1,223	—	—
17.0	Dehydrated alfalfa meal	—	—	100	—
50.0	Meat and bone meal[3]	—	550	400	—
61.0	Fish meal, menhaden	—	—	—	200
	Calcium carbonate (38% Ca)	90	45	55	80
	Dicalcium phosphate (26% Ca, 18.5% P)	160	60	85	140
	Iodized salt	50	50	50	50
	Trace mineral premix	12	12	12	12
	Vitamin premix	60	60	60	60
	Feed additives[4]	—	—	—	—
	Total	2,000	2,000	2,000	2,000

CALCULATED ANALYSIS

		1	2	3	4
Protein	%	39.48	40.66	40.87	41.46
Calcium	%	3.95	4.02	3.96	3.98
Phosphorus	%	2.01	2.05	2.02	2.05
Salt added	%	2.50	2.50	2.50	2.50
Lysine	%	2.67	2.55	2.60	2.86
Methionine	%	0.55	0.56	0.56	0.66
Cystine	%	0.59	0.48	0.52	0.58
Tryptophan	%	0.55	0.46	0.49	0.56
Metabolizable energy	kcal./lb.	1,123	968	1,047	1,122

Feeding Directions

[1]These supplements can be used to make growing-finishing, gestation, or lactation rations.

[2]Supplements 2 and 3 may be self-fed free choice with shelled corn.

[3]The meat and bone meal was considered to have 8.10 percent calcium and 4.10 percent phosphorus. If meat and bone meal with a higher concentration of calcium and phosphorus is used, the amount of dicalcium phosphate should be reduced accordingly.

[4]The level of feed additives will depend on the type of ration in which the supplement is going to be used, but should be four to six times higher than desired in the complete ration.

Source: *Life Cycle Swine Nutrition*, Iowa State University, 1973.

TABLE 8-9   COMPLETE RATIONS
(Using Supplements in Tables 8-7 and 8-8)

	Percent Protein in Complete Ration							
	13		14		15		16	
Ground yellow corn	1,725	1,675	1,650	1,600	1,600	1,525	1,550	1,450
40% supplement	275	—	350	—	400	—	450	—
35% supplement	—	325	—	400	—	475	—	550
	2,000	2,000	2,000	2,000	2,000	2,000	2,000	2,000

CALCULATED ANALYSIS FOR COMPLETE RATIONS (using an average analysis for the 40% or 35% supplements)

		13		14		15		16	
Protein	%	13.2	13.1	14.3	14.1	15.1	15.1	15.9	16.1
Calcium	%	0.56	0.58	0.71	0.71	0.81	0.84	0.91	0.97
Phosphorus	%	0.49	0.49	0.56	0.55	0.60	0.61	0.64	0.66

	Suggested Uses							
Growing Finishing	x	x	x	x	x	x	(Ca High)	
	(after 125 lbs.)							

Source: *Life Cycle Swine Nutrition*, Iowa State University, 1973.

vitamin-antibiotic premixes. A good vitamin-mineral mixture should be self-fed free choice if the pigs are fed a balanced supplement with the corn.

When the prices for corn or soybean meal are unusually high, it is sometimes advisable to feed rations slightly lower in protein content than those fed for optimal gain and feed efficiency. The savings in feed cost may have greater economic value than performance gains when optimal rations are fed. Presented in Figure 8-9 are three charts, one for each of three weight groups, indicating the relationship between the price of corn and the price of 44 percent soybean meal in ration formulation. The lines of the charts represent break-even points, or divisions between suggested protein levels. The spaces between the lines represent areas in which a specific protein level is recommended. The Nebraska tests that provided the information

on which these three charts are based are summarized in Table 8-3, page 186.

The suggested protein levels shown on the charts represent the levels that produce the lowest cost of gains. After determining protein level, the producer can refer to Tables 8-1 and 8-2 to find the number of pounds of corn and soybean meal to be included in the ration.

In using the charts, let us assume that corn is selling for $2.50 a bushel and soybean meal for $300 per ton, and that a least-cost ration is needed for pigs weighing 90 pounds. A vertical line is drawn from the price of soybean meal to a point where it intercepts a horizontal line drawn from the price of corn. A 14 percent protein ration is found to be desired. If the price of corn remains at $2.50 a bushel, but the price of soybean meal drops to $150 per ton, a 16 percent ration would be recommended.

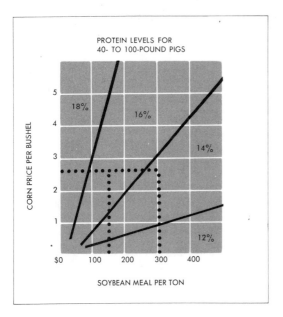

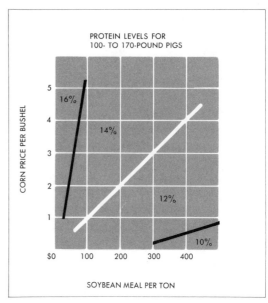

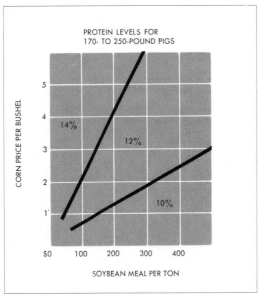

FIGURE 8-9. *Above left:* Chart for use in determining most economical protein level for 40-to 100-pound pigs. *Above right:* Chart for use in determining most economical protein level for 100-to 170-pound pigs. *Left:* Chart for use in determining most economical protein level for 170-to 250-pound pigs. (*Courtesy University of Nebraska.*)

Conditioner and Special Rations

Conditioner rations are often necessary when pigs have been transported to new locations, intermingled with pigs from other herds, and when weather conditions have been unfavorable. Boars in swine testing stations are also fed rations somewhat different from those fed by the average farm producer.

In Table 8-10 are the rations fed at the Central Iowa Swine Testing Station.

Starting Feeder Pigs

The stress put on feeder pigs at time of movement necessitates special care. The National Feeder-Pig Dealers Association and the American Association of Swine Practi-

TABLE 8-10 ISU RATIONS FOR IOWA CENTRAL BOAR TESTING STATIONS

Ingredients	Conditioner[1] Ration	Test Ration
Ground yellow corn (8.9% protein) lbs.	1102	1347
Wheat midds (15.5% protein) lbs.	200	—
Soybean meal (44% protein) lbs.	450	500
Whey (12% protein) lbs.	100	—
Binder (for pelleting) lbs.[2]	50	50
Molasses lbs.[2]	25	25
Dicalcium phosphate (26% Ca, 18.5% P) lbs.	25	35
Calcium carbonate (38% Ca) lbs.	15	10
Iodized salt lbs.	10	10
Trace mineral premix (7.0% Fe, 0.45% Cu, 5.5% Mn, 8.0% Zn) lbs.[3]	3	3
Vitamin premix (6 million I.U. Vit. A, 0.8 million I.U. Vit. D, 4.0 gm. riboflavin, 16 gm pantothenic acid, 30 gm. niacin, 20 mg. B_{12}, 4 gm menadione) lbs.[4]	20	20
Antibiotics		
Oxytetracycline	100 gm	—
Neomycin base	140 gm	—
Chlortetracycline	—	40 gm
Total pounds	2000	2000
CALCULATED ANALYSIS		
Protein %	17.0	17.0
Calcium %	0.73	0.72
Phosphorus %	0.62	0.64
Lysine %	0.85	0.94
Methionine %	0.28	0.28
Cystine %	0.29	0.28
Tryptophan %	0.21	0.20
Metabolizable energy kcal./lb.	1240	1255

[1] Twenty-five pounds of the conditioner ration is fed to each pig when he enters the station.

[2] Binder and molasses level may be varied according to needs for proper pelleting.

[3] Amount of trace mineral premix may vary but should provide approximately 100 ppm Fe, 6 ppm Cu, 75 ppm Mn, and 120 ppm Zn.

[4] Amount of vitamin premix may vary depending on the amount of carrier.

tioners have developed a guide to aid swine producers in starting newly purchased pigs. The following practices are recommended:

1. Check pigs carefully at time of arrival for signs of diseased or poorly doing pigs. Check ear tags of any found and notify the dealer. Call a veterinarian if necessary.

2. Remove any diseased or poorly doing pigs and give special medication and feed.

3. Keep newly purchased pigs isolated from other pigs on the farm. Do not mix new pigs and old pigs.

4. House pigs in clean, dry, draft-free pens. Observe them several times each day. Have separate pens for isolation of pigs that are off feed. Provide at least 4 square feet of floor space per pig, and one uncovered feeder space for each 4 pigs. Allow no more than 25 pigs per waterer or water space.

5. Provide dry bedding, preferably straw. Keep temperature at 75 degrees F.

6. Keep pigs grouped by source and subgroup them, if necessary, by size.

7. Feed a specially formulated starting ration for the first 10 to 14 days. It should contain 12 to 16 percent protein, provide a high level of vitamins and antibiotics, and have some extra fiber to decrease stress.

8. The addition of 2 pounds of ferrous sulfate and 1 pound of copper sulfate per ton of feed will help reduce stress and correct borderline anemia conditions.

9. Use high levels of feed additives. Check with the veterinarian or feed consultant. The following levels have been approved:

	Grams per ton
ASP 250	250
CSP 250	250
Tylan	100
and sulfamethazine	100
Neomycin	200
and Terramycin	100
Mecadox	50
Furazolidone	300
Furazolidone	150
and bacitracin	100
Nitrofurazone	500
Terramycin	150

10. Provide medication in water as recommended by a veterinarian or swine specialist.

11. Worm pigs upon arrival using a broad spectrum wormer. Delay worming if pigs show signs of stress. Use malathion or Ciodrin to control lice and mange.

12. Some pigs may not be used to mechanical feeders and waterers. Feed water in open troughs when it appears advisable.

13. Vaccinate for erysipelas.

14. The following ration is recommended by the University of Illinois:

Corn	1,400 lb.
Oats or barley	350 lb.
Soybean meal	200 lb.

Dicalcium phosphate or bone meal	20 lb.
Feeding limestone	20 lb.
Swine trace-mineralized salt	10 lb.
Ferrous sulfate	2 lb.
Riboflavin	2 gm.
Pantothenic acid	10 gm.
Niacin	30 gm.
Choline	100 gm.
Vitamin A	6 million IU
Vitamin D	600 thousand IU
Vitamin E	20 thousand IU
Vitamin B_{12}	32 mg.
Antibiotics	
ASP 250	250 gm.
or	
Tylan sulfa	200 gm.

METHODS OF FEEDING

Pigs may be hand-fed or self-fed. Self-feeding saves labor and usually produces more rapid and economical gains than those produced by hand-feeding.

FIGURE 8-10. Equipment for feeding paste or gruel rations to growing-finishing pigs. (*Frank Cooper photograph. Courtesy Ritchie Manufacturing Co.*)

In tests conducted at the University of Illinois, pigs weighing 95 pounds gained 1.59 pounds per day when self-fed in confinement while hand-fed pigs gained 1.42 pounds. The self-fed pigs used 4.14 pounds of feed to produce a pound of gain, while the hand-fed pigs needed 4.25 pounds of feed to produce the same gain.

Self-feeding also promotes sanitation and provides large feed storage space. Self-feeders, however, are expensive and are sometimes neglected by irresponsible caretakers. Sometimes the use of the self-feeders decreases the effect of good pasture because the appetites of the pigs are satisfied while they are at the feeders. Some farmers, who raise only a few pigs, prefer to hand-feed their pigs. They can watch them more closely and can feed types of feeds that cannot be self-fed, such as ear corn and mash feeds.

If pigs are to be hand-fed, feed at one time only the amount of feed that they will clean up in an hour or two. This is especially important if paste or wet mash feeding is practiced.

Self-feeders

A good self-feeder should make available to the pigs a supply of clean, dry feed at all times. It should be solidly constructed and protect the feed from wind or rain. It should be large enough to store amounts of feed that will last the pigs for several days. It should have an adjustable throat or opening, and some type of agitator, so that various types of feeds will feed down in desired amounts. The trough should be constructed in such a manner that little feed will be wasted. Doors or lids over the troughs are desirable. If the ration is fed as a complete mixed ration, only one feeder will be needed. If feeds, such as shelled corn, ground oats, and supplement, are fed separately, separate feeders will be needed, or separate compartments in a large feeder may be used.

FIGURE 8-11. Portable grinder-mixers permit ease in feeding large numbers of pigs either in confinement or on pasture. (*Courtesy Speed King Manufacturing Co.*)

FIGURE 8-12. A large confinement feeding facility with slatted floors, mechanical feed handling, automatic waterers, and controlled ventilation. (*Courtesy Butler Manufacturing Co.*)

The use of large feeders which will handle 250 bushels of shelled corn and 1,500 pounds of ration supplement will cut down the cost of feeding equipment and will save labor.

There should be one foot of feeder space for each three to four hogs. If supplement is fed separately, from 15 to 20 percent of the total feeder space should be reserved for supplement. Heavier hogs require less supplement and less feeder space.

Liquid or Paste Feeding

Tests conducted in England and at Iowa State University indicated that pigs fed a restricted ration gained faster on wetted feed than on dry feed. In Iowa tests pigs fed limited rations of wetted feeds gained 7.8 percent faster than pigs fed dry rations. Pigs on full feed of the wetted ration gained only 1.4 percent faster than pigs full-fed dry rations.

The wetting of the feed resulted in a 7 percent improvement in feed conversion for the limited ration pigs. Pigs self-fed wetted rations required 5.4 percent more feed than the pigs fed dry rations. The paste or liquid rations had no effect on carcass quality.

Liquid feeding involves a water to feed ratio of about 3 to 1. Paste feeding ratios are between 1.5 to 1 and 2 to 1. Paste feeding and equipment is considered to be simpler and less expensive than automatic dry feeding. There is no dust problem nor is there any pumping of excess water or air as is done in slurry or pneumatic systems. Paste feed is not splashed or blown away.

Paste and liquid feeding systems must be protected from freezing temperatures and may require more power than other mechanical systems. Grain must be finely ground to insure proper blending and distribution. There may be spoilage of feed mixture when sanitation practices are not followed.

FIGURE 8-13. A confinement facility with walls and ceilings covered with KEMPLY paneling. Gruel feeding equipment is in use. (*Courtesy Kemlite Corporation.*)

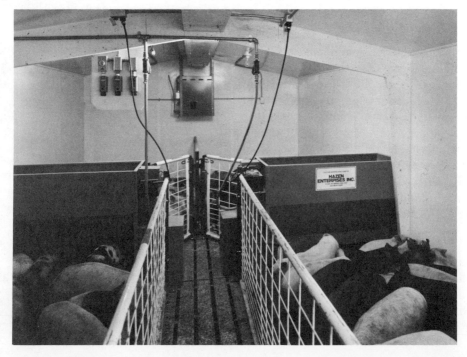

MANAGEMENT PRACTICES

Water

For pigs to make fast and economical gains, they must have plenty of fresh water available at all times. A pig needs from a half gallon to 1½ gallons of water daily for each 100 pounds of live weight. Eighty pigs weighing 150 pounds on pasture in July may consume as much as 80 to 150 gallons of water in one day. Young pigs and sows nursing litters have high water requirements.

When only a few pigs are involved, hand-feeding of water in clean troughs is satisfactory if done often enough, and if the pigs can be kept out of the troughs. Slats may be nailed across the trough to accomplish this purpose. The labor necessary to provide a sufficient amount of water by hand is usually not available. The use of automatic waterers is recommended.

Recent tests at Iowa State University indicated that pigs that were supplied heated water in automatic waterers from December 12 to January 23 gained approximately 10 pounds more than did pigs that were hand-watered. The heating of the water did not affect the quantity of feed required to produce a pound of gain. The use of waterers with heating units is recommended during cold winter months in areas where freezing temperatures prevail.

One automatic watering cup should be provided for each 20 pigs. The waterer should have the capacity to provide 25 gallons of water for each 10 pigs in summer and 15 gallons in winter. The temperature should not fall below 40 to 50 degrees in winter, or above 70 degrees in summer.

Shade

The use of rotated pastures during the summer months makes it necessary to provide some form of shade to protect the pigs

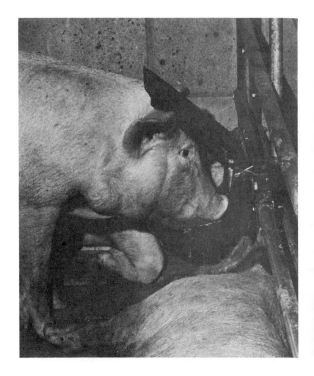

FIGURE 8-14. Cup waterers installed on partitions over slatted floor are recommended. Note attached lid on waterer. (*Agricultural Associates photograph. Courtesy Starcraft Swine Equipment.*)

from the hot sun. Pigs in wooded areas or those which have access to the farm buildings usually need no other shade unless the pasture area is at some distance from the buildings.

The shades should be from 4 to 5 feet above the ground to permit circulation of air and should be located on high ground. A 10 by 12 foot shade will accommodate 10 to 20 pigs, depending upon their size.

It pays to keep hogs cool during the hot summer months. Tests conducted at Davis, California, showed that 100-pound pigs did best when the temperature was around 70 degrees. Hogs kept in a room at 70 degrees used 250 pounds of feed to produce 100 pounds of gain. When the temperature was 90 degrees, 450 pounds of feed were used to

FIGURE 8-15. These Hampshires have adequate shades to protect them in hot weather. (*Courtesy Hampshire Swine Registry.*)

FIGURE 8-16. Pigs enjoy nozzle cooling device during hot weather. (*Courtesy American Yorkshire Club, Inc.*)

produce the same gain. Hogs weighing 200 pounds did best at temperatures of about 60 degrees.

Some hog producers have found it profitable to install cooling devices in central-type hog houses used during the summer months.

Hog Wallows

Sanitary wallows may be used during the summer months to aid in keeping the pigs cool and in maintaining health. Portable or permanent wallows may be constructed. The portable wallow is more adaptable to the needs of the farmer who feeds his pigs on rotated pastures.

Shoats fed to weights of 100 pounds in hot weather in Purdue University tests gained 2.09 pounds per day when they had access to a spray cooler system and 2.04 pounds per day when they had access to a portable wallow. Pigs that had only shade gained 1.89 pounds per day. In a Louisiana test the use of hog wallows saved 35 pounds of feed for every 100 pounds of gain.

The use of sprinklers or wallows is recommended during the summer months if pigs are not confined in an air-conditioned house. One-hundred square feet of wallow will accommodate up to 50 pigs if shade or shelter is nearby. A sprinkler that will provide 1 to 2 gallons of water per hour for each 10 to 12 pigs and cover one-fourth to one-half of the floor area is desirable. Time clocks should be used to operate sprayers 1 to 2 minutes in each 10-minute period. They can be set to start when temperatures are above 80 degrees.

Hogging Down Corn

The practice of hogging down corn is less popular today than it was a few years ago. With the multiple-litter system of breeding, the spring pigs are farrowed in February and March and are marketed before the new

FIGURE 8-17. Hogging down corn in Iowa. This practice is becoming uncommon. (*Courtesy U.S.D.A. Bureau of Animal Industry.*)

crop of corn is ready for hogging down. The fall pig crop may be large enough to glean a field following the corn picker, but the pigs are usually too small to break down the stalks in the hogging down process.

In hogging down corn, best results are obtained by using thin, active pigs weighing from 90 to 150 pounds. Heavier hogs waste more corn and do not make economical gains.

An early maturing variety of corn is best for hogging down, and the pigs should be turned in after the corn is well dented. A protein-mineral supplement and water should be available at all times.

Toxins produced by molds in corn may be present. These molds may cause abortion in gestating gilts but should not cause problems with growing-finishing pigs unless large amounts are consumed. If molds appear to be present in large amounts, it might be well to turn hogs into corn fields on alternate days or every third day.

Garbage Feeding

Garbage feeding accounted for nearly 40 percent of the hog production in California in the late 1960's. One of each 35 hogs produced in the nation at that time was finished on garbage. In Los Angeles County, California, about 1,700 tons of garbage are collected daily. The greater share of this garbage is still used in swine feeding. It is purchased for $2.00 to $3.00 per ton and it costs about $2.00 per ton to transport it to the farm. About two tons of garbage are needed to produce 100 pounds of pork in the Los Angeles area.

There were 516,292 hogs being fed garbage in the mainland states in October, 1973. Garbage-fed hogs were found on 7,677 premises. More than 4,800 hogs on 150 premises were being fed raw garbage even though all states involved had laws prohibiting the feeding of uncooked garbage.

The cooking of garbage kills disease organisms, but it is costly, and the cooked garbage is less palatable to the pigs. All states now have laws requiring heat treatment of garbage before it is fed to hogs.

Iowa in November, 1973, was one of nine states that prohibited the feeding of garbage to swine. The number of hogs fed garbage decreased from 764,000 in 1969 to 516,292 in 1973.

A new process has been developed in Georgia that converts garbage into a feed product containing 10 percent fat, 17 percent protein, and 1.5 percent fiber. The product has been proven useful in poultry feeding.

Garbage-fed hogs have been responsible for the rapid spread of cholera during the past few years, and many cases of trichinosis have occurred in some communities where raw garbage is fed to hogs. In feeding garbage extreme care should be taken to maintain sanitary quarters. Concrete feeding floors are recommended, and the yards should be thoroughly cleaned and disinfected regularly. Thorough cooking of pork products is necessary to kill any disease organisms which may be prevalent in the meat.

SUMMARY

About two-thirds of the total cost of growing out a pig is incurred during the weaning to market period. The feeding and management program during this period can greatly affect the profits from the enterprise.

Weaned pigs should be wormed with dichlorvos, hygromycin, piperazine compounds, Tramisol, or some other broad spectrum wormer when 6 to 8 weeks of age. Pigs can be raised profitably in confinement if adequate rations and sanitation are provided. Pastures, however, save feed, aid in controlling diseases and parasites, save labor, and may produce better carcasses.

Farm grains must be supplemented in feeding growing-finishing hogs. Protein, mineral, vitamin, and antibiotic feeds are essential. Corn in the Corn Belt and grain sorghum in the Southwest are usually the best grains. A mixed protein is considered better than a single protein feed. However, by feeding vitamin B_{12}, the pigs can make more effective use of single protein feeds. Pigs weighing from 40 to 120 pounds require rations containing 14 to 16 percent protein. Heavier hogs need rations with 12 to 14 percent protein.

Pigs weighing from 100 to 200 pounds will show an increase of about 8 percent in gains when antibiotics are fed. Pigs doing poorly will show greater increases. A combination of antibiotics may be better than a single antibiotic.

Growing-finishing hogs require thiamine, riboflavin, niacin, pantothenic acid, pyridoxine, biotin, folic acid, choline, vitamin B_{12}, and vitamins A, D, and E. These vitamins can best be provided in the form of farm grains, forages, and vitamin premixes.

Growing-finishing hogs require minerals. A simple mineral mixture composed of ground limestone, dicalcium phosphate, bonemeal, and iodized salt should be supplemented with a trace mineral premix.

Copper sulfate has proven to be effective as a growth stimulant and feed conserver.

Arsenicals are now considered standard ingredients of growing-finishing swine rations. Arsenicals and copper sulfate supplement antibiotics as growth stimulants and feed conservers in feeding out market hogs.

Forty- to 120-pound pigs should be fed a grower ration containing 14 to 16 percent protein, 0.65 to 0.75 percent lysine, 0.4 percent methionine and cystine, 0.1 percent tryptophan, 0.5 percent calcium, and 0.5 percent phosphorus.

The growing and finishing rations should also provide 1,000 units of vitamin A, 100 units of vitamin D, 5 micrograms of vitamin B_{12}, 1 milligram of riboflavin, 4 milligrams of pantothenic acid, and 7.5 milligrams of niacin per pound of ration. The growing ration should have 50 to 100 grams of antibiotics per ton of feed, and the finishing ration should have 25 to 50 grams.

It pays to self-feed a complete ration to growing-finishing pigs. A "least-cost" ration should be fed when a stable market is anticipated. When a lower market is anticipated, it may pay to feed a "least-time" ration in order to market the hogs before the break in price.

Shades, wallows, and sprinklers should be used during hot weather. Pigs do better when the temperature is from 60 to 70 degrees. Fresh water should be available at all times; each pig will consume one-half gallon to 1½ gallons per day.

Feeder pigs should be provided special care. They should be grouped by source and size with no more than 25 in a pen. Pigs showing stress or disease should be removed and fed medicated rations.

The pens should be well bedded and provide dry, clean, and well-ventillated quarters. There should be no drafts. Each pig should have 4 or 5 square feet of floor space. One space at the feeder should be available for each 4 pigs, and there should be a waterer space for each 25 pigs.

The temperature should be maintained at 70 to 75 degrees. A highly fortified starter ration should be fed during the first 10 to 14 days. It should contain 12 to 16 percent protein. Stressed pigs should receive 12 percent protein rations. Tylan sulfa, ASP 250, or equivalent antibiotics are recommended and should be fed at high levels.

It is a good policy to worm pigs with a broad spectrum wormer and treat pigs for external parasites 7 to 10 days after weaning.

QUESTIONS

1   Which is the better system of feeding hogs on your home farm, in confinement or on pasture? Why?

2   What is the best grain to feed the hogs on your farm? Why?

3   What should be the protein content of rations for 40- to 125-pound pigs? 120- to 240-pound pigs?

4   Outline a good protein mixture for use in balancing the rations for the growing-finishing pigs on your farm.

5   What is the place of antibiotics in feeding growing-finishing pigs? Explain.

6   Which of the vitamins needed by growing-finishing pigs must be provided in the form of vitamin premixes?

7   Outline a simple mineral mixture which will meet the mineral requirements of growing-finishing pigs.

8   What trace minerals are needed by pigs from weaning until market time?

9   Is it better to feed antibiotics or other growth stimulation additives such as arsenicals? Explain.

10   Which is the better system of feeding growing-finishing pigs on your farm, self-feeding grain and supplement separately, or feeding a complete mixed ration? Explain.

11   Describe what is meant by a "least-time" ration and indicate when it should be fed.

12   Explain a "least-cost" ration and indicate the conditions under which it should be fed.

13   Outline a complete balanced ration which you could use as a "least-time" ration on your farm.

14 How can you provide cool quarters for the pigs on your farm during the hot summer months?

15 What precautions must be taken in feeding garbage to hogs? Explain.

16 Of what value is a broad spectrum wormer in swine production?

17 How do corn, oats, wheat, barley, and grain sorghum compare as swine feeds?

18 What feeds can you use in place of soybean meal in formulating a ration for growing-finishing pigs?

19 Outline a plan for starting a group of 200 feeder pigs on your farm.

REFERENCES

Carlisle, G. R., and H. G. Russell, *Your Hog Business—Ration Suggestions*, AS-377. Urbana, Illinois: University of Illinois, 1969.

————, *Your Hog Business—Management Suggestions*, AS-377a. Urbana, Illinois: University of Illinois, 1969.

Ensminger, M. E., *Swine Science*, 4th ed. Danville, Illinois: Interstate Printers and Publishers, 1971.

Holden, Palmer, Vaugan Speer, and E. J. Stevermer, *Life Cycle Swine Nutrition*, Pm. 489. Ames, Iowa: Iowa State University, 1974.

Jensen, A. H., B. G. Harmon, G. R. Carlisle, and J. H. Muehling, *Management and Housing for Confinement Swine Production*, Cir. 1064. Urbana, Ill.: University of Illinois, 1972.

Jones, J. R., and A. J. Clawson, *Raising Hogs in North Carolina*, Cir. 505. Raleigh. N.C.: North Carolina State University, 1972.

Kortan, L. J., Wallace Aanderud, Louis Lubinus, and J. H. Bailey, *South Dakota Feeder Pig Production*, EMC 617. Brookings, S. D.: South Dakota State University, 1971.

Krider, J. L., and W. E. Carroll, *Swine Production*, 4th ed. New York, N.Y.: McGraw-Hill Book Company, 1971.

9 MANAGING SWINE IN PASTURE AND CONFINEMENT

The development of the McClean County swine production program in Illinois several decades ago was founded on good pasture management. Before nutritionists improved swine rations, and before programs to control swine diseases and parasites were developed, it was almost impossible to raise hogs profitably in confinement. Swine specialists now have solved many of the problems of confinement hog production.

PASTURE VERSUS CONFINEMENT FEEDING

The use of pasture in feeding bred sows and gilts is highly recommended. The extent that pasture should be used in feeding pigs from weaning to market time depends upon the following factors: (1) price of land, (2) kind of pasture available, (3) housing and equipment available, (4) labor and management available, (5) size of the enterprise, and (6) disease level on the farm.

Each farmer must evaluate his facilities and resources and decide whether he should grow his pigs on pasture or in confinement. Many farmers use pastures in their production programs. Confinement feeding, however, is increasing.

A study in 1970 indicated that about 10 percent of all U.S. hog farmers finished their hogs in total confinement. At that time 20 percent of the producers of 1,000 head, or more, used a total confinement system. Nearly 65 percent of all producers used pasture until the last month or so of the growing-finishing period. They then moved the swine to confinement quarters. About 20 percent marketed their hogs directly from pasture. With the rapid increase in the number of large producers and the decrease in the total number of swine producers since 1970, it is estimated that nearly 25 percent of the hogs in the nation were produced in confinement in 1974.

Following are the advantages of the two systems.

Advantages of Pasture

1. Pork can be produced on clean hog pastures with 15 to 30 percent less concentrates than can be done in confinement feeding when only corn and soybean meal are fed.
2. In Illinois tests an acre of legume pasture carried 25 pigs and saved $90 worth of protein supplement and concentrates over confinement feeding when corn and soybean meal were fed.

   In Minnesota tests pigs grown on pasture and fed a complete ration gained 1.52 pounds per day on 3.1 pounds of feed per pound of gain. The feed cost was $8.96 per 100 pounds of pork produced. Pigs fed in confinement gained 1.56 pounds per day on 3.2 pounds of feed per pound of gain, at a cost of $9.26 per hundredweight of pork produced.

   There is evidence that pasture feeding of growing pigs, under some conditions, may not save sufficient feed to pay for the cost of pasture when complete rations are fed.

3. The protein supplement feed bill can be cut one-third to one-half by using good legume pasture.
4. There is less chance of loss caused by disease when pigs are on "clean" legume pasture.
5. The equipment necessary to feed and manage hogs under sanitary conditions is less when they are on pasture than when they are confined.
6. There is no bedding problem during the summer months.
7. Hogs on pasture will spread their own manure. Less labor is involved.
8. Many tenants must rely upon use of pastures in growing out hogs. Many landowners will not provide desirable facilities for confinement feeding programs.
9. It takes less know-how to grow pigs on pasture.
10. Land can be used that is not suitable for cropping.

Mature sows and gilts make excellent use of green pasture. Pasture feeding will increase the quantity and improve the quality of proteins, provide necessary vitamins, and improve the mineral content of the ration.

FIGURE 9-1. A common sight in the Corn Belt. Many farmers have sows farrow in late spring in individual houses on clean-ground pasture. (*Courtesy American Yorkshire Club.*)

Feed costs may be reduced and desirable exercise for sows during gestation provided. In South Dakota tests an acre of alfalfa or clover carried 8 to 10 sows. A feed saving of $50 to $70 per acre was possible. Pasture feeding of sows resulted in (1) increased litter size, (2) increased vigor of pigs at birth, (3) increased number of pigs weaned, and (4) fewer breeding failures.

In Illinois tests bred sows that received 2 pounds daily of a 12 percent protein ration while on good legume pasture produced litters as large and as strong as those sows that received 4 pounds of the same ration in confinement.

It was found in Illinois that for maximum use of pasture, protein can be reduced 2 percent in rations fed to growing pigs on pasture, as compared to rations fed pigs in confinement. An acre of legume pasture saved $44.00 worth of feed in these tests.

Tests conducted in Ohio to determine the value of legume pasture in feeding pigs from 40 to 210 pounds are summarized in Table 9-1. The acre of legume pasture saved $28.05 worth of feed. Using 1974 feed prices instead of the 1964 to 1967 prices, the savings would be about $70.00 an acre.

In another Ohio test an acre of good alfalfa-ladino clover pasture resulted in a feed saving of 35.5 pounds per pig. Based upon 1974 feed prices, this would amount to about $1.80 per pig.

Advantages of Confinement Feeding

1. It may not be economical to use land valued at $600 an acre or more as hog pasture.
2. More rapid gains can be produced when pigs are in confinement if adequate rations and good management are provided.
3. Confinement facilities can be used during more months of the year than can pasture facilities.
4. There is usually less labor used in hauling and handling feed when pigs are confined.
5. There is less labor and difficulty in providing an adequate supply of water.

FIGURE 9-2. These Yorkshire pigs have an abundance of ladino, red clover, and brome mixed pasture. (*Courtesy American Yorkshire Club.*)

T A B L E 9-1   VALUE OF LEGUME PASTURE FOR FINISHING SWINE

	Pasture	Confinement
Average daily gain (lbs.)	1.45	1.44
Feed per 100 lbs. gain (lbs.)	313.00	335.00
Feed cost per 100 lbs. gain	$9.39	$10.05
Value of feed saved by pasture (25 pigs per acre)	$28.05	

Source: Ohio Agricultural Research and Development Center, Wooster, 1964-1967.

6. Fence problems will be minor as compared to pasture feeding.
7. The confinement method is best for the man who wishes to grow out a large number of hogs.
8. Better use can be made of mechanization in feeding and watering.
9. Equipment can be used during the entire year, thus reducing the cost per litter.
10. Death loss of young pigs caused by chilling and mechanical injury can be reduced. Also fewer growing pigs will be lost because of heat and weed poisoning.
11. Large numbers of pigs can be produced with limited manpower and few acres of land.

Presented in Table 9-2 is a summary of pasture versus confinement feeding trials at the University of Nebraska. Pigs gained more slowly but required less feed to produce 100 pounds gain on properly rotated pastures than on concrete. An acre of pasture saved $13.00 worth of feed.

According to a Purdue economist, the system, whether pasture or confinement, is not the most important factor in swine production. The use that is made of the system adopted is the factor that has the greatest influence on profit. The Purdue study, summarized in Table 9-3, indicated that under some conditions each system studied might be best for an individual farm operator.

The pasture system involving individual farrowing houses, portable shelter, and pasture was compared with a low-investment confinement system based on the use of a solid-floored central farrowing house and open-front nursery, finishing, and gestation buildings. Also included in the study was a high-investment system involving completely enclosed, mechanically ventilated buildings with slatted floors.

The study revealed that the pasture system with two farrowings a year produced 100 pounds of pork for $19.25, the lowest of the three systems. The cost with two farrowings a year with the inexpensive confinement system was $19.45, whereas the cost in using the high-investment system was $20.50. When four farrowings were made each year, the pasture system had the highest cost, and the inexpensive system had the lowest cost per 100 pounds of pork produced. When there were six farrowings per year, the pasture system had the highest cost. The other two systems had identical costs.

Much of the research that has been done concerning the value of pasture to growing-finishing pigs was done previous to 1965. The emphasis in hog production during the past ten years has been on confinement systems. Some studies have been made of the value of pasture to the reproductive efficiency of breeding herd. In Virginia it was found that total confinement of the breeding herd decreased reproductive efficiency in sows about 15 percent compared with sows kept on dry lot with no forage or pasture. The sows did not conceive when bred.

Pasture management of the breeding herd previous to breeding and during at least a part of the gestation period has been proven to be effective in increasing percentage of conception on first service, increased litter size, increased number of pigs weaned, and fewer feet and leg problems among the breeding animals. Presented in Table 9-4 is a summary of tests conducted in Indiana in comparing productivity of sows on pasture and in partial and complete confinement.

T A B L E 9-2   PASTURE VERSUS CONFINEMENT FEEDING

	Confinement	Pasture
Number of test comparisons	69	69
Number of hogs	633	640
Average daily gain, lbs.	1.48	1.40
Concentrates per 100 lbs. gain	324	315

Source: University of Nebraska, 1959.

T A B L E 9-3   PASTURE AND CONFINEMENT SYSTEMS COMPARISON

Variable Expense	Per Sow Farrow-to-Finish Production Costs		
	Controlled Confinement	Open Confinement	Pasture System
Total feed	$373.45	$372.35	$371.35
Corn	223.45	234.35	234.35
Pasture rental			10.00
Supplement	150.00	138.00	127.00
Health and veterinary	8.00	13.00	15.00
Breeding	3.00	3.00	3.00
Marketing	22.00	22.00	22.00
Power, fuel, and equipment repair	32.00	21.00	27.00
Misc. (bedding, supplies)	10.00	14.00	30.00
Total variable expense (excluding labor)	448.45	445.35	468.35
Nonlabor variable expense per cwt. pork produced	13.25	13.15	13.75
Direct labor	55.00	82.50	87.50
Variable expense per cwt. pork produced (including labor)	14.80	15.50	15.90
Feed per 100 pounds pork (lbs.)	414	424	398
Overhead cost 2 litters/year	194.00	134.00	99.00
Overhead cost 4 litters/year	117.00	87.00	85.00

Source: Purdue University, 1973

T A B L E 9-4   PERFORMANCE OF SOWS ON PASTURE, IN PARTIAL CONFINEMENT, AND FULLY CONFINED

	No Confinement Pasture-Drylot	Partial Confinement Open-Front House	Complete Confinement Enclosed House
Pigs farrowed per sow	11.8	10.4	11.0
Pigs weaned per sow at 42 days	9.2	8.4	8.0
Conception rate %	91	82	82

Source: Purdue University.

KINDS OF PASTURE CROPS

A number of pasture crops have been tested and their value for swine has been determined. Shown in Table 9-5 is a summary of tests conducted by the Missouri Agricultural Experiment Station.

According to the Missouri tests, alfalfa proved superior to all other crops in pounds of pork produced per acre and in value of pork produced. Red clover and rape ranked second and third.

Other pasture crops that were not included in the Missouri tests are ladino clover, Balboa rye, lespedeza, Sudan grass, sweet clover, bromegrass, and bird's-foot trefoil.

Pasture crops for swine may be classified as (1) legume pastures, (2) winter and early spring pastures, or (3) emergency pastures.

Legume Pastures

Ladino clover. This comparatively new clover, which resembles white Dutch clover, is rated in several states as the best pasture crop for swine. In Purdue tests it was found superior to alfalfa as a pasture for full-fed hogs, as indicated by a more rapid gain and a 40 percent saving in protein supplement.

The feed cost of 100 pounds of gain was $9.35 on the ladino pasture and $10.42 on the alfalfa.

Ladino has a shallow root system and is best suited to moist soils, but it is not as winter hardy as is alfalfa and may winter kill in extreme northern states.

On a dry-matter basis, ladino clover may contain from 22 to 24 percent protein. It produces lots of feed per acre, is very palatable, and is commonly seeded with alfalfa or with bromegrass. The rate of seeding ladino is one-half to 1 pound with 7 to 10 pounds of alfalfa per acre.

Alfalfa. This is the best pasture crop for hogs on dry soils and in areas where winter killing is a problem. It withstands dry weather and severe winters better than does ladino clover. Tests conducted at the Illinois and Michigan experimental stations indicated that an acre of alfalfa used as hog pasture saved 13 bushels of corn and 1,200 pounds of protein supplement compared to dry-lot feeding. With corn at $1.50 per bushel and supplement at $5.50 per hundred pounds, the pasture saved about $85 in feed. It should be noted that the hogs were not fed complete rations. When they were fed complete rations, the value of pasture decreased.

T A B L E 9-5   VALUE OF VARIOUS PASTURE CROPS FOR SWINE*

Crop	Number of Days Pastured	Number of Hogs per Acre	Pounds of Pork per Acre of Forage	Value per Acre with Pork @ $20 per Cwt.
Alfalfa	163	10	592	$118.40
Red Clover	130	12	449	89.80
Rape (Dwarf Essex)	82	23	395	79.00
Sorghums	87	15	275	55.00
Bluegrass	136	12	274	54.80
Soybeans	25	17	175	35.00
Cowpeas	32	13	149	29.80

*The rations of the pigs were limited to one-half to two-thirds of a full feed.
Source: *Pork Production in Missouri*, University of Missouri, Bulletin 587, Columbia, Mo., 1952.

In some areas alfalfa may be seeded in the fall without a nurse crop. It is usually seeded in the spring with a nurse crop. Ten to 12 pounds of seed per acre is recommended.

Red clover. This clover does not yield quite as heavily as alfalfa and is not as palatable. Like alfalfa, it provides an early pasture and grows well during the hot summer weather. It is usually seeded in the spring with a nurse crop. Eight to 10 pounds of seed per acre is recommended.

In Missouri tests an acre of red clover produced 449 pounds of gain when pigs were fed only corn and protein supplement. At $18 per hundredweight the acre of this clover was worth about $80. Red clover does not withstand grazing as well as alfalfa does.

Lespedeza. This crop is rather popular as a legume pasture in the southern states. It is less palatable than the other legume crops described in this section. Lespedeza is an annual but reseeds itself each year. It does not stand early grazing.

Bird's-foot trefoil. This crop ranks with alfalfa in palatability, in feed value, and in ability to stand drought. It does well seeded with bluegrass as a permanent pasture crop. Often two to three years are required to get a stand, but it will last for many years. The usual rate of seeding is from four to five pounds of inoculated seed per acre.

Sweet clover. This clover is fairly satisfactory as a hog pasture during the first year but becomes coarse and woody during the second year. It is seeded with a nurse crop in the spring and can be grazed during the late summer and fall of the same year. Ten to 12 pounds of seed per acre are recommended.

Winter and Early Spring Pastures

Balboa rye. Winter or Balboa rye makes excellent pasture during the winter and early spring. It is seeded in the fall at the rate of six pecks per acre and can usually be pastured from November through April.

FIGURE 9-3. Feed may be transported to self-feeders for pigs on pasture or in confinement. These pigs have access to pasture. (*Larry Day photograph. Courtesy Fairall and Co.*)

FIGURE 9-4. A litter of Poland China pigs on excellent pasture. (*Courtesy Poland China Record Association.*)

Pigs on Balboa rye pasture at the University of Kentucky made average daily gains of 2.18 pounds as compared to a gain of 1.9 pounds made by pigs in a control group in dry lot.

An acre of rye will pasture four to six sows or ten to twelve fall pigs per acre.

Bluegrass. This grass furnishes excellent grazing for hogs during the fall, early winter, and early spring but dries up during the hot summer months of July and August.

Emergency Pastures

The perennial legume pastures are preferred for summer pasture, but they sometimes winter kill, or there is difficulty in getting a stand. It is then necessary to rely upon other pastures. A number of crops may be used as emergency pastures.

Shown in Table 9-6 is a summary of some of the most commonly used emergency pasture crops for swine.

Rape. At the Minnesota Station rape produced slightly larger gains than alfalfa. This pasture is on a par with the clovers for hogs. It may be seeded in the spring and can be grazed after about six weeks. It has a high carrying capacity and a long growing season.

The Dwarf Essex is more palatable for hogs than other varieties of rape. From 20 to 25 growing-finishing hogs can be grazed on one acre. Rape sometimes causes the skin to blister and become sore. A treatment of carbolated Vaseline will usually clear up the trouble.

Oats and rape. A common practice in providing emergency pastures is to use oats and rape for late spring and early summer pasture, follow with Sudan grass during late summer and early fall, and use rape as a late fall pasture. Shown in Table 9-7 is a summary of a test made at the Purdue Station compar-

TABLE 9-6   EMERGENCY PASTURES

Crop	Date of Seeding	Rate of Seeding (per Acre)	Availability for Grazing
Rape	April 1 to July 1	Drill: 4 lbs. Broadcast: 7-9 lbs.	4 to 6 weeks
Oats and rape	April 1 to May 1	Drill: 1 bu. oats Broadcast: 6-8 lbs. rape	6 weeks
Oats, peas, and rape	April 1 to May 1	Drill: 1 bu. oats, 1 bu. peas Broadcast: 5-7 lbs. rape	6 weeks
Sudan grass	May 20 to June 20	Drill or broadcast: 30 to 35 lbs.	6 weeks

T A B L E 9-7   COMPARISON OF OATS, RAPE, AND SUDAN GRASS PASTURE WITH ALFALFA

	Lot 1	Lot 2
Kind of pasture	Oats, rape, and sudan grass	Alfalfa
Number of pigs	40	40
Av. initial wt.	39.8 lbs.	39.9 lbs.
Av. final wt.	169.8 lbs.	173.7 lbs.
Av. gain	130.0 lbs.	133.8 lbs.
Av. daily gain	1.55 lbs.	1.59 lbs.
Feed for 100 lbs. gain:		
Corn	287.5 lbs.	282.3 lbs.
Supplement	39.9 lbs.	33.1 lbs.
Mineral mixture	.2 lbs.	.2 lbs.
Salt	.5 lbs.	.7 lbs.
Total	328.1 lbs.	316.3 lbs.
Feed cost per 100 lbs. gain including pasture	$13.78	$13.18

Source: *Emergency Pasture Systems for Hogs*, Mimeo. No. 36, Purdue University, Lafayette, Ind., 1948.

FIGURE 9-5. This growing-finishing setup has bulk tank self-feeders that hold several tons of feed. The wooded area provides shade. (*Courtesy Allied Mills, Inc.*)

ing alfalfa with an emergency pasture of oats, rape, and Sudan grass. It should be noted that these pigs were not fed a complete ration.

The pigs on alfalfa pasture gained slightly faster and used less feed per 100 pounds of gain than did the pigs grazed on the oats, rape, and Sudan grass pasture.

The oats and rape were sown together at the rate of 5 pounds of rape and 1½ bushels of oats per acre. The Sudan grass was broadcast at the rate of 30 pounds per acre.

Oats, peas, and rape. In southern states field peas are seeded with oats and rape. This forage outyields oats and rape, and the peas produce pasture high in protein. The usual rate of seeding is 1 bushel of oats, 1 bushel of peas, and 7 pounds of rape per acre.

Sudan grass. This crop has a heavy carrying capacity during a short grazing period. It is an excellent crop to supplement oats and rape or bluegrass pasture. It furnishes green grazing during the hot months of July and August.

Sudan grass must be seeded thickly to prevent the growth from becoming coarse and woody. Thirty to 35 pounds of seed per acre is recommended.

There is little danger of prussic acid poisoning if the Sudan grass is not grazed until it is at least 15 inches in height. An acre of Sudan grass will carry 20 pigs per acre, but for a shorter grazing period than will ladino or alfalfa.

In tests conducted at Purdue, alfalfa pasture increased the gain of hogs 8.7 percent over those fed in dry lot. The increase with Sudan grass was 2.1 percent.

Bluegrass. Although listed as an emergency pasture, bluegrass is a permanent pasture, but it does not compare with ladino and alfalfa as a hog pasture. It should be used only when the preferred types of pasture are not available.

With white Dutch clover or bird's-foot trefoil, bluegrass produces excellent early pasture in the spring and late pasture in the fall. It is especially good for sows and litters. The grass dries up during July and August when heavy grazing is needed. The clover and trefoil provide good grazing during the hot dry periods. One acre can carry 15 pigs during the early spring and late fall grazing periods.

Bromegrass. This crop is used in combination with alfalfa, ladino, or red clover as a hog pasture. It should not be used alone unless as an emergency pasture. It produces abundantly, but since it is a grass rather than a legume, the forage is not as high in protein as ladino or alfalfa.

PASTURE MIXTURES

A mixture of grasses and legumes may provide more desirable pasture during all stages of the growing season, and on varying

FIGURE 9-6. These Hampshire pigs have made excellent growth on rotated pasture. (*Courtesy Hampshire Swine Registry.*)

T A B L E 9-8   RECOMMENDED FORAGE CROPS FOR SWINE

Crop (Optimum pH)	Rate of Seeding (Per Care)	Date of Seeding	Fertilizer (Per Acre)	Grazing Management
Ladino Clover (6.5)	5 lbs.	Aug.-Sept.— Piedmont Sept.-Oct.— C. Plain	Seeding: 400-700 lbs. 0-14-14 Annual Maintenance: 400-600 lbs. 0-10-20 applied Aug.-Sept.	Do not graze below 2 inches.
Small Grain (5.5-6.5)	3 to 4 bu.	September	Seeding: 600-800 lbs. 6-12-12 Topdress: Fall 30-60 lbs. N Spring 30-60 lbs. N	When grain is 6" to 10", graze back to 2". Allow for regrowth.
Ryegrass Crimson Clover (6.0-6.5)	30 to 40 lbs. 20 lbs. c. clover	Sept.1-Oct. 15	Seeding: 600-800 lbs. 6-12-12 Topdress: Fall 30-60 lbs. N Spring 30-60 lbs. N	Graze when 6" to 8" as long as forage is available.
Soybeans (Biloxi) (6.0-6.5)	1 bu. in rows	May-June	Seeding: 400 lbs. 0-10-20	Graze when 8" to 10" as long as forage is available.
Millet (Gahi-1) Millet (Starr) (5.6-6.5)	40 lbs. broadcast or 15 to 20 lbs. in 2-3 ft. rows	April-June (Successive plantings)	Seeding: 600-800 lbs. 6-12-12 Topdress: 60-100 lbs. N	Graze 8" to 10" and not below 3".

Source: Clemson University.

soil conditions. The following mixtures are recommended:

1. Alfalfa—6–8 lbs.
   Ladino clover—½–1 lb.
   Bromegrass—5 lbs.
2. Ladino clover—1–2 lbs.
   Bromegrass—5 lbs.
3. Alfalfa—5 lbs.
   Red clover—4 lbs.
   Bromegrass—5 lbs.
4. Sweet clover—4 lbs.
   Red clover—2 lbs.
   Alsike clover—2 lbs.
   Ladino clover—½ lb.
   Bromegrass—5 lbs.

PASTURE MANAGEMENT

The management given the pastures and the hogs that graze them may greatly influ-ence the effectiveness of the pastures in hog production programs. Many good pastures are not managed properly.

Data indicate that the average swine producer in southern states markets fewer hogs than do those in the Corn Belt, and more of them are using pastures in their pro-duction programs. The lower cost of farm land and the high cost of using confinement systems when small numbers of pigs are in-volved are probably contributing factors.

Recommendations concerning forage crops and grazing management for swine in South Carolina are presented in Table 9-8.

Rotated Pastures

In order to prevent disease and parasite outbreaks, it is desirable to provide a new pasture area each year. Farmers who make

FIGURE 9-7. These pigs have access to large feeders which are filled mechanically. Their pasture appears to be very limited. (*Courtesy Hawkeye Steel Products, Inc.*)

FIGURE 9-8. These Yorkshire pigs have an abundance of good alfalfa pasture. Removing a hay crop will improve the quality of pasture when excessive growth exists. (*Courtesy American Yorkshire Club.*)

new seedings each year have a new pasture area for each crop of pigs.

Farmers who use permanent or semi-permanent pastures accomplish the same result by dividing the pasture area into three or four parts, and then rotating the use of the divided areas.

Pigs per Acre

The number of pigs or hogs per acre depends upon the kind of pasture, fertility of the soil, weather conditions, and whether a full or limited feeding program is followed. An acre of legume or legume-grass pasture should provide adequate pasture for 20 to 30 growing-finishing pigs on full feed. If the pigs are on a limited ration, the number should be reduced to 10 to 15 per acre.

Clipping

Frequent clipping of pastures during the summer months is recommended if the pigs do not keep the growth down.

Location of Feeders and Waterers

Considerable labor may be saved by locating the feeders and waterers near the fence so that they may be filled by use of power equipment without driving into the field. Large feeders which are protected from the weather are recommended. They do not need to be filled as often as other feeders, and if they are of the walk-in type, they provide shade on at least one side which encourages pigs to consume feed during hot weather.

Tests conducted at the South Dakota Station indicate that the feeders and waterers should be located close together. When located more than 300 feet apart the average daily gain decreased.

FIGURE 9-9. These Berkshire sows are on a grass pasture. (*Courtesy American Berkshire Association.*)

Ringing Pigs

It is sometimes necessary to nose-ring pigs to keep them from rooting the sod in the pasture. The feeding of balanced rations containing animal proteins and minerals tends to reduce rooting. Care should be taken in ringing pigs to avoid injury to the bone structure of the nose.

CONFINEMENT MANAGEMENT

Three types of confinement systems are in use: (1) semi-confinement using an open-front building and outside concrete slab, (2) complete confinement in enclosed building but no environmental control, and (3) complete confinement in an environmentally controlled building. The advantages and disadvantages of these systems are described below.

Semi-confinement (open-front building and concrete slab)

1. Most inexpensive of the three systems.
2. Provides for flexible feeding and watering systems.
3. Decreases ventilation problems.
4. Manure can be handled by tractor scoop and spreader.
5. More labor required.
6. Bedding is required.
7. Snow and ice problems in the north.
8. Less operator comfort.
9. Less feed efficiency.

Complete confinement (enclosed building with no environmental control)

1. Intermediate first cost.
2. Manure handled by gutter cleaner or tractor and scoop.
3. Provides for mechanical feeding and watering systems.
4. Bedding is required.
5. Some ventilation problems with wall openings.
6. Minimum snow and ice problems.

7. Intermediate in operator comfort.
8. Intermediate in labor requirements.
9. Good feed efficiency.

Complete confinement (enclosed, environmentally controlled building)

1. High first cost.
2. Provides for mechanical feeding and watering.
3. Provides flexible waste handling, slats, gutter cleaner, or liquid.
4. Controlled heat and ventilation possible.
5. No bedding needed.
6. No snow and ice problems.
7. Comfortable quarters for operator.
8. Least labor requirement.
9. Best feed efficiency.

Details concerning buildings and equipment recommended for use of the three systems are presented in Chapter 10. Practices recommended for use in starting feeder pigs were listed in Chapter 8. Many of them should be continued during the entire finishing period. Following is a list of suggestions

for managing growing-finishing pigs in partial or complete confinement.

Stress

1. Pen pigs by size.
2. Provide adequate space (See Chapter 10).
3. Avoid mixing pigs from various sources.
4. Provide adequate feeders and waterers.
5. Feed adequate rations—keep feed and water before pigs at all times.
6. Limit the number of pigs in each pen to 20.

Sanitation

1. Provide partial or fully slatted floors.
2. Use liquid manure handling methods.
3. Use lagoons, oxidation ditches, or haul liquid directly to field.
4. Slope floors ½ to ¾ inch per foot if solid floors are used.
5. Maintain recommended temperature and ventilation.
6. Locate waterers over slatted floor area and feeders away from slatted floor areas.
7. Isolate pigs showing signs of sickness.
8. Control flies.

FIGURE 9-10. A large-scale confinement feeding facility. The dunging area is at the lower level. (*Courtesy Confinement Builders, Inc.*)

9. Control odors.
10. Train pigs to sleep in one area and dung in another area.

Tail and Ear Biting

1. Best control of tail biting is to avoid it by docking tails of pigs at birth.
2. Provide adequate space in pen.
3. Provide adequate feeder and waterer space.
4. If few pigs are involved, isolate the tail biters.
5. Treat affected tails and ears with commercial solution of methyl violet.
6. Feed for short time free-choice a mixture of ⅓ Glauber's salts, ⅓ salt, and ⅓ mineral.
7. Add 2 to 4 pounds of magnesium oxide to each ton of complete feed.
8. Suspend a chain, rope, or tire for the pigs to play with.

Feeding

1. Feed the complete ration recommended in Chapter 8.
2. Keep feeders filled.
3. Adjust feeders to avoid feed waste.
4. If paste or gruel feeding is done, make certain that all equipment is clean and stale feed removed.
5. Keep feeder troughs about one-third filled.
6. Use automatic feeding system when economically sound.
7. Use pelleted feeds to increase consumption and decrease spillage losses.
8. On solid floor systems, locate feeders away from dunging area.
9. Use feeders with partitioned troughs and lids over each feeder space.
10. Observe withdrawal times of rations containing feed additives requiring withdrawal before marketing the pigs.

Temperature

1. Keep temperature between 50 and 60 degrees in winter and under 75 degrees in summer.
2. Use sprinkling system set automatically to provide 1/10 gallon of water per pig per hour.
3. Use supplemental heat in winter if necessary.

Ventilation

1. Close part of front of open-front confinement houses in northern areas during extremely cold weather.
2. Use ceiling inlets with baffled slots and a fan system in inclosed buildings.
3. Avoid drafts.
4. Provide about 15 cfm (cubic feet per minute) of air per pig weighing 50 pounds, 25 cfm per pig weighing 125 pounds, and 200 cfm per pig weighing 200 pounds during the winter. During the summer the number of cfm should be tripled for all three weight classifications.
5. Sidewall openings may be desirable in some houses during the summer months.

Water

1. Provide adequate waterers, either nipple or cup waterers.
2. Keep waterers properly adjusted.
3. Clean waterers regularly.
4. If slats are used, place waterer over slatted area.
5. Keep water ice-free during winter.

Bedding

1. Bedding is not needed in slatted floor confinement houses.
2. Bedding is needed in sleeping areas of open-front and other solid floor confinement houses.
3. Provide more bedding in cold weather than during warm weather.
4. Keep bedding dry. Clean as often as necessary and replace with new bedding materials.
5. Use dust-free bedding when possible.
6. Wheat and oat straw, shavings, or wood and bark chips are preferred.

Health

1. Worm with broad spectrum wormer when 6 to 8 weeks of age.
2. Vaccinate for erysipelas (when prevalence warrants) when 6 to 8 weeks of age.
3. Spray pigs for lice and mange at 8 weeks and routinely treat as necessary.
4. Avoid stress by providing adequate rations, water, feeder and waterer space, temperature, ventilation, and space.

5. Keep quarters clean and dry.
6. Isolate and treat sick animals.
7. Isolate pigs brought on to farm for 3 weeks before moving them into the confinement house.
8. Keep outsiders out of the hog house.

SUMMARY

Pastures are used by most hog producers. They save 15 to 30 percent of concentrates used by pigs in confinement and may reduce the protein supplement feed bill by one-third to one-half, unless carefully planned complete rations are fed. Rotated pastures aid in controlling diseases and parasites.

Confinement feeding may be more economical on high-priced land and when large numbers of hogs are produced. Rations for pigs fed in confinement must be more highly fortified than for pasture feeding. Breeding stock can make more economical use of pasture than can growing-finishing pigs.

Ladino clover and alfalfa are the two best legume pastures. Ladino is superior to alfalfa. In Missouri tests, alfalfa pasture produced $118.40 worth of pork per acre. Red clover produced $89.80, and an acre of rape produced pork worth $79.

A mixture of ladino, alfalfa, and brome grass is recommended where soil variations exist and irregular rainfall is expected.

Rye makes an excellent late fall, early winter, and early spring pasture.

Bluegrass provides excellent early spring and late fall pasture but dries up during July and August, when heavy grazing is usually needed.

Rape is an excellent emergency pasture during the summer months. It is seeded in the spring and can be grazed after about six weeks. Rape, oats, and Sudan grass may be

FIGURE 9-11. An open-front finishing house in Gates County, North Carolina. It can hold 1,000 head of swine. (*Courtesy J. Ray Woodard, North Carolina State University.*)

used to provide pasture throughout the season.

Sudan grass has a heavy carrying capacity for a short season during July and August.

Hog pastures should be rotated. New ground should be available for each crop of pigs. One acre of legume pasture will graze from 20 to 30 growing-finishing pigs on full feed.

Frequent clippings of pastures during the summer are recommended if the growth becomes heavy.

Feeders and waterers should have large capacities and be located close together near the fence.

Pigs should be nose-ringed if they root up the sod. Shades, open buildings, or trees should provide protection from hot sun in summer.

Nearly 25 percent of all hogs marketed in 1973 were reared in confinement. The pasture system of swine production results in lowest costs per 100 pounds of pork when only two farrowings are made each year. Lowest costs when four or more farrowings are had each year result from the use of a high initial cost, environmentally controlled, complete confinement system. The open-front building, partial confinement system provides lowest costs in southern sections of the U.S., and about equal costs to the use of complete confinement systems in the north.

Growing and finishing pigs in confinement requires better management than in using pastures. Pigs should be grouped by size and source (if purchased). No more than 20 pigs should be in a pen. Adequate space is a must. When possible, slatted floors should be used in entire pen. Bedding must be used in solid-floor buildings. It must be kept clean and dry.

Mechanical feeders should be used. Self-feeders should be adjusted to avoid feed loss. Individual cup or nipple waterers may be used.

Complete rations should be fed. The feeders should be kept full. Either dry or gruel feeding systems may be used. Pelleted feeds are more palatable and there is less waste than with the use of dry, ground feeds. Withdrawal times in feeding rations containing feed additives must be observed.

The temperature should be kept 50 to 60 degrees F in winter and under 75 degrees in summer. Supplementary heat may be necessary in winter. Sprinkling systems and fans are necessary in the summer. Adequate ventilation must be provided during both winter and summer to remove odors and humidity, and to keep fresh air in house at all times.

There should be no drafts. Open fronts may be closed, or partially closed in winter. Sidewall openings may be used in confinement houses during the summer.

Pigs should be wormed and treated for lice and mange at 6 to 8 weeks. Vaccination for erysipelas should be done if conditions demand it. Sick pigs should be removed and isolated. New pigs should not be put into the confinement house until isolated on the farm for at least 3 weeks. Keep outsiders out of the hog house.

QUESTIONS

1 What are the advantages of pasture feeding over confinement feeding?

2 In what ways is confinement feeding better than pasture feeding?

3 Which system is best for your farm? Why?

4 Compare ladino clover, alfalfa, and red clover as pastures for swine.

5 How do the legume pastures compare with rape as hog pastures? Explain.

6   What is the place of Sudan grass as a hog pasture?

7   Outline the pasture crops which might be used to substitute for legumes in case they are winter killed.

8   What crops are recommended for late fall, winter, and early spring pasture for hogs?

9   List the number of pounds of seed recommended in seeding an acre of (1) ladino, (2) alfalfa, (3) red clover, (4) rape, and (5) Sudan grass.

10   How many acres of legume pasture are needed to provide adequate hog pasture on your farm? Explain.

11   What management practices should be followed in making effective use of hog pastures?

12   Outline a program which will provide adequate hog pasture for the swine enterprise on your home farm.

13   Under what conditions would it be desirable to use a combination of pasture and confinement feeding systems?

14   How can stress be avoided in confinement feeding?

15   List some methods of controlling nose and tail biting among confined pigs.

16   Outline sanitation practices necessary in confinement systems of growing-finishing pigs.

17   Which feeding method would be most economical on your farm, (1) dry, ground feed, (2) pelleted, (3) paste or gruel? Why?

18   What temperatures should be maintained in the confinement house during the summer and in winter? How can these temperatures be maintained?

19   Under what conditions is it necessary to use bedding in confinement feeding? List desirable bedding materials.

20   How can health problems be prevented or controlled in confinement feeding during the growing-finishing period?

21   Select from among the following the best system of growing-finishing pigs on your farm: (1) pasture, (2) combination pasture and confinement, (3) partial confinement, (4) complete confinement. Give reasons for your choice.

REFERENCES

Aanerud, Wallace G., *Practical Hog Production for S. D. Farmers*, EC 649. Brookings, S.D.: South Dakota State University, 1973.

Ensminger, M. E., *Swine Science*, 4th ed. Danville, Ill.: The Interstate Printers and Publishers, 1970.

Granite City Steel Company, *Farmstead Planning Swine Manual*, Book 3. Granite City, Ill., 1973.

Jensen, A. H., B. G. Harmon, G. R. Carlisle, and A. J. Muehling, *Management and Housing for Confinement Swine Production*, Cir. 1064. Urbana, Ill.: University of Illinois, 1972.

Jones, J. R., and A. J. Clawson, *Raising Hogs in North Carolina,* Cir. 505. Raleigh, N.C.: North Carolina State University, 1972.

Krider, J. L., and W. E. Carroll, *Swine Production,* 4th ed. New York, N.Y.: McGraw-Hill Book Company, 1971.

Moyer, W. A., and B. A. Koch, *Feeding Hogs for Profit.* Manhattan, Kans.: Kansas State University, 1971.

Whiteker, M. D., R. Edwards, and D. O. Liptrap, *Swine: Feeding and Management, Growing and Finishing Pigs,* ASC 2. Lexington, Ky.: University of Kentucky, 1972.

Younkin, Dwight, *Managing the Growing-Finishing Pig,* Sp. Cir. 156. University Park, Pa.: Pennsylvania State University.

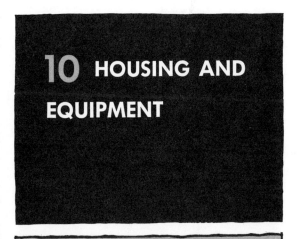

10 HOUSING AND EQUIPMENT

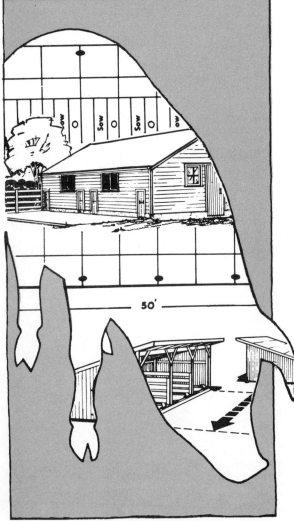

It is the opinion of many swine producers that improvement in the housing of hogs has not kept pace with developments in nutrition and breeding. Those improved practices in swine housing which have come about have been introduced largely because of the interest in confinement swine production programs, in multiple-farrowing programs, and in disease and parasite control.

Considerable emphasis has been given to the development of swine production equipment. New developments in pig brooders, farrowing stalls, self-feeders, mechanical feeding equipment, hog waterers, shades, sprinklers, and manure handling equipment have been common.

Pork production is much more competitive and complicated today than it was 20 years ago. Feed efficiency and economical use of labor are important factors in pork production. Both are affected to quite an extent by the housing facilities and equipment used. Disease problems are serious, and their control is usually dependent, at least in part, upon housing conditions and equipment available. Pork producers must give attention to housing and equipment if they are to make effective use of new developments in nutrition and breeding. The best of breeding stock and the most nutritive rations cannot take the place of adequate housing and equipment.

We present in this chapter principles and recommendations concerning (1) housing the breeding herd, (2) providing farrowing and nursery quarters, (3) providing shelter for pigs on pasture, (4) providing housing for pigs fed out in confinement, and (5) providing swine-production equipment.

HOUSING THE BREEDING HERD

Breeding animals do not require the exacting housing conditions required by sows and litters at farrowing time, or by pigs being fed out in confinement.

Summer Housing

Brood sows and boars need shade to protect them from the sun during hot summer weather and need protection from winds and cool weather which may occur especially during the early and late summer months. An open cattle shed or sheep barn makes an excellent shelter for breeding animals.

Portable houses may also be used. Houses which are open on one side, or which can be converted into shades are recommended. Portable houses with tight walls should not be used unless they are located under trees or in wooded areas.

Winter Housing

It is important that sows and gilts be given adequate housing during the winter months. In some areas there is danger of animals getting the "flu" as a result of cold, drafty, or wet quarters. Heated or insulated houses with concrete floors are not necessary.

In most areas an open shed, pole barn, or lean-to to the barn will provide adequate protection. Deep, dry bedding is recommended. Movable houses which may later be used as farrowing houses may also be used. Care should be taken to avoid crowding the animals. Sows on pasture and others that have an outside run should have 12 to 15 feet of housing. Confined sows and gilts should have 16 to 20 square feet per animal.

FARROWING QUARTERS

Adequate housing is very important at farrowing time. Good housing facilities will save one or several more pigs per litter.

There are two general methods of housing sows and litters at farrowing time: (1) using portable houses and (2) using central-type houses. Each of the systems provides certain advantages and has some disadvantages. Each hog producer must analyze production program and finances in deciding which system of housing to use.

Portable Houses

Advantages of portable houses. The use of movable hog houses in the past was common in the Corn Belt. Nearly 40 percent of the farms in some states were tenanted.

FIGURE 10-1. This confinement finishing facility will accommodate 560 head. Three or four groups of pigs can be finished each year. (*Courtesy Confinement Builders, Inc.*)

Landowners in general did not see fit to provide and maintain adequate central farrowing houses for their tenants. As a result many tenants relied upon portable houses.

The number of farmers using portable houses for farrowing is decreasing, but many pigs in the southern Corn Belt and in the South are still farrowed on pasture in movable houses. Even in Iowa and Illinois, the leading swine producing states, many pigs are farrowed in the late spring or early fall on pasture.

The prevalence of disease and parasitic organisms around the central-type hog houses and farmsteads has forced some farmers to move their hogs out to clean legume pasture. These farmers must have portable housing. While some of them have central farrowing houses, many of them use portable hog houses for farrowing. The portable houses are moved up to the building site where water and electricity are available, and it is convenient to look after the pigs. In some cases concrete slabs or feeding floors are provided, and the movable houses are lined up around them.

From the standpoint of cost it is much more economical to acquire and maintain portable houses than a central-type farrowing house unless multiple farrowing is done. The investment per litter is smaller because the house may be used for a longer period of time each year.

It may be better to use portable houses than pens in a central farrowing house when farrowing litters from new breeding stock. New stock can be isolated from the herd by housing it in a portable house in a separate lot.

Disadvantages of portable houses. One handler can care for more sows and litters at farrowing time if they are in a central-type house than if they are housed in portable houses. It is usually easier to keep the houses clean since the floors are of concrete and water may be piped into the building. It is much easier to provide heat and properly ventilate central farrowing houses. They pro-

FIGURE 10-2. This building may be used for either farrowing or finishing. Fresh air is brought in from the top, circulated through the slatted floors, and exhausted at the sides of the building just below the slatted floors. (*Courtesy Behlen Co.*)

FIGURE 10-3. A modern environmentally controlled farrowing house. Floors are partially slatted. Feed is automatically delivered to each feed hopper. The solid panels around creep area prevent drafts and conserve heat for pigs. (*Courtesy A. R. Wood Mfg. Co.*)

vide for more ease in feeding both the sow and the pigs during the creep feeding period.

Much more labor is involved in using portable houses than in central farrowing houses, especially if more than two pig crops are reared each year. Central houses provide greater comfort for both the animals and for the herdsman.

Types of movable houses. A large number of plans for portable houses are available from the various agricultural extension services and lumber companies. Some of them are for individual sows and litters. Others will accommodate two or more litters. Shown in Figures 10-4 and 10-5 are examples of movable houses.

The single-pen houses are commonly used for farrowing when the sows are permitted to farrow in the field or timber lot with little assistance provided. These houses are not usually equipped with a farrowing stall, but they may have guard rails.

While some plans for portable houses were made before farrowing stalls were introduced, it is not difficult to change the plans to make possible the use of stalls. The house shown in Figure 10-5 is a popular building in the Corn Belt, but it was not planned for the use of farrowing stalls. It is possible, however, to use the same plan in building a house 15 or 16 feet long which will contain three farrowing stalls. The doors on the side and roof must be changed.

The most popular portable house is a multi-use or "sunshine" house which has been recommended highly by Purdue University, University of Illinois, and Iowa State University specialists. The Doane Agricultural Service, Inc., of St. Louis provided the picture for the building shown in Figure 10-7.

These houses are made in pairs with the long side of the house left open or partly open. By pulling the two houses together, a large house is provided which may be used

FIGURE 10-4 (*Above*). The Purdue University two-unit hog house may be used to provide farrowing quarters or shelter for hogs on pasture. The units pictured have been grouped and wired with electricity for use as farrowing quarters. (*Courtesy J. C. Allen and Son.*)   FIGURE 10-5 (*Below*). Movable houses may be grouped in housing pigs on pasture.

for farrowing or for shelter during the winter. Two Doane houses drawn together will provide space for 12 farrowing stalls and a 6-foot alley. Two 8- by 14-foot houses when put together will provide space for six farrowing stalls and a 6-foot alley.

It is possible to build the houses with removable floor panels that are removed when the houses are taken to the pasture for summer shelter.

The advantages of twin "sunshine" or multi-use houses over other types of movable houses are as follows:

1. They can be used the year around.
2. They combine the advantages of the individual portable and the central-type hog houses.
3. They can be moved through standard farm gates.
4. One man can take care of several sows and litters at one time.
5. The houses are easy to clean.
6. Heat can be easily provided.

Materials. Three types of materials are used in building portable houses. They are lumber, corrugated steel, and weatherproof panel board. Corrugated steel and panel board will make the tightest building. Steel is a good conductor of heat. It will be hotter in the summer and colder in the winter than wood. Painting the corrugated steel with aluminum paint is recommended. Price often determines the material that should be used.

Central-type Houses

The multi-use houses have taken the place of central-type farrowing houses on

the farms of most small operators. The man who raises a large number of pigs and follows a multiple-farrowing program will find a central-type farrowing house a good investment.

Advantages of central-type houses. One man can feed and care for a larger number of sows when they are housed in a central farrowing house than when they are housed in individual houses. It is usually easier to keep the pens clean, and heat can be provided more easily. The wiring and water system can be permanent installations. It is easier to move pigs from large litters to small litters and easier to group the pigs according to size at weaning time.

Disadvantages of central-type houses. Houses of this type are costly. One farmer invested $10,000 in a hog house which would accommodate 16 sows and litters at one time. Assuming that two crops of pigs were produced annually, the investment per litter was $313. With four crops produced each year, this investment would be reduced to $156 per sow. Two or more crops of pigs must

FIGURE 10-6 (*Above*). These portable houses permit raising the roof for cleaning, and the back for summer ventilation.   FIGURE 10-7 (*Below*). The Doane designed multi-use 12-sow hog house. (*Plan No. 52. Courtesy Doane Agricultural Service, Inc., St. Louis, Missouri.*)

usually be produced each year to justify the building of a central-type farrowing house.

Disease can spread very rapidly from one litter to another when the litters are housed under the same roof.

A duplication of investment in housing is often necessary if the pigs are moved from a central-type house to clean-ground pasture.

Types of central farrowing houses. We have almost as many types and floor plans as we have hog houses. Rarely do we find two alike. A large number of central farrowing houses are remodeled barns, machine sheds, or cattle feeding sheds.

The system to be followed in the management of the litters and the availability of portable housing facilities determines to a large extent the type of farrowing house which will be needed. Shown in Figure 10-10 is a 16-sow farrowing house. With the use of portable houses after the pigs are two to four weeks of age, if the pigs are to be raised on pasture, or the use of a finishing house if raised in confinement, this house could be used in farrowing 24 to 32 litters during each farrowing season.

The plan provides for eight farrowing stalls and four pens 8 by 10 feet. When the litters are four to five days old, the sows and litters are moved to the pens. Two sows and litters are placed in each pen.

This plan provides for concrete feeding floors on each side of the house with separate pens and feeders for each two sows. Since all feeding is done outside the hog house, a 4-foot service alley is sufficient.

The use of slatted floors is not provided for in the plan. It is assumed that cleaning will be done by scoop. Cleaning should be minimal since the sows will be fed in the outside pens.

Note that the floor is sloped toward the middle for drainage. A slope of one-quarter inch per foot is recommended. Drains are provided in the service alley. Each pen has an automatic waterer. A solid wall partition

FIGURE 10-8. This facility is used for farrowing, suckling, and finishing. (*Courtesy Big Dutchman.*)

FIGURE 10-9. Note that these farrowing crates are elevated several inches above floor level, and that the automatic waterers drain into a drain at the lower level. (*Courtesy Big Dutchman.*)

FIGURE 10-10. *Above:* The outside of a 16-sow farrowing house equipped with farrowing stalls. *Below:* The floor plan for the same building. (*Midwest Plan Service Plan No. 72668.*)

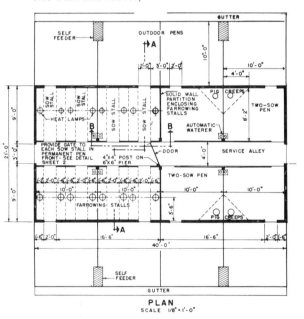

divides the farrowing stall section of the house from the remainder of the house.

It is a good idea to insulate the walls and ceiling of the house. A vapor barrier should also be provided. Mechanical ventilators are preferred, but louvers may be used in each end.

This building may be made of lumber or tile. Price may be a determining factor. The concrete floor should be underlaid with gravel, and a layer of waterproof material should separate the concrete from the gravel base. An inch of foam glass separating the foundation from the floor will aid materially in keeping the floor warm during winter months.

A comparison of seven different farrowing systems was conducted at Purdue University. The study involved the use of (1) 6- by 8-foot modified "A" houses with slatted platform outside, (2) a central farrowing house with 7- by 8-foot pens with guard rails and heat lamps, (3) a central house with six conventional farrowing crates and floors equipped with heat units, (4) six convential farrowing crates in a central house with a slatted floor, a central manure pit, and electric heat lamps, (5) a central house with six farrowing stalls with heat lamps and the sides of the stalls and pig creep covered with plywood, (6) a central farrowing house with six farrowing stalls on a raised centrally-located concrete base equipped with electrical heating units, and (7) a central house with six farrowing stalls, raised 18 inches above a partially slatted floor, and equipped with heat lamps.

A summary of the calculated annual use costs of the different systems per year per sow and litter is presented in Table 10-1.

There was an overlay loss of 1.5 more pigs per litter in individual houses and in pens in central houses. The costs per sow and litter for the first 21 days were similar for the various systems, but the slatted-floor house reduced labor 30 to 40 percent compared to the other systems. The building investment was slightly higher in the slatted-floor unit than in the other central-house systems.

The individual portable houses required $100 to $200 less investment per sow capacity but resulted in higher labor requirements than the central-house systems. Tests at Purdue indicated little economic advantage of farrowing capacities over 32 sows in the farrowing house.

Shown in Figure 10-11 are end and floor plan views of a 24 by 48 foot farrowing house

built in North Carolina for approximately $10,000. Note that it has two 8-foot manure pits and an air tunnel under the center alley. The 16 farrowing crates are over a completely slatted floor. The building is heated with a furnace and is zone cooled. Air enters from the attic in the winter and directly from the outside in the summer. All air can be discharged from the pits. Thermostatically controlled fans are used.

A similar design recommended by the University of Illinois, is shown in Figure 10-12. The Illinois plan makes provision for 24 farrowing stalls with individual feeders and waterers so that the sows do not need to be let out of the stalls until the pigs are weaned, or the sows and litters moved to nursery quarters. Note the insulation on the walls and ceiling.

The sows will be hand-fed and should face the center for convenience in feeding. The outside aisles are only 2½ feet wide, but provide access to the back of the stalls. The floors may be completely slatted, or slats on only the back one-third of the stall and the outside aisles.

T A B L E 10-1   CALCULATED ANNUAL USE COSTS OF DIFFERENT FARROWING SYSTEMS PER YEAR PER SOW AND LITTER

System (32-Sow Capacity)	Number of Farrowings per Year			
	2	4	6	8
SOWS FED INSIDE				
Crates, concrete floor	$41.64	$31.33	$27.89	$26.18
Crates, concrete floor, plywood floor under pig creep	43.24	32.63	29.09	27.33
Crates, raised slotted floor	43.01	31.70	27.93	26.05
Crates, slotted floor with pit	43.50	31.38	27.34	25.32
SOWS FED OUTSIDE				
Crates, concrete floor	39.80	31.42	27.96	26.48
Pens	43.03	32.58	29.10	27.36
Individual houses	41.87	32.58	29.48	27.94

Source: *Productivity and Cost of Swine Farrowing and Nursery Systems,* Purdue University, 1967.

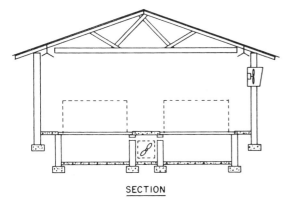

SECTION

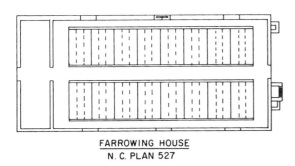

FARROWING HOUSE
N. C. PLAN 527

FIGURE 10-11. End and side view of a 16-crate farrowing house in use at the North Carolina Swine Development Center, Rocky Mount, North Carolina. (*Courtesy North Carolina State University.*)

Standard *farrowing stalls* are 5 feet wide with a 2-foot wide sow stall and 18-inch brooders for pigs on each side. Stalls normally have adjustment arrangements to widen for large sows and narrow for small sows. The length can also be adjusted.

The use of *free-stall pens* for farrowing is increasing rapidly. These stalls vary greatly in size ranging from 5 to 6 feet wide and 10 to 14 feet long. They usually provide guard rails, a creep area and a slatted-floor area at the end of the pen for the sow to get water, be fed, and dung. Sows learn rapidly to use the back of the pen for dunging. The front of the pen is kept clean. There is very little

labor involved in the use of free-stall pens. The sows and litters usually remain in the pens until the pigs are weaned. Sows adjust to the free-stall pens much more readily than to stalls and there is less stress at farrowing time. It is much easier for the attendant to assist the sow and pigs at farrowing time than when crates or stalls are used.

An alley 2½ to 3 feet wide should be available at the back of the pen to facilitate feeding the sow. Some users of free stalls prefer to feed the sows in an alley in front of the pens.

A comparison of a standard farrowing stall and two types of free-stall pens is shown in Figure 10-13.

FARROW-TO-MARKET HOUSING SYSTEMS

Developments in swine nutrition and in disease control now make it possible to farrow and grow out pigs to market weight profitably under one roof. This system is used extensively in some sections of several European countries.

Labor is an important item in feeding and managing hogs fed in confinement. The hog house shown in Figure 10-10 may be used in feeding pigs in confinement. The feeders are on the outside of the building on the feeding floors and can be filled easily from a truck or trailer. The labor problem in using this type of house arises, however, in connection with the removal of manure. With very young pigs the problem is simple. It becomes complicated as the pigs reach an average weight of 100 pounds and above.

Some swine producers using a two-litter system, farrowing two litters per sow with only one group of sows, use one building in caring for the pigs from birth to market. They use a combination of crate and free-stall arrangements. In Figure 10-14 is a plan recommended in South Dakota. The building is 30 feet wide and 50 feet long with a 4½

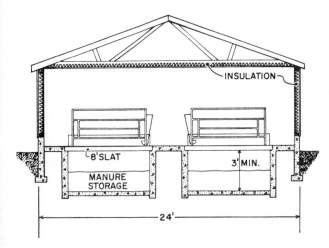

FIGURE 10-12 (*Left and below*). The cross section and floor plan of a slatted-floor farrowing house recommended by the University of Illinois. (*Courtesy University of Illinois.*) FIGURE 10-13 (*Bottom*). (*A*) Standard farrowing stall with partially slatted floor. Stall 2 by 7 feet with 1-foot creep for pigs on each side. (*B*) Farrowing stall 2 by 7 feet with 1½-foot creep on each side and feeder in end. (*C*) and (*D*) Two types of free-stall farrowing pens. Type *C* is 5 by 13 feet and has pig creep in end. Type *D* is 6 by 10 feet and has creep at one side. It also has slatted floor, feeder, and waterer at one end for the sow.

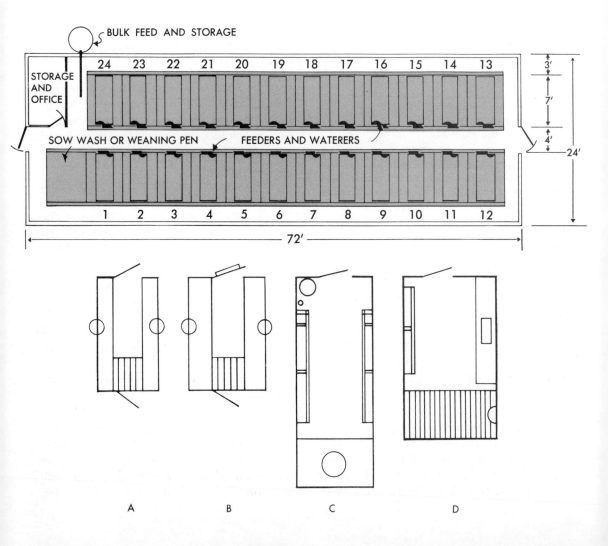

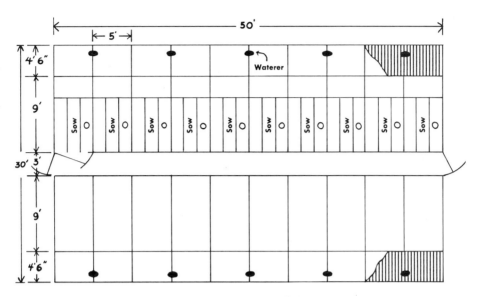

FIGURE 10-14. Free-stall farrowing house recommended by the South Dakota State University. (*Courtesy South Dakota State University.*)

foot slatted area in the back of the 20 pens, each 5 feet wide and 13½ feet long. Water is provided over the slats and the sows are fed in the crate, or in the slatted area.

When the litters are weaned, the crates are removed, and the 20 pens are used in growing out and finishing the pigs. The pigs may be penned by size, sex, litter, or by a combination of these methods.

Some swine producers have developed farrow-to-finish confinement systems with the farrowing, nursery, and finishing units under one roof. The three units are separated only by plywood walls. Such a system permits ease in the movement of pigs, more effective use of mechanized feed-handling equipment, economy and efficiency in removal of manure, and some decrease in construction costs.

Presented in Figure 10-15 is a layout for a production-line system recommended by the University of Illinois. The building is a 32-foot wide clear-span building with slatted floors. The 8-foot slats are supported on continuous foundations which form four separate manure storage tanks. The house is designed to handle six 30-sow farrowings each year, or about 1,500 pigs from farrowing to market.

The farrowing unit of the production-line housing system is presented in Figure 10-16. There are two rows of farrowing crates over fully slatted floors. The slats under the crates are 3 to 4 inches wide with a ⅜ inch spacing. The spacings between slats in a 2-foot square area directly behind the sow are widened to 1 inch to assume adequate cleaning. A metal grate is used to cover the wider spacings during and for a few days after farrowing. A space heater is used to maintain a temperature of 80°F at farrowing.

The slats may be of concrete, metal, or wood. Usually the slats are of concrete and are poured in place by use of metal forms. The pens are used to hold sows and to acclimate them to the new quarters before farrowing. The slats in the holding pen area are 4 to 5 inches wide with 1 inch spacings. Feeding stalls are provided.

The nursery unit in the Illinois production-line housing system is presented in Fig-

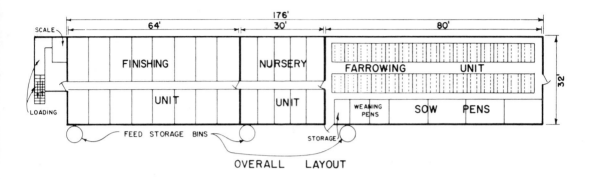

OVERALL LAYOUT

FIGURE 10-15 (*Above*). Overall layout for a production-line swine housing system recommended by the University of Illinois.　　**FIGURE 10-16** (*Below*). Farrowing unit of the Illinois production-line housing system. (*Figures 10-15 and 10-16 Courtesy University of Illinois.*)

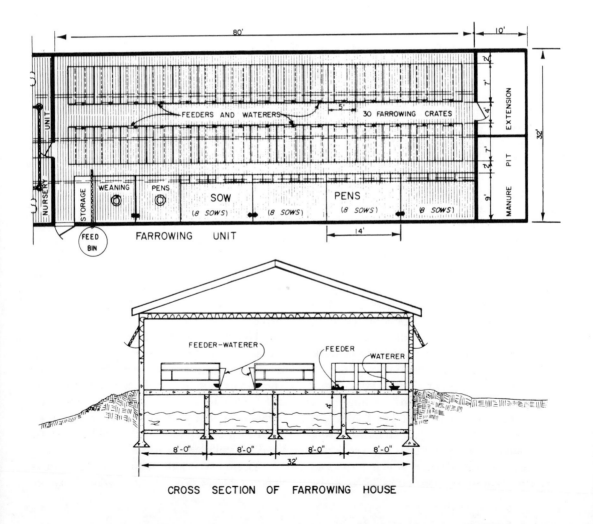

CROSS SECTION OF FARROWING HOUSE

ure 10-17. This unit consists of 5 pens 6 by 12 feet and of 5 pens 6 by 16 feet. Each pen accommodates about 3 litters or 24 pigs. The pigs remain in small pens about 30 days then move to the larger pens. The 6 by 12 foot pens provide 3 square feet of space per pig, whereas the 6 by 16 foot pens provide 4 square feet per pig.

The slats in the floor are 4 to 5 inches wide and have 1 inch spacings. A space heater is provided and the temperature is maintained at 65 to 75 degrees. Automatic time regulated feed conveyers may be used.

The finishing unit is shown in Figure 10-18. The pigs move from the nursery unit to the smaller pens of the finishing unit after about 60 days. Each pen in the finishing unit accommodates about 16 pigs or two litters. After about 30 days the pigs move to the 8 by 16 foot pens. The 8 by 12 foot pens provide 6 square feet of floor space per pig. The 8 by 16 foot pens provide 8 square feet. The floors are completely slatted and both feed and water are provided automatically.

Manure may be hauled away with a tank wagon, may be directed to a lagoon, or may be handled through the use of an oxidation ditch.

No nursery facilities are necessary if the sows and litters are kept in the farrowing crates or free-stall pens until weaned. Most growing-finishing houses have pens of 2 or 3 sizes and the pigs are started in the smallest pens and moved to larger pens as they need more space. Pigs weaned at three weeks of age need warmer housing and different feeds and feeders than larger pigs. Large producers quite often have separate nursery buildings.

The nursery facility constructed at the North Carolina Swine Development Center and recommended in that state, has partially slatted floors in pens 10 feet wide and 17 feet long. Each pen will accommodate three sows and their litters. Creep areas and heat lamps are provided. The roof is insulated

and there are shutters on one side and plastic curtains on the other.

Open-front finishing houses are very popular in the South and have a place in the Corn Belt. They are easier and less costly to construct than temperature-controlled buildings. They are best suited to housing pigs weighing 60 pounds or more.

Slats can be used in the South but freezing may be a problem in the North. Concrete floors are recommended. An epoxy-resin surface will aid in cleaning and decrease injuries caused by slipping and falling.

Most open-front houses have a concrete apron area equal to the width of the house. The houses open to the south and have a door, or vent, in the rear to permit ventilation in the winter. In southern states both sides may be open, but shutters or curtains are used to keep them warm in the winter.

Some houses have steam or electrically heated floors to keep them warm and dry in the winter. Bedding is used in the North, but not on slatted floors in the South. When partial slats are used they should be at the back of the house.

The pens should be 8 to 10 feet wide and the partitions should be solid to the ceiling, or to a height of at least 6 feet to prevent drafts.

The roof should be solid and well insulated. An open ridge will aid ventilation. Provision must be made in northern states to control drafts in cold weather. Doors, curtains or shutters should be provided. It is a good idea also to make provision for the partitions to extend to ceiling in cold weather.

Some producers have the front part of the apron area sloped to the front and an 8-foot section fenced off as a dunging area. Much of the manure is worked into a manure pit by the pigs. Where floors are heated, no bedding may be needed and there is little labor required in cleaning.

In Figure 10-19, page 245, is the plan used by the North Carolina Swine Develop-

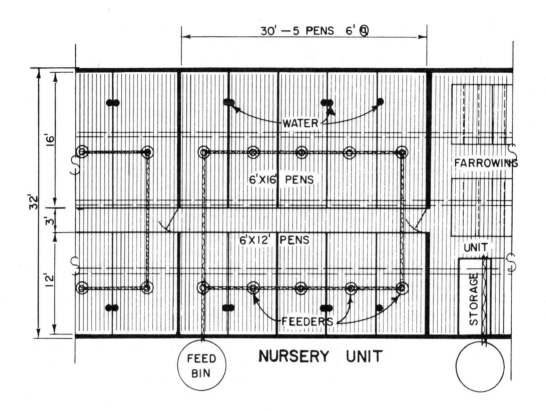

NURSERY UNIT

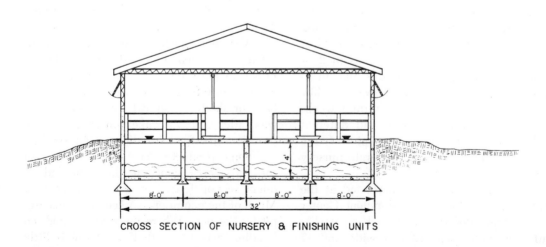

CROSS SECTION OF NURSERY & FINISHING UNITS

FIGURE 10-17. Nursery unit in the Illinois production-line swine housing system. (*Courtesy University of Illinois.*)

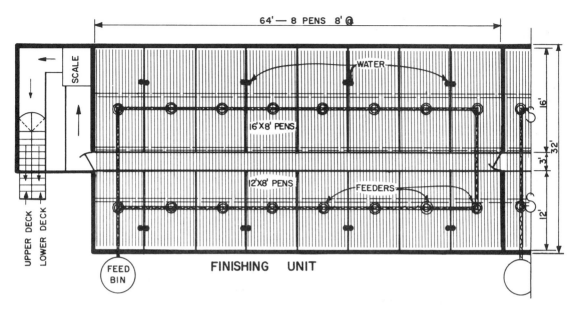

FIGURE 10-18. The finishing unit of the Illinois production-line housing system. (*Courtesy University of Illinois.*)

ment Center in building their finishing house. The floor is completely slatted. Pens 7 by 14 feet are on each side of a 4-foot alley. The sides of the house are open but roll-up plastic curtains are provided for use in cold weather. Ventilating fans are in the ceiling over the alley, and there is a thermostatically controlled fogging system for summer cooling. Use is not made of a concrete apron.

Planning for Confinement Housing

Some swine producers who feed out their pigs in confinement use two houses—a farrowing house and a finishing house. The finishing building, while relatively simple in construction, must be planned carefully. It is estimated that a finishing house adequate to feed about 500 pigs per year will cost about $12,000 to $15,000, providing multiple farrowing is practiced.

Presented in Tables 10-2 and 10-3 are the space needs of growing-finishing swine.

Most growers who feed in confinement follow systems of multiple farrowing and anticipate marketing the pigs at about 5 months. The pigs are kept in the farrowing stalls or free-stall pens with sows until weaned (21 to 42 days). They are then moved to the nursery unit. They remain there for 50 to 60 days. The pigs are then placed in the finishing unit. They are usually ready for market in 45 to 60 days.

Presented in Table 10-4 is a time-table showing the use of a growing-finishing building assuming that groups of sows were farrowing each two months. Two farrowing groups will be on the finishing floor at the same time. To illustrate, Group 1 consisting of 50 pigs from six sows farrowed on January 1, and reached Pen 1 in the growing-finishing building 60 days later, March 1. They remained in Pen 1 for a period of 40 days—until April 10. They then, according to schedule, weighed an average of 115 pounds and were moved to Pen 2 where they were kept 47 days, or until marketed at a weight of about 210 pounds about May 27. After Group 1 had been marketed, Group 2 was moved to the pen vacated by Group 1, and Group 3 was moved to the pen vacated by Group 2.

By use of the above timetable, a two-pen building would adequately house a total of 300 pigs during the growing-finishing period in one year. The pens would be vacated a total of 38 days for cleaning or could be used to finish out slow doers. Eighteen sows, each farrowing two litters per year on a multiple farrowing system, would be needed if each sow saved an average of 8⅓ pigs per litter.

Feeding and Feed Handling

Large volumes of feed may be required in finishing hogs. A feed cart that can be rolled or a feed carrier mounted on an overhead track in the service alley may be used in filling self-feeders. Self-feeders can be filled with self-unloading wagons or trucks or by use of systems of automation using either augers or blowers. Feed can be blown 200 to 300 feet from the feed-processing center. Feed can be augered up to 100 feet in a single run and longer distances by use of multiple augers.

Commercial hog-feeding systems are available that provide for storage of ingredients, the grinding and mixing of the feed materials, and the delivery of the complete

T A B L E 10-3 EQUIPMENT NEEDS OF SWINE

Feeder Space:

Complete ration — 1 feeder to 4 to 6 pigs
Supplement fed alone — 1 feeder space to 12 to 15 pigs
Grain fed alone — 1 feeder space to 5 to 7 pigs

Watering Space:

Gravity cup — 1 cup per 20 to 25 head
Pressure cup or nozzle — 1 space per 20 to 30 head

T A B L E 10-2 FLOOR SPACE NEEDS OF SWINE

Weight (Pounds)	Total Floor Space per Pig	
	Slatted Floors, Partial or Total (Square Feet)	Solid Floors (Square Feet)
PIGS		
25 to 40	3	4
41 to 100	4	6
101 to 150	5	8.5
151 to 210	8 (9 during hot weather)	11.5
SOWS	15	20

Source: University of Illinois.

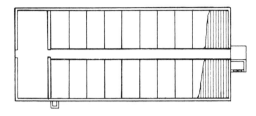

FINISHING HOUSE

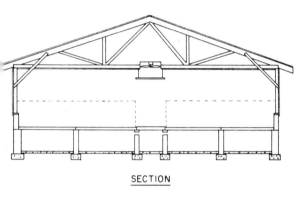

SECTION

FIGURE 10-19. End view and floor plan of the finishing house in use at the North Carolina Swine Development Center, Rocky Mount, North Carolina. (*Courtesy North Carolina State University.*)

ration to the unloading wagon or truck, or directly by automation to the self-feeders. These systems are efficient but expensive. Volume is necessary for economical use of this type of equipment. Shown in Figure 10-20 is a facility with a modern automatic feeding system.

Nearly 3,000 U.S. farms produce and market 2,500 or more pigs each year. About 1,000 of them raise 5,000 or more pigs annually. A large percentage of these producers are located in the Corn Belt states. Many of them have developed feed processing, storage, and delivery systems that permit one man to handle the processing and feeding of large volumes of feed mechanically. The processing center may involve auger or pneumatic transfer of feed. Feed can be blown through pipes above or underground. It is possible to deliver 2 tons per hour as far as 800 feet through 1¼-inch heavy-duty conduit underground.

It is advisable for swine producers to figure rather carefully alternate systems of grain and feed handling in planning their enterprises. Investments in mechanical feed handling equipment must reduce cost of production sufficiently to repay the cost of

T A B L E   10-4   TIMETABLE SHOWING USE OF GROWING-FINISHING BUILDING

Farrowing Group	Farrowing Day	Growing-Finishing Building		
		50 Pounds (Day Number)	117 Pounds (Day Number)	210 Pounds (Day Number)
1	1	60	100	147
2	61	120	160	207
3	121	180	220	267
4	181	240	280	327

Source: Adapted from *Housing and Equipment for Growing-Finishing Hogs*, Circular 799, University of Illinois, 1958.

FIGURE 10-20. Finishing house on Dean Luhman farm, Radcliffe, Iowa. The 40 x 160 foot steel building has slatted floors, convection tube ventilation, concrete pen dividers, and automatic feeding system. (*Courtesy Butler Mfg. Co.*)

FIGURE 10-21. Layout for an automatic electric mill system with pneumatic conveyer and provisions for expansion. (*Courtesy Oklahoma State University.*)

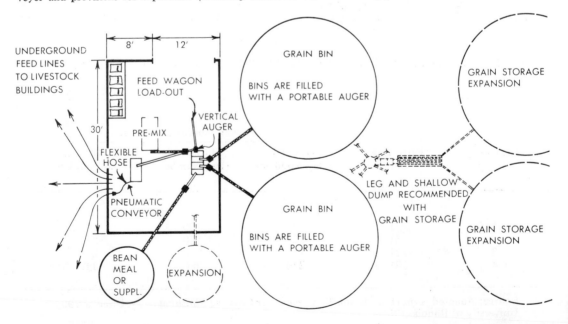

original equipment and annual cost for its use. The use of such equipment should reduce labor costs and permit sufficient increase in volume to net a larger profit to the owner.

A layout for an automatic electric mill system with a pneumatic conveyer and provisions for expansion is shown in Figure 10-21.

Slatted (or Slotted) Floors

The use of slatted floors has made it possible for large numbers of farmers to move to confinement systems of swine production. The self-cleaning feature of slatted floors eliminates much of the hand labor previously required in cleaning. They in addition provide drier, cleaner floors, and eliminate the use of bedding.

Some operators prefer to use slats on a part of the floors. Others use slats on the entire floor. A large percentage of the finishing buildings are completely slatted, whereas, most of the farrowing houses are partially slatted. When slats are used in farrowing houses, the spacings between slats must be covered for a few weeks after farrowing, and the spacings must be narrower than in finishing houses.

Slats may be of wood, concrete, steel, aluminum, or plastic. Wood slats have the lowest initial cost, but last only 2 to 5 years depending upon quality of wood. Pigs are inclined to chew them even if treated. Costs vary from $1.00 to $2.00 per square foot.

Concrete slats are durable and well suited to heavier hogs. The slats may be purchased in precast form, or forms may be used in casting them in place. Concrete slats should have a smooth, steel-troweled finish so that small pigs will not get knee and leg abrasions. It costs about $1.00 to $2.00 per square foot to cast the slats in place. Precast slats cost $2.00 to $3.00 per square foot.

Untreated steel slats should not be used. Corrosion will limit the life of the slat to

FIGURE 10-22 (*Above*). Interior of a growing-finishing confinement house with slatted floors. (*Courtesy Moorman Manufacturing Co.*)   FIGURE 10-23 (*Below*). Porcelainized steel slats are easy to clean and they do not corrode. (*Agricultural Services photograph. Courtesy Starcraft Swine Equipment.*)

about 2 to 3 years. Most steel slats sold are treated to prevent corrosion. Porcelainized steel slats 1 to 1½ inches wide and ³⁄₁₆ to ¼ inch thick have worked quite well in farrowing quarters.

Aluminum slats are light weight and easy to handle. They have a smooth surface and wear well. There is little corrosion and little damage to the legs of pigs due to abrasions. They are noisy and quite expensive. Low profile slats are about 1½ inches high; high profile slats are 2½ to 3 inches high. The low profile slats need to be supported every 4 feet, whereas high profile slats need supports only 8 feet apart. The slats cost from $2.00 to $4.00 per square foot.

Plastic slats are available and early indications of their usefulness are favorable. They have smooth surfaces and there is little danger of leg abrasions. They require more support than other types of slats and are costly. They have proven especially desirable in farrowing and nursery quarters.

The width of the slat and the space between slats varies with the size of animal. Larger animals are more efficient in working the manure through the spaces between slats than small pigs. Slats 3 to 4 inches wide spaced ⅜ to ½ inch apart are recommended for farrowing quarters, and slats 5 to 7 inches wide and 1 to 1¼ inches apart are desired in finishing houses. Recommendations concerning slat width and spacing are presented in Table 10-5.

Dunging Alley

While the hog is sometimes considered a "dirty" animal, he actually is very careful in eliminating body wastes. If given a chance, a hog will eliminate wastes away from the sleeping and feeding areas. Europeans have found that when an alley is provided the pigs will use it for dunging.

Shown in Figure 10-24 is a plan for a hog house containing dunging alleys. The alleys are 3 to 4 feet wide. A gate opens into the alley at the back of each pen. The gates are made so that when open they close the alley. Each pen has its separate alley.

To clean the alleys, the gates to the pens are closed, thus confining the pigs. A small tractor, equipped with a scoop the width of the alley, can be used, and the alleys are cleaned in minutes.

Manure Storage

Pits under the slatted floors may be used to store the manure from three to six months depending on the animals housed. The pits should be 2½ to 3 feet deep in farrowing houses and 4 or 5 feet deep in finishing houses. They should be constructed of high-quality concrete. Provision must be made to empty and clean the pits periodically. The pits may drain directly into a lagoon, or into a larger storage pit located some distance away at a lower level, or the liquid may be pumped out and spread directly on to a field. The pit must be able to hold the manure produced over a three- or four-month period in the northern states since it cannot be spread during much of the winter due to the freezing temperatures.

Mechanical Cleaners

Manure cannot be handled as a liquid in houses where bedding is used. Power driven scoops or mechanical cleaners are usually used in large buildings. The cleaners could be installed in the house shown in Figure 10-10. The gutter for the cleaner would be placed just inside the pens near the alley. It is usually 14 inches wide and 12 inches deep and is covered with sections of 2-inch plank that may be removed as needed when the mechanism is used.

Barn cleaners are expensive. A system for a 20-pen house may cost about $2,000. The building would have to be used during

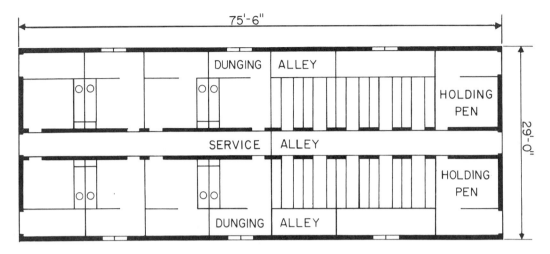

FIGURE 10-24. Floor plan of a hog house with dunging alleys designed for use when pigs are grown in confinement. (*Drawing by David Opheim.*)

T A B L E 10-5   SLAT WIDTH AND SPACING RECOMMENDATIONS FOR TOTALLY SLATTED FLOORS

Housing Unit	Slat Width (Inches)	Spacing Between Slats[1] (Inches)
Farrowing[2]	3 to 5	3/8 to 1/2
Nursery[2]	3 to 5	1
	1 to 2	1/2
Growing-finishing and gestation	7	1-1/4
	4 to 6	1 to 1-1/4

[1] A wider space adjacent to the partition or wall will reduce buildup of manure along the back or sides of the pen. One-inch slots in the rear 18 inches of the farrowing crate area will improve cleaning efficiency.

[2] No. 9-11 gauge, 3/4-inch, flattened, expanded, and galvanized metal is excellent for the front and rear of the farrowing crate floor and in the nursery.

Source: Cir. 1064, University of Illinois.

the entire year if the investment in the cleaner is to be considered economical.

Manure Disposal

Manure disposal is no problem when hogs are on pasture. It may be a serious problem in confinement feeding. Two methods of disposal are possible. Manure may be handled as a solid or as a liquid. Bedding may be used to absorb liquids. The material may be scraped by hand or by small tractor into a gutter or a mechanical barn cleaner, or directly into a storage pit. It later can be picked up by a tractor loader.

A liquid manure handling system involves the use of water under pressure to remove the manure to a storage tank or to

FIGURE 10-25. A mechanical barn cleaner is a great aid to the pork producer. (*Courtesy The Calumet Company.*)

FIGURE 10-26 (*Above*). This "Bob Cat" tractor scoop is especially designed for use in cleaning swine feeding facilities. (*Courtesy Clark Equipment Co.*)   FIGURE 10-27 (*Below*). Many producers using confinement facilities store the liquid manure until it accumulates, then use spreaders like the one below to spread it on fields. (*Courtesy Hawkeye Steel Products, Inc.*)

a lagoon. If a storage tank is used, it should hold at least three to four month's accumulation. From two to four gallons of liquid manure will be produced per hog per day, depending upon the amount of water used in cleaning the pens.

The liquid must be pumped from the storage tank into a tank wagon and spread on the field. Usually a three-inch diaphragm pump that delivers 175 to 450 gallons per minute, depending upon lift, will be satisfactory.

Lagoons

The use of oxidation ditches and aerated lagoons in disposal of swine wastes is common. These plans involve the supply of oxygen to aerobic bacteria which break down the waste materials. Carbon dioxide is produced which is odorless. About 50 percent of the organic solids will decompose in three to six months.

In oxidation ditches, the flow should be at 1.5 to 2 feet per second to prevent settling. The ditches should be 2 to 3 feet deep and 4 to 10 feet wide. Aeration rotors are used to supply oxygen and circulate wastes in both lagoons and in oxidation ditches.

The lagoon may be located adjacent to or 200 to 300 feet from the feeding floor or finishing house. Six- to eight-inch tiles are used to carry the liquid to the lagoon. The liquid in the lagoon should be at least three feet deep. The lagoon should at least be of a size equal to the floor area.

While odor is usually not a problem, the lagoon should be located at least 500 feet from the dwelling and in a direction other than that of prevailing winds.

An anaerobic lagoon (one using no oxygen) should be 8 to 10 feet deep and not more than 60 feet wide. Hog wastes will accumulate about a foot a year in this type of lagoon and provision must be made to remove the sludge after several years of use.

FIGURE 10-28. The lagoon used by a feeder-pig market in North Carolina. (*Courtesy Farmers Cooperative Service, U.S.D.A., New Bern, North Carolina.*)

The bank of the lagoon should be at least 2 to 3 feet higher than the maximum water level.

The size of the lagoon needed is based upon the weight of the animals to be housed. With growing-finishing pigs, there needs to be 1½ to 2 cubic feet of lagoon for each pound of animal. A lagoon of this type for 500 pigs weighing 200 pounds, assuming that the water is 7 feet deep, would require 200,000 cubic feet. A lagoon 60 feet wide and 500 feet long would meet the needs.

Aerobic lagoons and oxidations ditches (those using oxygen to enhance oxidation) are usually 3 to 5 feet deep. There should be a large surface area in aerobic lagoons because 2½ square feet of surface area is needed for each pound of animal. A lagoon of this type, with no special oxidation equipment, would have to have 250,000 square feet of surface area for 500 finishing pigs weighing a total of 100,000 pounds.

If the lagoon is to be used in processing the manure from other livestock enterprises on the farm, additional capacity should be provided.

Air and Water Pollution Regulations

Swine producers must meet federal and state laws in operating livestock waste removal programs, and in establishing new facilities. Any producer who has had, or is planning to have, 2,500 pigs weighing 55 pounds or more, is required by law to apply for a National Pollutant Discharge Elimination System (NPDES) permit. Each state has appropriate channels that individual producers can use in making application for permits.

In order to obtain an NPDES permit, the farm operator must have or develop, a water runoff control system and a manure-handling program that eliminates any discharge of manure, or polluted runoff, in any open ditch, stream, pond, or other body of water, except what might result from a very exceptional cloudburst.

While the law at present does not require producers with fewer than 2,500 hogs, or fewer than 1,000 animal units of livestock, to obtain a permit, these producers are subject to legal action if polluted water from their premises does enter tiles, streams, rivers, and ponds, or is carried with surface water to off-farm areas.

Swine producers should check carefully to make certain that both air and water regulations are being adhered to. Any operator planning new facilities should check to make certain that they will meet the established standards, or standards likely to be established by the NPDES or state control agency.

Information concerning these regulations may be obtained from the local or district extension director, the state geological survey agency, the Soil Conservation Service, the county ASCS office, and the Water Pollution Control Division of the State Department of Health. Additional information and permit application forms are available from the Permit Branch, Region V, U.S. Environmental Protection Agency, in Chicago, or by writing the EPA branch office in your region.

Insulation

Insulation is needed in central-type hog houses (1) to lower the temperature during the summer, (2) to increase temperature during the winter, and (3) to control condensation of moisture. In well-insulated buildings animal body heat may provide much of the heat needed for ventilation and for physical comfort. Space heaters are needed in the winter. Insulation will decrease the need for heat and reduce fuel bills.

Multiple farrowing results in the use of housing facilities continuously during the year. Farrowing takes place both during the hot summer and cold winter months. Moisture condensation may be a major problem during winter months. It is important that walls and ceilings be insulated.

Relative insulation values of some of the common construction materials are presented in Table 10-6. The recommended insulation value needed in Illinois and Iowa is 10 in the sidewalls and 14 in the ceilings. The building material will provide a part of the insulation value. The remainder must be supplied through the use of special materials.

Presented in Figure 10-29 are the insulation values of different walls. Note the need for insulation when walls are of concrete. Buildings with plywood walls and 2 inches of batt insulation will meet the wall insulation requirements. Three inches of insulation will be needed in the ceiling. It is important that a moisture barrier be provided in using insulation materials.

TABLE 10-6 INSULATION VALUES (R) FOR COMMON CONSTRUCTION MATERIALS

Common Construction Materials	Resistance (R)	
	Per Inch	For Thickness Listed
INSULATION MATERIALS		
Batts or blankets		
fiberglass, mineral rock, rock wool	3.70	—
wood fiber	4.00	—
Boards and slabs		
1/2-inch insulation board	2.63	1.32
expanded polystyrene	3.70-5.00[1]	—
urethane foam	6.67[1]	—
Loose fill		
fiberglass, mineral wool, rock wool	3.70	—
vermiculite (expanded)	2.08	—
sawdust and shavings	2.22	—
BUILDING MATERIALS		
25/32-inch wood siding	—	0.98
3/8-inch plywood	1.25	0.47
1/2-inch plywood	1.25	0.63
3/4-inch wood siding	1.25	0.94
1/4-inch hardboard	0.72	0.18
1/8-inch cement-asbestos board	0.25	0.03
4-inch poured concrete	0.08	0.32
WINDOWS		
single, glazed	—	0.89
double, glazed (storm windows)	—	2.22
double-pane insulating glass	1.50	1.75
AIR SPACE		
3/4-inch or more space	—	0.91
with reflective lining on one side	—	2.17
SURFACE CONDITIONS		
inside surface	—	0.61
outside surface (15 m.p.h. wind)	—	0.17

[1] Varies slightly with different products.

Source: *Farrowing Houses For Swine*, Cir. 973, University of Illinois.

Ventilation and Cooling Equipment

Ventilation is not usually a problem in growing-finishing buildings that are open on one side. Ventilation is a serious problem in farrowing houses and in buildings which house sows and pigs through the suckling period. Professional assistance should be obtained in planning the ventilation system for

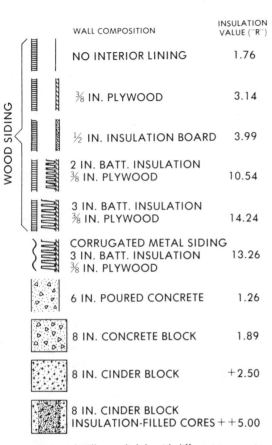

WALL COMPOSITION	INSULATION VALUE ("R")
NO INTERIOR LINING	1.76
⅜ IN. PLYWOOD	3.14
½ IN. INSULATION BOARD	3.99
2 IN. BATT. INSULATION ⅜ IN. PLYWOOD	10.54
3 IN. BATT. INSULATION ⅜ IN. PLYWOOD	14.24
CORRUGATED METAL SIDING 3 IN. BATT. INSULATION ⅜ IN. PLYWOOD	13.26
6 IN. POURED CONCRETE	1.26
8 IN. CONCRETE BLOCK	1.89
8 IN. CINDER BLOCK	+2.50
8 IN. CINDER BLOCK INSULATION-FILLED CORES	++5.00

+Will vary slightly with different aggregate
++Varies with different insulation

FIGURE 10-29. Insulation values of different walls. (*Courtesy University of Illinois.*)

buildings that are used for farrowing and as nurseries.

The ventilating system should be capable of removing from four to six pounds of moisture daily for each hog from 100 to 300 pounds, or 10 pounds or more daily for each sow and litter. Recommended ventilation rates are presented in Table 10-7.

A complete ventilating system has four parts: (1) fans to move the air, (2) inlets to provide proper distribution of incoming air with no drafts, (3) outlets for stale and moisture-laden air, and (4) controls to make the system nearly automatic.

In winter incoming air should enter the building slowly to avoid drafts. The inlets should be near the ceiling and located so that the entire building will be uniformly ventilated. Baffles may be used to direct the air along the ceiling.

Exhaust air should be moved rapidly as it contains large amounts of moisture. The outlets should be such that the water cannot get into walls or loft. The freezing and thawing of moisture around windows and under the roof can present moisture problems.

The humidity should be kept under 80 percent. It is better, however, to permit humidity to rise above 80 percent than to lower the building temperature to 40 degrees or below.

During the summer, the main problem is to keep the pigs cool. As shown by data in Table 10-8, 75 cfm per pig is required for a 125-pound pig when the temperature is 90 degrees, whereas only 25 cfm are required when the temperature is 50 degrees.

The fan is the outlet in the exhaust system. The high velocity exhaust fan carries moisture outside the building and in doing so creates a vacuum which brings air into the building. To obtain good distribution of air, proper inlets must be provided. Presented in Figure 10-30 are examples of slot inlets when the fans are located on one side and on both sides of the building. Presented in Table 10-9 are fan specifications.

The use of windows and doors as air inlets are usually unsatisfactory as drafts result. Air ducts properly spaced down the length of the building will provide more uniform ventilation with less draft in any one area. The inlets must be constructed to permit opening and closing as needed. Many ventilating systems have inlets that are thermostatically controlled to open and close mechanically. Shown in Figure 10-31 is a

T A B L E 10-7   MINIMUM VENTILATION RATES FOR SWINE

	Ventilation (rate, CFM)[1]	
	Winter	Summer
Farrowing unit (sow and litter)	80	350
Nursery unit (30- to 50-pound pig)	20	100
Growing-finishing unit (50- to 225-pound pig)	35	150
Breeding and gestation unit (gilt, sow, or boar)	50	350

[1]Cubic feet per minute. These values are considered minimum rates under normal conditions.
Source: University of Illinois, Cir. 1064.

T A B L E 10-8   AMOUNT OF AIR NEEDED ACCORDING TO OUTSIDE TEMPERATURE AND SIZE OF HOG

Outside Temperature (°F)	CFM Per Hog	
	125 Lbs.	200 Lbs.
100	75	100
90	75	100
80	75	100
70	75	100
60	30	40
50	25	35
40	20	28
30	12	17
20	9	12
10	8	11
0	7½	10½
−10	7	10

Source: *Ventilate Your Swine Finishing House*, Pm-443, Iowa State University, 1968.

diagram of a motorized inlet developed by Cornell University.

The ventilation system illustrated in Figure 10-32 is quite popular in Oklahoma. The combination system uses a fan mounted at ceiling height away from the wall which pressurizes a plastic distribution duct. The duct runs the length of building and has evenly spaced holes for air distribution. The fan runs continuously. Motorized shutters are mounted on the wall near the fan. When the shutters are closed, the air inside the building is recirculated. When they are open, an exhaust fan draws air from the building. This system can also be used in forced-air evaporative cooling.

Ventilation of farrowing houses with slatted floors is more difficult than ventilation of finishing houses. There is more heat and moisture in the building and there can be less movement of air. Many swine producers

T A B L E 10-9   PROPELLER FAN SPECIFICATIONS
(1/8-inch static pressure)

Fan Diameter (Inches)	Revolutions per Minute	Fan Horse Power	Cubic Feet of Air per Minute
13	1,550	1/35	500
13	1,725	1/8	1,500
19	1,740	1/6	2,500
26	1,140	1/3	4,340

Source: *Ventilation for Swine*, Cir. 862, University of Illinois.

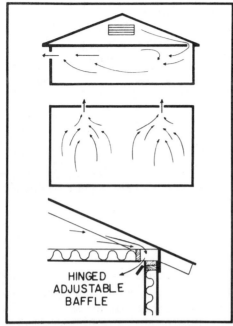

Slot inlets with fans on one side
of the building.

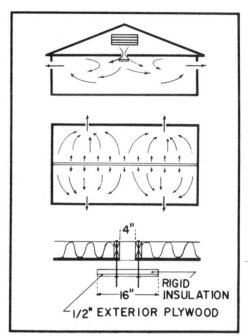

Slot inlet in center of building with fans on
both sides.

FIGURE 10-30 (*Above*). **Slot inlet and fan arrangements in swine housing.** (*Courtesy Iowa State University.*)   **FIGURE 10-31** (*Below*). **Motorized inlet for use in ventilating houses for swine.** (*Courtesy Cornell University.*)

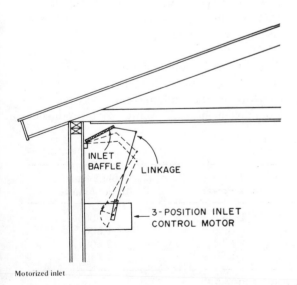

Motorized inlet

with slatted floors are using under-slat ventilating systems. They indicate that under-floor ventilation decreases the odor problem from the pit by exhausting the gases before they enter the house atmosphere. Drier floors are therefore maintained, drafts are minimized, and the temperature and moisture conditions favorable to good pig performance are maintained.

Under-slat ventilation is also recommended for finishing houses. In finishing houses one method is to duct a fan or fans to the pit just under the slats. In farrowing houses the two manure pits may be connected in the middle and fans ducted at the ends to remove the air and gases. Shown in Figure 10-33 is a diagram of an under-slat ventilation system for a totally enclosed farrowing house in North Carolina.

Cooling Equipment

Air conditioning of farrowing houses is costly. Indiana tests indicate the most eco-

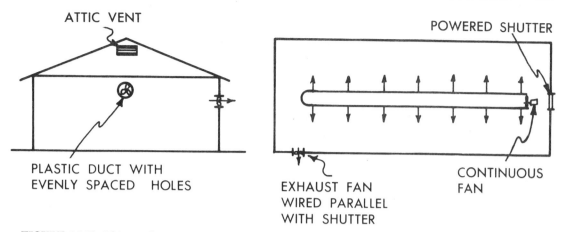

FIGURE 10-32. This combination ventilating system is recommended by Oklahoma State University. Air velocities and temperatures are nearly uniform and constant throughout the building. (*Courtesy Oklahoma State University.*)

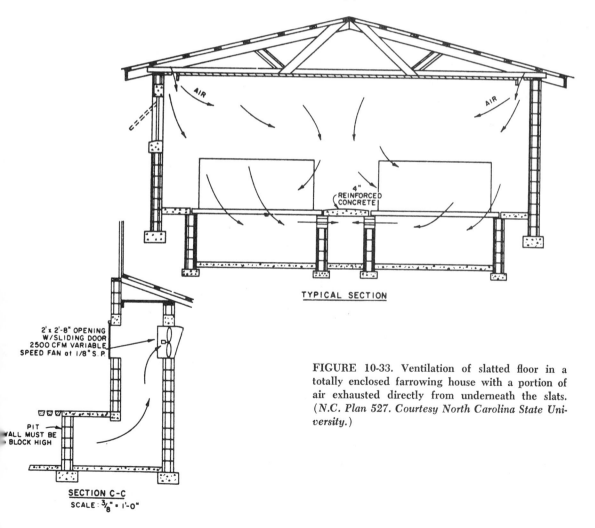

FIGURE 10-33. Ventilation of slatted floor in a totally enclosed farrowing house with a portion of air exhausted directly from underneath the slats. (*N.C. Plan 527. Courtesy North Carolina State University.*)

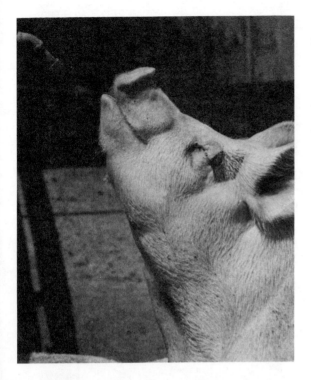

FIGURE 10-34 (*Above*). Sprinkler for cooling hogs in finishing house. (*Courtesy Moorman Manufacturing Co.*)   FIGURE 10-35 (*Below*). The multi-use house provides excellent shelter during summer as well as during winter months. (*Courtesy University of Illinois.*)

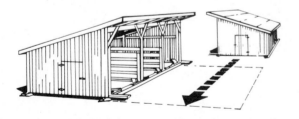

nomical method of reducing stress in sows in hot weather is to supply mechanically cooled, dry air to the sow for breathing purposes at the rate of eight cubic feet per minute. Less than one-tenth ton of refrigeration per sow is theoretically required to condition this amount of air when outside temperatures are at 100 degrees and 45 percent relative humidity.

Spray cooling systems are recommended for hogs in confinement quarters having concrete floors. Pigs weighing 100 to 200 pounds make most rapid gains at an air temperature of 70 to 75 degrees.

Hogs differ from other animals in that they sweat very little. They dissipate body heat through evaporation from the lungs. Water sprinkled on pigs in hot weather evaporates and the animal is cooled. Fogging is not effective. The use of sprinkling systems in the Corn Belt is common in confinement houses.

Evaporative cooling systems are often used in the South where humidity is low and evaporation is rapid. The coolers operate on the principle that as a stream of air passes over the cooling device it becomes saturated with moisture. Heat is given off as the water evaporates and the air is cooled.

In spray-cooling pigs one nozzle is used for 20 to 30 pigs depending upon the size of the pigs. The size of nozzle will vary with the number of pigs in the pen. About 0.09 gallons of water should be sprinkled over each pig in an hour. The nozzles should be placed about eight feet apart and four to six feet from the floor. They should be set to shutoff at about 78 degrees F.

RANGE SHELTER

Most farmers move the portable houses to the pastures to provide summer shelter. They are usually supplemented with shades if trees are not available.

The multi-use or "sunshine" types of houses provide protection for more pigs than do other types of houses, and, since they are usually taller buildings, they are cooler in hot weather. Shown in Figure 10-35 is an example of a desirable range shelter.

FEEDING FLOORS

A large percentage of the spring pig crop produced in the nation is fed out while on pasture. The fall and winter pigs, however, are largely fed out in dry lot. Where dry-lot feeding is practiced a concrete feeding floor will pay dividends.

Some farmers lay concrete strips to use in getting the little pigs started before moving them to pasture. The movable houses are pulled up to the edge of the paving.

Size of Floor

Most hog producers like feeding floors or strips of concrete at least 10 feet wide. A 150- to 200-pound hog requires 10 to 15 square feet of area. A feeding floor 10 by 60 feet will provide space for 40 to 60 hogs weighing 150 to 200 pounds, or for about 80 younger pigs.

A strip of concrete the width of the portable house and 10 to 12 feet deep is usually considered adequate for a sow and litter.

Materials

A mix of one part Portland cement, two and one-quarter parts sand, and three parts gravel or crushed rock is recommended in making a feeding floor. Five gallons of water should be used per sack of cement if the sand is in average moist condition. A mushy rather than a soupy mix is desired.

Laying Concrete

The area should be graded and leveled before placing the concrete. A slope of one-quarter inch per foot is recommended. Forms may be made of two-by-fours or two-by-

sixes. If heavy equipment is to be driven over the slab the concrete should be six inches thick. It is usually better to lay the concrete in sections ten feet square. One section should be laid at a time. A level but rough surface is desired.

Concrete should be cured properly by covering it with straw or earth as soon as it is hardened. It should be kept moist for several days.

Gutter

It is a good practice to make a gutter two feet wide and one foot deep at the edge of the floor. Gates which can be raised will aid in cleaning the feeding floor.

Grates for Confinement Feeding

Iron grids are being used to make grates over gutters in confinement housing. Manure is pushed into the gutters and washed down. If round grids are used, the hogs will not cross the grates if they are 30 inches wide.

Steel Mesh Floors

The use of steel mesh or slatted floors in growing-finishing buildings facilitates cleaning and keeping the quarters dry. According to Illinois tests, pigs stay cleaner on steel mesh floors and make faster gains than pigs on concrete. Mesh floors are especially desirable in hot weather.

PRODUCTION EQUIPMENT

Plans for farrowing stalls were presented in Chapter 7. On the following pages plans or pictures of other swine equipment are shown.

SUMMARY

Open sheds and shades provide adequate shelter for sows during the summer. For winter housing open sheds are satisfac-

FIGURE 10-36 (*Above*). Large 60-bushel capacity self-feeders in use in a pasture finishing facility. Each feeder has 12 doors. (*Agricultural Associates photograph. Courtesy Starcraft Swine Equipment.*)   FIGURE 10-37 (*Below left*). This automated finishing facility features the feeding of high moisture corn. (*Courtesy A. O. Smith Harvestore Co.*)   FIGURE 10-38 (*Below right*). Feeders must be kept in adjustment to avoid feed losses. (*Courtesy Big Dutchman.*)

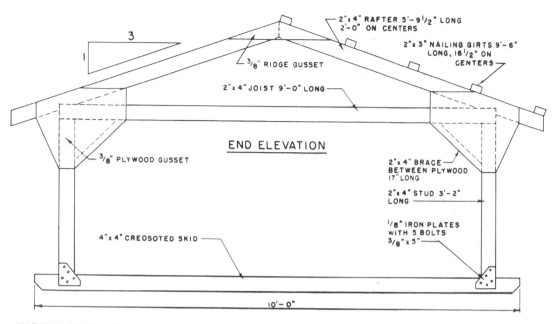

2"x 4" RAFTER 5'-9 1/2" LONG
2'-0" ON CENTERS

2"x 3" NAILING GIRTS 9'-6"
LONG, 16 1/2" ON
CENTERS

3/8" RIDGE GUSSET

2"x 4" JOIST 9'-0" LONG

END ELEVATION

3/8" PLYWOOD GUSSET

2"x 4" BRACE
BETWEEN PLYWOOD
17" LONG

2"x 4" STUD 3'-2"
LONG

1/8" IRON PLATES
WITH 5 BOLTS
3/8" x 5"

4"x 4" CREOSOTED SKID

10'-0"

FIGURE 10-39 (*Above*). A portable hog shade. (*Midwest Plan Service Plan No. 72690.*)   FIGURE 10-40 (*Below left*). A portable stairstep loading chute. (*Midwest Plan Service Plan No. 87341.*)   FIGURE 10-41 (*Below right*). The nipple hog waterer in use. (*Courtesy Systems Engineering.*)

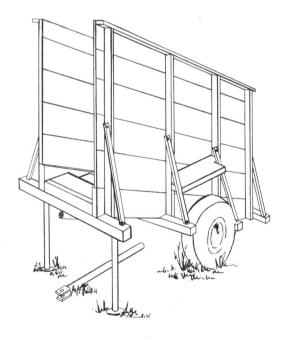

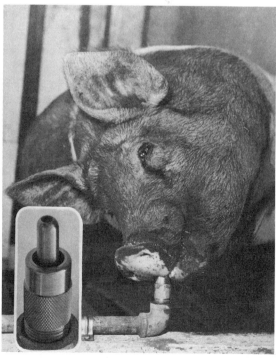

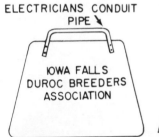

ELECTRICIANS CONDUIT
PIPE

IOWA FALLS
DUROC BREEDERS
ASSOCIATION

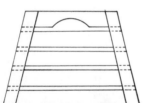

FIGURE 10-42 (*Above*). An automatic hog waterer that can be used winter or summer. (*Cooper photograph. Courtesy Ritchie Manufacturing Co.*) FIGURE 10-43 (*Left*). Hand hurdles. (*Drawings by David C. Opheim.*)   FIGURE 10-44 (*Below left*). Automatic waterer on a concrete lot. (*Courtesy Moorman Manufacturing Co.*)   FIGURE 10-45 (*Below right*). Manure in this facility is scraped into a gutter equipped with a manure loader. (*Courtesy Patz Material Handling Equipment Co.*)

tory in central and southern states. In far northern states buildings closed on all sides are necessary.

From one to several pigs per litter may be saved by providing improved housing and farrowing equipment. The use of farrowing stalls, or free-stall pens, equipped with heat lamps or heated floors is a must.

Farrowing stalls may be installed in portable or central-type houses. Pork producers who farrow only one crop of pigs per year, or who keep less than eight to ten sows, usually cannot afford to maintain a central-type house. A central-type house is a good investment when large numbers of pigs are raised and where multiple farrowing is being practiced.

There is less investment in buildings, and there may be better control of disease when portable houses are used. More labor, however, is involved in caring for the hogs and in maintaining adequate housing than when a central-type house is used.

Central-type houses will save labor. One man can care for more sows and litters, and it is easier to clean a central-type house than a large number of portable houses.

The multi-use or "sunshine" house is a highly recommended portable house. The houses are used in pairs at farrowing time. They are built with open sides. Two buildings pulled together will provide farrowing stalls for six to fifteen sows. The same houses can be used as range shelters and as fall and winter shelters for the breeding herd.

Central-type houses should have insulated walls and ceilings. The floors may be of concrete, partially slatted, or fully slatted. Fully or partially slatted floors save labor, but may make the house colder and more drafty unless proper ventilation is provided.

Farrowing houses should have the stalls or free–stall pens facing the center of the house with access aisles back of the pens. A 3- to 4-foot alley in the middle is satisfactory. The area in the back of the stalls and free-stall pens should be slatted. If the entire pen is slatted, the area for brooding and nursing should be covered for a few weeks after farrowing. Under-slat ventilation is recommended if slats are used.

Open-front finishing houses are more economical than completely closed finishing houses. Provision should be made to close the side or sides in cold weather. Slatted floors may be used in southern states.

Slatted floors are recommended in completely closed houses in the Corn Belt. Concrete, aluminum, steel, wood, or plastic may be used. Slats 6 to 7 inches wide with 1 to 1¼ inch spacing between slats are recommended.

Slats in the farrowing house should be 3 to 5 inches wide and spaced ⅜ to ½ inch apart. Surfaces should be smooth, but not slick when wet.

Pigs weighing 25 to 40 pounds should have 3 square feet of floor space per pig, pigs weighing 41 to 100 pounds should have 4 square feet, pigs 101 to 150 pounds should have 5 square feet and pigs above 150 pounds should have 8 to 9 square feet when housed on partially or totally slatted floors. The floor area in solid-floor houses should be increased 33 percent.

Manure pits 2½ to 3 feet deep should be used in farrowing houses with slat floors. Pits in finishing houses should be 4 to 5 feet deep. They may drain into a lagoon, oxidation ditch, or a manure storage pit.

State and federal air and water pollution regulations must be adhered to in building and operating a waste removal system for a swine operation.

Insulation is needed in farrowing and confinement finishing houses. An insulation value of 10 to 12 is needed on side walls and 14 to 16 in the ceiling.

A 125-pound pig needs 75 cubic feet of air per minute (cfm) during 80 degree temperatures and only 25 cfm when the temperature is 50 degrees.

Dunging alleys, 3 to 4 feet in width and located at the back of the pens, will save considerable labor when hogs are confined on solid floors.

Ventilation is very important in a confinement house. Door and window ventilation is not satisfactory. Fans and ducts evenly spaced near the ceiling on the sides should be used. Under-slat ventilation should be provided. Air should be ducted out of the building using fans drawing air, water, and gases from the pit. The fan speed should be adjusted to both outside and inside temperatures.

Automatic feeding and watering equipment should be used in confinement feeding. Manure disposal can best be done by use of liquid manure tanks or lagoons.

Finishing hogs make most rapid gains at temperatures of 70° to 75°F.

Mechanical barn cleaners are expensive but will save labor.

Concrete feeding floors should be at least 10 feet wide and have a 1- by 2-foot gutter along the side to aid in cleaning. Each 150-pound hog should have from 10 to 15 square feet of floor space.

Large self-feeders and large automatic waterers are essential if labor is to be used efficiently in swine feeding.

Most farmers prefer having a concrete apron outside the house. Some prefer to confine the sows and litters in the house.

Building costs per sow and litter can be greatly reduced by using a multiple farrowing program. A farrow-to-finish confinement housing system provides for most efficient use of labor, mechanical feed handling, waste disposal, and some decrease in construction costs.

QUESTIONS

1 Outline a program which will provide desirable housing for the breeding herd on your home farm.

2 What are the advantages and disadvantages of portable hog houses?

3 What are the advantages and disadvantages of central-type hog houses?

4 Make a rough drawing of a central-type house for a producer who expects to keep 40 sows for both spring and fall litters.

5 Make a diagram of a movable hog house which would be adaptable for use on your home farm.

6 What are the advantages of the multi-use houses?

7 What systems of ventilation are preferred in a completely slatted finshing house?

8 Under what conditions should dunging alleys be provided in a hog house?

9 What are the advantages of using slatted floors in confinement houses?

10 Should you use wood, steel, aluminum, concrete, or plastic slats? Why?

11 What width and spacing between slats are recommended for farrowing houses, nurseries, and finishing houses?

12 Describe a desirable manure pit under the slatted floor in a finishing house.

13 Explain the difference between an anaerobic and an aerobic lagoon?

14 Explain the state and federal regulations affecting systems of handling wastes in confinement swine housing?

15 Plan a concrete feeding floor which will meet the needs on your home farm.

16 Describe the types of self-feeders best suited to your home farm situation.

17 What is the average investment per sow and litter on your home farm in swine housing and equipment?

18 Outline a program for the improvement of the swine housing and equipment on your home farm.

REFERENCES

Aanderud, W. G., *Practical Hog Production and Marketing for S.D. Farmers*, EC 649. Brookings, S.D.: South Dakota State University, 1973.

Jensen, A. H., B. G. Harmon, G. R. Carlisle, and A. J. Muehling, *Management and Housing for Confinement Swine Production*, Cir. 1064. Urbana, Ill.: University of Illinois, 1971.

Jones, J. R., C. R. Weathers, L. B. Diggers, and C. M. Stanislaw, *N.C. Swine Development Center*. Raleigh, N.C.: North Carolina State University, 1973.

Krider, J. L., and W. E. Carroll, *Swine Production*, 4th ed. New York, N.Y.: McGraw-Hill Book Company, 1971.

Meyer, V. M., *Ventilating Air Intakes for Swine Housing*. Ames, Ia.: Iowa State University.

———, *Swine-Farrowing Facilities*, Pm-520. Ames, Ia.: Iowa State University, 1973.

———, *Ventilating Fans for Swine Housing*, Pm-505. Ames, Ia.: Iowa State University, 1971.

———, and Larry Van Fossen, *Electric Floor Heat for Swine*, Pm-502. Ames, Ia.: Iowa State University, 1971.

———, *Ventilate Your Farrowing House*, Pm-394. Ames, Ia.: Iowa State University, 1973.

Midwest Plan Service, *Swine Handbook, Housing and Equipment*. Ames, Ia.: Iowa State University, 1972.

Muehling, A. J., *Confinement Housing Slotted Floors*, AE-875. Urbana, Ill.: University of Illinois, 1969.

———, and G. R. Carlisle, *Farrowing Houses for Swine*, Cir, 973. Urbana, Ill.: University of Illinois, 1967.

Van Fossen, Larry, and V. M. Meyer, *Open-Front Swine Facilities*, Pm-500. Ames, Ia.: Iowa State University, 1971.

Young, H. G., *Ventilation for Swine Buildings*, FS 360. Brookings, S.D.: South Dakota State University, 1967.

11 DISEASES AND PARASITES— PREVENTION AND CONTROL

Swine diseases and parasites are responsible for losses amounting to millions of dollars each year. It is estimated that only about one-third of the pigs farrowed are grown out as healthy pigs. On many farms the profit or loss from the enterprise is determined largely by the extent to which disease and parasite losses have been controlled.

PREVENTION IS BETTER THAN CURE

Most diseases, ailments, and parasitic conditions of hogs are preventable. The treating of diseased pigs is expensive because of medicine and veterinarian costs and because feeds are wasted when fed to unthrifty pigs. A stunted or runty pig requires a long feeding period and a large amount of feed to get it ready for market. It is cheaper to prevent the unhealthy condition than to remedy it.

Farmers who raise large numbers of hogs year after year, on the same lot, usually have disease and parasite losses. The disease germs and worm eggs live through the winter in the filthy lots and unsanitary houses. The use of disease- and parasite-free breeding stock, rotated legume pastures, clean and disinfected houses, and good balanced rations fortified with vitamins and antibiotics can do much to reduce losses. Some diseases and parasites, however, must be controlled by vaccination and medication.

SWINE DISEASES

There are a large number of infectious diseases of swine, and their prevalence varies from community to community and from year to year.

Anemia

This disease was discussed in Chapter 7.

African Swine Fever

This disease, closely resembling hog cholera but not related to it, is potentially the most devastating of all swine diseases. To date no cases have been diagnosed in the U.S. but a serious outbreak occurred in Cuba in the early 1970's. Nearly 10 percent of all hogs in Cuba were slaughtered in order to eliminate the disease. It has been a serious threat to swine production in Africa, Spain, Portugal and Greece.

Cause. The disease is caused by a virus quite different from the cholera virus. It can withstand heat, dryness, and putrefaction.

Symptoms. In acute form, the disease resembles hog cholera. The pigs run high fevers, lose their appetites, the skin reddens and they appear depressed. A high percentage of infected animals die. Those that live may carry a low fever for months, are retarded in growth, and are unthrifty. They may become carriers of the disease.

Treatment and Control. There is no treatment or vaccine. A veterinarian should be called if suspicious cases are found. In Africa, Cuba, and other countries where the disease has been found, control has involved depopulation of all infected animals, and all other animals on the premises.

The disease spreads by contact and by carriers. The tick is a carrier in Cuba. The disease has spread in many countries through feeding of raw garbage. Extreme care must be exercised in the handling of garbage from planes and ships coming to this country from other countries. The effort to eliminate the feeding of all garbage, or at least of uncooked garbage, in this country is perhaps the best means of avoiding the introduction of the African swine fever in the U.S.

Arthritis

Arthritis may result in heavy losses to producers in that pigs make inefficient use of feed and require additional time to reach market weight, or are inefficient as breeding animals. There may be some dockage of animals by the packer.

Cause. At least 10 bacteria may cause arthritis and the condition may result from injury. The most common causes of arthritis are streptococcus and erysipelas bacteria, and mycoplasma hyosynoviae. Damp, cold, and rough floors with little bedding may be contributing environmental factors.

Symptoms. Streptococcus arthritis affects baby pigs. Joints become swollen and soft at first, but become firm as the condition continues. The pigs will be inclined to avoid rising and putting weight on the limb, or limbs involved. The pigs may become stunted, or permanently lame.

Lameness and swollen joints caused by erysipelas have been described elsewhere. The condition usually becomes evident when pigs are 8 to 12 weeks old. The joints may become stiff and hard. The condition may continue until market time.

PPLO, or mycoplasmic arthritis, develops when pigs weigh 70 or more pounds and are housed under stress conditions. Affected animals become lame in one or more joints and are reluctant to move. Usually other signs of PPLO are present, inclination to stretch, high temperature, and difficult breathing.

Treatment and prevention. Streptococcus arthritis can be treated in early stages by intramuscular penicillin shots. Treating the navels of newborn pigs with iodine is recommended. The use of carpets, bedding, and nonabrasive floor surfaces will aid in decreasing a chance of leg and knee injury.

Penicillin and erysipelas antiserum are the best treatments for arthritis conditions resulting from erysipelas infections. Vaccination of sows before breeding, and vaccination of pigs when 8 to 10 weeks of age is the best prevention.

Mycoplasmic arthritis may be controlled by early injection of tylosin and lincocin. The feeding of the conditioner ration presented in Table 11-1 will aid in the control of all three types of arthritis. * Pharmaceuticals re-

T A B L E 11-1   SWINE CONDITIONER RATIONS[1]
(Corrective rations for stress conditions)

Percent Protein	Ingredients		1	2	3
8.9	Ground yellow corn		842	892	1,027
12.0	Ground oats		300	600	600
15.5	Wheat middlings		300	—	—
44.0	Solv. soybean meal		200	250	300
12.0	Dried whey		200	200	—
17.0	Dehydrated alfalfa meal		50	—	—
31.0	Fish solubles		50	—	—
	Calcium carbonate (38% Ca)		15	10	15
	Dicalcium phosphate (26% Ca, 18.5% P)		15	20	25
	Iodized salt		5	5	10
	Trace mineral mix		3	3	3
	Vitamin premix		20	20	20
	Feed additives[2] (gm/ton)		100-300	100-300	100-300
	Total		2,000	2,000	2,000
CALCULATED ANALYSIS					
	Protein	%	14.67	14.27	14.77
	Calcium	%	0.66	0.61	0.68
	Phosphorus	%	0.56	0.54	0.55
	Lysine	%	0.76	0.74	0.74
	Methionine	%	0.25	0.23	0.24
	Cystine	%	0.27	0.26	0.26
	Tryptophan	%	0.18	0.17	0.18
	Metabolizable energy	Cal./lb.	1,213	1,237	1,217

Feeding Directions

[1] These rations are recommended for the first ration fed to newly received feeder pigs, for stress periods and convalescence.

[2] Be certain that only approved feed additives and levels are used for therapy. The feed additive may be part of the vitamin premix, or if a separate premix, it should replace an equal amount of corn.

Source: *Life Cycle Swine Nutrition*, Iowa State University, 1972.

quiring a particular time for withdrawal before marketing are indicated with an * throughout this chapter. See Table 11-2.

Baby-Pig Hypoglycemia

This disease is also known as *acute hypoglycemia,* which means *not enough sugar in the blood.* The disease affects the baby pigs during the first three days after they they are farrowed.

Cause. This is unknown, but the disease is not contagious.

Symptoms. The pigs appear normal at birth but become listless, shiver, and have no desire to nurse. They usually die within 48 hours after the symptoms appear.

Treatment and prevention. The disease is related in some way to the milk production of the sow and to the lack of sugar in the blood. Feeding the sow balanced rations before and following farrowing will stimulate milk production. The injection in the pigs of sugar solution or the hand-feeding of equal

FIGURE 11-1. *Above:* Pigs with twisted and distorted snouts due to atrophic rhinitis. (*Courtesy M & M Livestock Products Co.*) *Below:* A pig infected with bull nose. (*Courtesy University of Illinois.*)

parts corn sirup and water at the rate of one to two tablespoonfuls each two or three hours is recommended. Keep pigs warm. Avoid chilling.

Bordetella Rhinitis

This disease has been prevalent for more than 20 years but has caused most losses since 1950. It is now widespread among herds of both purebred and commercial hogs. It is quite often confused with *bull nose*. Rhinitis is very infectious, whereas bull nose is **not** contagious.

Cause. The usual cause of atropic rhinitis is an infection of the nasal cavity by the bacterium *Bordetella bronchiseptica*. The organism gets into wounds or scratches in the mouth or nose of pigs. It may spread directly from sow to pig at birth.

It spreads from one pig to another through contaminated feed or water or by body contact. In tests conducted at Iowa State University, 80 percent of the infected pigs had trichomonads in the snout, whereas only 2.8 percent of those not affected had them. These findings may provide a clue to the control of this infection.

Symptoms. The infection seems to start when the pigs are a few days or a few weeks old. They first show signs of sneezing, which becomes more pronounced as they grow older. At four to ten weeks of age the skin on the snout begins to wrinkle, and the snout may bulge or thicken. Sometimes the snout becomes twisted or distorted. Some pigs recover and show no outward signs of the infection. However, they may be carriers.

The hair coat is usually rough, and infected pigs are poor doers. Quite often there is a discharge from the nose which may be clear and thick or puslike and bloody.

Treatment and prevention. It has been pointed out that some animals may be carriers of atrophic rhinitis yet show no outward signs. A study was made at Beltsville, Maryland, to determine the presence of infection in the nasal passages of pigs by use of an adapted otoscope, now called a *rhinoscope.* Specialists found that the rhinoscopic inspections were 75 percent accurate.

There is no known cure for the disease but several practices are recommended as control measures:

1. Have veterinarian make swab and culture of the nasal cavity of adult breeding stock, then eliminate carriers.
2. Feed sulfamethazine ° to infected pigs and isolate them from other hogs.
3. Gilts and boars from litters in which there are pigs with rhinitis should not be saved for breeding purposes.
4. The dams of infected pigs should be sold as soon as possible.
5. Pigs weaned from the sow when they are three to five days of age and removed to clean housing are less likely to contract rhinitis.
6. Bred gilts or sows showing outward signs of rhinitis should be sold for slaughter in advance of farrowing.
7. Tests indicate that rations high in vitamins and antibiotics may aid materially in rhinitis control. Aureomycin-Sulfamethazine-Penicillin (Aureo Sp 250), Chlortetracycline (CSP250), and Tylan plus Sulfa are recommended.
8. Boars, gilts, and bred sows purchased and brought on to the farm should come from herds that have been proven rhinitis free by swab tests. Even so, they should be kept in isolation from the herd for at least 30 days.
9. It is good practice to have sows farrow in groups and in separate quarters.
10. Litters or individual pigs showing signs of infection should be isolated immediately and fed a highly fortified ration.
11. Use SPF (specific pathogen free) breeding stock. Depopulate farm of hogs for 60-day period and give facilities complete cleanup before bringing new hogs to farm.
12. Cornell tests indicate a relationship between levels of dietary phosphorus and calcium and turbinate atrophy. Rations containing 1.2 percent calcium and 1 percent phosphorus are recommended for growing-finishing pigs. A sulfamethazine-antibiotic additive may aid in maintaining a desirable calcium balance in the pigs body.

Brucellosis or Bang's Disease

This disease has been known as *contagious abortion of swine* and as *Bang's disease.* Surveys indicate that 0.5 to 1 percent of the hogs in the United States are infected, and each year many persons become infected with brucellosis from cattle and hogs, or from dairy and pork products.

Tests being made in 1974 in Iowa indicated less than 0.3 percent of Iowa hogs were infected with brucellosis. Nearly 65 percent of the cases of brucellosis in humans have been traced to swine nationally.

California buys many live hogs and much pork from Iowa and other Corn Belt states. Since January 1, 1973, all imported hogs and pork must come from brucellosis-free farms.

Losses caused by brucellosis in swine herds are due to the abortion of pigs, pigs born weak, sterility or infertility in sows and gilts, and sterility in the boar when the reproductive tract becomes infected.

Cause. The disease is caused by an infectious organism, *Brucella suis,* and is spread from animal to animal through contact, and by contaminated feed, water, and afterbirth.

Symptoms. There are no symptoms which can be easily recognized with certainty. Many animals carry the disease but appear to be normal in every respect. Symptoms which may appear are premature abortion of litters, sterility, and inflammation of the uterus or womb, testicles, and joints.

The only sure way to determine whether or not your herd is infected is to have the entire herd blood-tested. A herd of swine may be considered free from infection when, on two tests made from 60 to 90 days apart, no animal reacts, no previous evidence of infection (such as abortion and weak pigs) exists, and no new stock has been added to the herd during the previous three months.

Herds are validated as brucellosis-free when all animals over 6 months of age pass two consecutive negative tests that are not less than 60 nor more than 90 days apart. The validation is for a 12-month period. Revalidation for an additional 12 months may be obtained upon completion of an annual negative herd test on all breeding animals 6 months of age or older.

A state becomes validated brucellosis-free when at least 90 percent of the sows and boars going to slaughter have been tested, and 90 percent of the reactors have been traced back to the farm of origin and brucellosis eliminated from those farms. Arkansas, Arizona, California, Montana, Nevada, Utah, Vermont, and Wyoming were officially validated free states by 1974. Seventeen other states had active swine brucellosis eradication programs underway.

Treatment and prevention. There is no treatment. The herd should be tested and all reactors marketed. A retest should be made within 30 to 90 days and all infected animals sold. The operation should be repeated until the herd is free from the disease. All swine seed stock herds should be tested and validated. All sows, boars, and stags should be tested at slaughter and reactors traced back to herd or origin.

Breeding and feeding stock should be selected from brucellosis-free herds, and all animals brought onto the farm should be isolated for at least four weeks.

Boars are the most common sources of Bang's infection.

Hog Cholera

Losses in the United States due to hog cholera in years past amounted to as much as 50 million dollars annually. Cholera 40 years ago was our most serious disease.

Cause. The disease is caused by a small virus. It is present in the blood and body tissues of an infected animal, and is spread through the urine and feces, as well as through nose and mouth secretions.

Young hogs are more susceptible to the disease than are older hogs, and the virus is usually introduced by new animals brought to the herd, or by farm attendants carrying in the virus on their shoes.

Symptoms. The incubation period is usually from three to seven days. Infected animals first show fever and a loss of appetite. Later the eyes become filled with a sticky discharge, and the hogs prefer dark quarters. They lose weight, and the underside of the neck and abdomen may show dark red or purple coloration. Infected animals cough and have difficulty in breathing.

Usually the first cases which appear are in acute form and the animals die in from

FIGURE 11-2. Hogs in advanced stage of cholera. (*Courtesy U.S.D.A. Bureau of Animal Industry.*)

three to seven days. Chronic cases of hog cholera, however, may last longer.

Treatment and prevention. There is no treatment for the disease. Infected animals must be slaughtered. Congress authorized the U.S. Department of Agriculture in 1961 to undertake a broad federal-state cholera eradication program. A four-phase program was begun and by June, 1974, all 50 states and Puerto Rico had been declared free of hog cholera. A state is declared free after a 12-month period with no cholera reported in the state. Should a state lose its cholera-free status, it could be reinstated after a 6-month period without a diagnosed case of cholera.

Mexico began a cholera eradiation program on January 1, 1974. Canada, Austria, Denmark, New Zealand, Northern Ireland, and the Republic of Ireland are free of hog cholera. It is estimated that the cholera eradication program has saved U.S. swine

producers about $125 million in vaccination costs. Iowa, the leading hog production state, has not had a case of cholera since 1970.

Clostridium Enteritis

This disease, sometimes called enterotoxemia and necrotic enteritis, affects pigs during the first week after farrowing. The pigs may die within ten days, or the disease may become chronic and they die later, or are killed as runts.

Cause. The disease is caused by *Clostridium perfringens* Type C bacteria. The organism can pass with the feces into the soil or housing area and cause recurring outbreaks.

Symptoms. The disease resembles scours. Diarrhea begins with a watery, yellow scours that may show traces of blood. Pigs become weak and want to lie under the heat lamp.

Vomiting may occur; pigs that live over the second day have dark brown, liquid feces.

Treatment and prevention. Since the disease resembles other diseases causing pigs to scour, a veterinarian's diagnosis is recommended. The treatments for other scours are ineffective. Usually vaccination of the sow before farrowing with toxoids and innoculating pigs at birth is recommended. General sanitation practices should be followed.

E. Coli Baby-Pig Scours

A study in the Corn Belt revealed that pig scours was considered to be a more serious disease to swine producers than erysipelas, pneumonia, or rhinitis. Of the many causes of scours, E. coli (enteric colibacillosis) is considered to be the most common killer of newborn pigs. It is estimated that about 75 percent of all cases of baby-pig scours are due to E. coli infection.

Cause. E. coli bacteria are present in the intestinal tracts of most species of animals and most strains of this bacterium are harmless. The E. coli strains that are enteropathogenic may cause scours depending upon the age and susceptibility of the pigs. Healthy pigs and sows carry the organisms but are unaffected.

Symptoms. Infected pigs develop diarrhea, usually when 2 to 8 days old. They become dehydrated, weak and inactive. Because of the weakened condition, they are more susceptible to death by being lain upon or by starvation. They may chill and become infected with other bacteria or viruses.

Treatment and Control. Sows may become immunized to E. coli by moving them to the farrowing quarters, or by exposing them to fecal material and bedding from the farrowing house 3 weeks before farrowing.

High humidity encourages the growth of E. coli. It is suggested that adequate ventilation be provided and that if bedding is used, it be kept clean and dry. Sanitation

FIGURE 11-3. A typical example of a pig with infectious scours. (*Courtesy Abbott Laboratories.*)

is very important. Slat floors back of the farrowing stalls or free stall pens are helpful.

Thorough cleaning of the sows and farrowing quarters should have high priority. Veterinarians can make slides of bacteria and determine presence of type of organism causing diarrhea and recommend treatment. A vaccine which can be fed to sows is being developed in Ohio.

Baby-pig scours can usually be avoided if (1) well balanced and highly fortified rations are fed, (2) dry and sanitary quarters are provided, and (3) abrupt changes in rations of the sow and pigs are avoided. Overfeeding of the sow and the feeding of a poorly balanced ration can cause pig scours. It is suggested that the rations recommended in Chapter 7 be fed to the sows and that pre-starter and starter creep rations be fed that have been fortified with arsenicals and nf-180 (furazolidone). Both of the additives should be included in the ration for the sows that are suckling litters. Directions should be followed.

Oral dosage of recommended antibiotic or antibiotic combinations may be given infected pigs. An injection of 3cc to 5cc of a 10 percent sulfadimethoxime solution in each pig is recommended. Antibiotics should also be included in the sow's ration. Strict sanitation is a must.

Erysipelas

This disease is prevalent in the Midwest and is especially serious in some areas. Hogs may contract the disease in the acute form and die quickly, or they may have the chronic form, which causes swelling of the joints in the legs and general stiffness.

Cause. The disease is produced by a microorganism or bacteria similar to those which cause arthritis in some animals. Diseased animals, and others which carry the disease, pass off the organism in the urine and feces, and other hogs pick it up by eating contaminated feed or water.

Symptoms. Erysipelas presents itself in many forms and is difficult to diagnose. Hogs with the acute form have a high body temperature and may have a redness of skin as in cholera. They may die within 24 hours to three or four days. Those with the less acute form run high temperatures for a couple of days; when the temperature drops, red diamond-shaped patches, which disappear in time, show on the skin.

Those that have the chronic form are stiff in the joints and lame. They become sluggish and are easily fatigued. Often they rest on their haunches or on their breastbones.

Treatment and prevention. No medicines have been effective in treating erysipelas, but the injection of antierysipelas serum in the early stages of the disease is helpful. Exceptionally good results have been obtained when penicillin or tylosin are injected with the serum in both the acute and chronic forms of erysipelas.

Sick animals should be removed from the herd and isolated and treated. As a pre-ventive, the entire herd may be vaccinated with serum and penicillin.

Recent tests indicate that the disease is spread by healthy animals that are carriers rather than by soil infestation. It was formerly thought that the organism could live over in the soil for long periods of time. Tests indicate that it will survive in the soil only 52 hours at temperatures above 86 degrees, and 32 days at 32 degrees F.

One means of preventing the disease in the past was to double-treat the pigs at five to seven days of age with live-culture erysipelas vaccine and antiswine-erysipelas serum. When bringing new animals to the herd, it was necessary that they came from disease-free herds or from herds which, as pigs, were double-treated for erysipelas. Pigs which had been double-treated for erysipelas could give the disease to unvaccinated animals. The use of live-culture vaccine has been discontinued.

Erysipelas bacterin. Killed vaccine bacterins have been developed which are used in controlling erysipelas without danger of spreading the disease or of humans contracting the disease. The new vaccine has completely replaced the double-treatment.

FIGURE 11-4. An animal infected with swine erysipelas. (*Courtesy U.S.D.A. Bureau of Animal Industry.*)

It has been found that the vaccination of the gilts before breeding, or three weeks before farrowing, will give the pigs immunity against erysipelas for as long as six weeks. It is much cheaper to vaccinate the sows than the baby pigs. The sow will retain immunity for as long as eight months. The pigs should be vaccinated with the erysipelas bacterin when about eight to ten weeks old. The vaccine produces active immunity in about 21 days. It is sometimes advisable to revaccinate pigs one to two months later. Breeding stock should be revaccinated every six to eight months—sows before breeding time.

Sanitation is important in controlling erysipelas infections. Rotated pastures, sanitary quarters, and the isolation of any sick pigs will help prevent losses.

Influenza or Flu

In a survey made in one state 16 percent of the farmers interviewed said that their hogs had had influenza, or flu, during the past year. Very few animals die of the flu, but it can take the profit out of the hog business.

Cause. Flu in hogs, much like flu in humans, is caused by a virus in combination with a bacterium, *Hemophilus suis*. The disease is much more severe when both agents are present. It is passed readily from one animal to another. It is most likely to show up when the resistance of hogs has been lowered by fatigue, exposure to temperature changes, changes in feeds, or movement to new quarters. Hogs confined in a poorly ventilated building may become overheated and may chill when exposed to the air. Drafty and extremely cold sleeping quarters can cause just as much trouble as buildings which are too hot.

The germ is harbored in the nasal cavity of hogs or in the larvae of lung worms or earth worms.

Symptoms. Infected pigs usually go off feed and become listless and inactive. They appear to be distressed, and breathing is difficult. There may be a discharge from the eyes and a cough. At later stages, the cough is deep and loud. Infected hogs have fever for a few days. The sickness usually lasts less than a week if permitted to run its course, but treatment may shorten this time.

Treatment and prevention. Veterinarians can inject a porcine bacterin or supply water soluble or feed-additive medication to prevent death and weight loss. There is no effective vaccine or serum for the disease. Hogs should be provided with warm, clean, well-ventilated quarters and plenty of fresh water. Healthy pigs fed good rations are less likely to get the flu. In the fall, hogs should be kept away from straw stack bottoms and other areas where they can root for worms.

Jowl Abscess

Federal meat inspectors condemn parts of over 4 million swine carcasses each year because of abscesses in the neck and other areas of the carcass. It is estimated that because of such abscesses $12 million dollars worth of pork is lost. In addition there is some loss in feed efficiency on infected farms.

Cause. Jowl abscesses are caused by streptococcus bacteria which invade the tonsils where they grow, multiply, and spread. The infection spreads through feed, water, and pig-to-pig contact.

Symptoms. Localized abscesses appear in the throat, neck, and head. They commonly appear at one side of the jowl. Suckling pigs appear to be more resistant to the infection than weaned pigs.

Treatment and prevention. Breeding animals showing signs of abscesses should be culled. The breeding herd should be put on a ration containing 50 grams of chlortetracycline (Aureomycin) per ton of complete ration for a week before farrowing and during the suckling period. The pigs should be given a starter feed containing 100 grams of chlortetracycline per ton and continue on this antibiotic level through weaning.

At 10 to 11 weeks of age the pigs should be vaccinated with Jowl-Vac. The oral vaccine should be given with no feed or water for two hours before or after treatment. Strict sanitation should be followed.

Leptospirosis

We have heard little about this disease until recent years, but it is of great concern to all livestock farmers because it affects cattle, swine, sheep, horses, and man. In Illinois tests involving 2,153 herds of cattle and 575 droves of hogs, it was found that nearly 30 percent of dairy and beef herds and nearly 29 percent of swine herds had infected animals.

Cause. The disease is caused by several species of bacteria known as *Leptospira.* The organism may be found in the kidney or urinary tract. It spreads rapidly through the urine.

Symptoms. Diagnosis of the disease is difficult because the symptoms resemble those of cholera and erysipelas. The disease causes fever, loss of appetite, loss in weight, jaundice, anemia, abortion, and reduced milk flow. Abortions are common as well as a high percentage of dead pigs. Pigs farrowed alive may show signs of anemia and scouring. Many may die during the first two weeks.

Treatment and prevention. As with other swine diseases, a veterinarian should be called in as soon as symptoms are apparent. Blood and tissue tests should be made.

All animals showing symptoms of leptospirosis should be isolated from the herd, and care should be taken in feeding to avoid spreading the disease.

Only disease-free stock should be brought onto the farm, and such stock should be isolated from the herd for at least 30 days.

All contaminated bedding and aborted fetuses should be burned. Milk from infected cows should not be fed. The vaccination for leptospirosis is recommended. Pigs should be vaccinated at eight to ten weeks of age. Breeding stock should be vaccinated two to three weeks before they are to be bred.

Care should be taken in working with animals showing symptoms of this disease. Rubber gloves should be worn, and a disinfectant should be used to avoid contracting the infection.

Infected animals usually respond to antibiotic therapy. Several antibiotics may be used. Streptomycin, chlortetracycline and oxytetracycline* have been most effective.

All animals should be kept away from streams or surface water that is accessible to other animals. Rat and other rodent populations should be eliminated.

MMA

The MMA syndrome (metritis-mastitis-agalactia) has become a major swine disease. The disease takes a heavy toll of baby pigs. The infection is especially serious in herds that are kept in confinement.

Cause. The agent that causes MMA has not been determined. The syndrome represents a number of disease symptoms and there may be several causes. MMA is really a group of diseases. It may be transmitted by boars at breeding time or may be due to contamination of birth canal with manure at farrowing.

Symptoms. Metritis is an inflammation of the milk producing glands in the udder. Agalactia has to do with poor milk flow.

MMA usually appears a day or two after farrowing. Sows lie on their bellies and pigs cannot nurse. Constipation may be present. The udder may be swollen, hard and hot to the touch. There is little milk. A yellow or white discharge is present from the vulva of the sow.

FIGURE 11-5. There was no MMA problem with this sow and litter. (*Courtesy Abbott Laboratories.*)

The pigs appear hungry and weak from lack of milk. Many will die unless treatment of both sow and pigs is started early.

Treatment and prevention. Treatment consists of practices to overcome constipation, stop infection, and start milk flow. Injections are recommended as follows:

1. 3 to 5cc post-pituitary hormone
2. 10cc of a combination antibiotic
3. 4 to 6cc of cortisone product (5cc B_{12} complex)

The injections should be repeated in 12 hours and daily until satisfactory results are obtained. A veterinarian is necessary in control of MMA.

Other recommended treatments are furazolidone* and oxytetracycline, mixed bacterin shots, streptomycin, tylosin, and corticosteroids.*

An abundance of water and liquid feeding are recommended. Pigs should receive the colostrum milk if possible. Pigs should be fed a milk replacer, or fed clean cow's milk at body temperature. A tablespoon of corn sirup and an antibacterial agent should be added to each pint of milk.

Parakeratosis

This skin disease is seldom fatal but has caused hog farmers much concern during the past few years.

Cause. The cause is not known but it appears to be related to an imbalance in the amounts of zinc and calcium in the ration.

Symptoms. First symptoms are a skin reddening and pimple-like formations. The shoulders, backs of front legs, underside of belly, and hocks are first affected. In later stages the skin becomes thick and crusted. A dark serum is secreted. The entire body may be affected.

Treatment and prevention. University of Wisconsin studies show that the feeding of trace minerals containing zinc helps to prevent and cure the disease. One-half pound of zinc sulphate in one ton of ration is recommended. On farms where the disease is an annual problem, the zinc should be provided in the creep ration. Care should be taken that the zinc compound is finely ground and well mixed with the feed.

The disease resembles mange. Mange oils and remedies, however, tend to aggravate rather than cure the disease. The zinc treatment should be stopped a few weeks before the pigs are of market weight.

Pseudorabies

This disease is thought to be quite new in the U.S., but may have been present and confused with hog cholera. With the eradication of cholera, its existence may have become apparent. It is quite deadly with young pigs, but also affects older animals.

Cause. Pseudorabies is caused by a virus and has been associated with hogs fed garbage. Recently, however, it has been found in the Corn Belt among grain-fed hogs. The disease is quite common in Europe and Ireland.

Symptoms. There is some variation in the affect of the disease on animals, which may indicate more than one strain of the virus. Nearly 90 percent of baby pigs that become infected die within a few days. Pigs run a high fever, become depressed, show signs of nervous system disorder, a lack of coordination, and paddling with their feet. They may cough and have convulsions. Pigs four to six weeks old show many of the same signs, but death losses may not exceed 40 to 50 percent.

In older hogs, symptoms are coughing, loss of appetite, some fever, and trembling, or convulsions. Few older animals die, but about 50 percent of the bred females abort.

Treatment and prevention. There is no vaccine approved for use in the U.S., and there is no treatment. A veterinarian should be called to assist with control. Only a few cases have been reported in any state. Elimination of garbage feeding, and strict sanitation practices are essential. Infected animals should be isolated, and dead animals destroyed.

Shaky Pig Syndrome

This disease is sometimes called congenital tremors. It occurs more frequently with first litter gilts than with mature sows.

Cause. The actual cause of the disease is not known, but is thought to be caused by a virus which infects the animal early in the gestation period.

Symptoms. The disease apparently damages the spinal cord and as a result the pigs are shaky and lack coordination. The shaky condition is identifiable shortly after birth. They may continue to lack control and be susceptible to overlay. Some may recover.

Treatment and prevention. There is no treatment. The use of farrowing stalls and heat lamps will aid in keeping the pigs from being laid upon. There is no vaccine.

Control of the disease is through prevention. It is thought that infection may be carried by the boar as well as by other females. Boars should be brought onto the farm and kept in isolation at least a month before the breeding season starts. During this time the sheath of the boar should be irrigated with a nonirritating disinfectant several times. It is a good idea to have a nasal swab made for rhinitis, and the boar tested for brucellosis and leptospirosis also at this time.

The boar and females should be brought together at least 30 days before breeding so that immunity for any infection carried would be attained before time to breed.

SMEDI Syndrome

SMEDI is a term used to label a number of symptoms of irregularities in swine reproduction. They include stillborn pigs, mummified pigs, embryonic death of pigs, and infertility of sows and gilts.

Cause. There are at least nine viruses that may cause the disease. The stage of reproduction determines the nature of the effect on the reproductive system.

Symptoms. Infected animals show no signs of infection. The only visible signs of the disease are the resulting problems in reproduction. Infection at breeding time results in a low percentage of females conceiving. The first 30 days of the gestation period are critical times for infection to take place. It is at this time that the embryo pigs may be killed. As a result small litters, and mummified pigs may result.

Treatment and prevention. At present, prevention is the only method of control. There is no vaccine. There is no treatment. It has been found that infected animals will become immune to the virus within 30 days after infection. As a result the best control is to give the breeding animals an opportunity to infect others at least 30 days before the breeding season starts. Boars and gilts should be brought together and have at least fence-line contact. It is preferable to place all females in the same pen.

When it is impossible to bring animals together, the manure and bedding from one pen or house can be brought to other pens. The animals can pick up the infection if it is present. No new animals should be brought onto the farm 30 days before breeding until after pigs are weaned. They then should be isolated for a 3-week period. The herd should be managed as a closed herd. Outsiders should not be permitted in the house or pens.

Stress Syndrome

Porcine stress syndrome (PSS) has become a problem on many farms, especially when meaty, well-muscled pigs are grown in confinement in large numbers.

Cause. Stress may be caused by a number of factors, some of them are genetic. It has been found that extremely meaty animals with firm muscling are more inclined to suffer stress than other pigs. The temperament or disposition of the pig is also a factor.

Environmental factors affecting stress are housing, temperature, crowding, confinement, noise and forced movement. It has been estimated that environmental factors may be responsible for 50 to 80 percent of stress, whereas heredity factors are responsible for 20 to 50 percent.

Symptoms. Stress may be reflected in extreme muscularity, muscle tremors, reddening of the skin, anxiety, and pale, soft, and exudative carcasses when slaughtered. Stressed pigs are usually short legged, compact and tight skinned. They may have leg weaknesses. Surveys indicate that more than 40 percent of swine herds have PSS losses.

Often pigs die when being moved into new quarters or when being transported to market.

Treatment and prevention. Stress can be controlled by improvement in environmental conditions, and by testing pigs with a 2 to 6 percent halothane gas to determine susceptible pigs and providing them more ideal growing conditions. Pigs showing susceptibility to stress, and their littermates and parent stock should not be kept for breeding.

It is a good idea to follow-up pigs marketed as they are slaughtered in the packing plant to determine the extent of PSE (pale, soft, and exudative) condition in the carcasses. Strains of breeding stock producing these characteristics should be culled from the herd.

Crowding of pigs in pens should be avoided and they should be moved with as little excitement as possible. Hot and cold temperatures should be avoided. Clean, dry and comfortable quarters with as little distraction from noise are desirable.

Swine Vesicular Disease

There were no known cases of swine vesicular disease in the U.S. in 1973, but a serious outbreak occurred in Great Britain. The disease is prevalent in several European countries, but only Great Britain and Austria are conducting large scale eradication programs.

An outbreak of swine vesicular exanthema occurred in the U.S. in 1952. It resembles closely the disease now present in European countries.

In June of that year a load of raw garbage containing uncooked pork chops was carried out of California on the diner of a train and set off at Cheyenne, Wyoming. A hog raiser at Cheyenne fed the raw garbage to pigs which were later shipped to a hog-cholera serum firm in Nebraska. Those Wyoming pigs apparently became infected with

SVD (vesicular exanthema) from the garbage and carried it to Nebraska. From the serum firm it was carried to the Omaha stockyards. By the close of 1953 vesicular exanthema had shown up in 43 states.

Cause. Vesicular swine disease, like foot-and-mouth disease, is caused by a submicroscopic germ of the virus group. It is extremely contagious and spreads rapidly from animal to animal. It also may be spread by contaminated carriers or objects.

Symptoms. The chief symptom of SVD is the formation of blisters on the lips, tongue, snout, and above and between the claws of the feet. There is usually a rise in body temperature and a loss of appetite. The blisters often break and ulcers form. Pigs may show lameness and they may lose their hoofs.

Treatment and prevention. There is no known vaccination for SVD, and infected animals are quarantined and later slaughtered. No treatment is available.

All states now have enacted legislation prohibiting the feeding of raw garbage to hogs. Connecticut was the last state to enact legislation. In 1957 the Secretary of Agriculture announced the eradication of SVD after the completion of a seven-year state-federal cooperative SVD eradication program.

Care in buying breeding and feeding stock, isolation and new animals brought onto the farm and immediate isolation of sick animals, and sanitation in transporting, housing, and feeding hogs, as recommended for other swine diseases, should be practiced to prevent losses due to swine vesicular diseases. A veterinarian should be called immediately if symptoms show up.

Swine Vibrionic Dysentery

This disease is sometimes called *black scours, bloody scours,* or *swine dysentery,* and may be confused with necrotic enteritis and with trichinosis.

Cause. The cause is unknown, but it is recognized as a specific disease. Circumstances suggest that a Vibrio bacterium and a spirochete are necessary to produce the disease. Healthy pigs fed feeds contaminated with materials from the intestines of diseased animals become infected.

Symptoms. This disease is very acute and infectious, and the main symptom is a bloody diarrhea or black feces. The animals may or may not go off feed. They have some fever, but it is not high. Some pigs die within a few days, others linger for several weeks or longer. Fifty percent or more of the young pigs in an infected herd may die.

The disease should be diagnosed and infected herds quarantined by a veterinarian. Arsenicals, tylosin, furacins, Carbodox, tetracyclines, and other antibiotics are the most commonly used treatments.

Treatment and prevention. Many remedies have been used in the control of swine dysentery, and some with favorable results. Prompt treatment of infected pigs with sulfamethazine is quite effective. One or two grains per pound of body weight has been recommended as the initial dose. This should be followed with about half this amount each 24 hours for two or three days.

The isolation of sick animals and the moving of healthy pigs to clean ground are recommended, as in the control of necro.

The contaminated quarters should be cleaned and the lots cultivated, if possible. Avoid bringing to the farm pigs which have come from infected herds, or have been transported in infected trucks, or have passed through sales barns or stockyards.

Tests indicate that rations containing arsenilic acid or 3-Nitro will aid in the control of this disease. These materials also stimulate growth and improve feed conversion. The nitrofuran nf-180 fed to the sows two weeks before farrowing, three weeks after farrowing, and while the pigs were suckling also reduces scours. Mecadox* has been helpful as a preventive. The directions of the veterinarian or feed manufacturer should be closely followed.

Tuberculosis

In 1971, the Federal Meat Inspection Service reported about 700,000 hog carcasses with TB lesions. Most of the lesions were in the head of the animals and the carcasses were approved for human consumption after the lesions were removed. New regulations now will not permit the use of the carcass for food if there are lesions in any two primary sites unless it is cooked to 165 degrees for a minimum of 30 minutes. The resulting decreased value of the carcass has caused much interest in eliminating the disease.

Cause. There are three types of TB bacteria, (1) *Mycobacterium tuberculosis*, which cause TB in humans, (2) *Mycobacterium bovis*, which cause the infection in cattle, and (3) *Mycobacterium avian*, the cause of TB in poultry. Swine may be infected with any of the three types, but about 90 percent of the cases reported are of the avian type.

Symptoms. TB infected swine usually cannot be identified until slaughtered. Swine that have TB lesions are identified at the slaughter plant, and when possible the producer is notified of the condition.

Treatment and prevention. There is no treatment. The herd should be tested and reactors sold. If the herd is heavily infected, it should be depopulated. The housing and lots should be cleaned and disinfected. The organism can live in the ground for 3 or 4 years under favorable conditions. It is more easily controlled in confinement feeding than in pasture feeding, unless new pastures and lots are available.

New stock brought onto the farm should pass a negative test for TB. All poultry and birds should be kept away from the swine enterprise. Poultry litter should not be spread on land used for hog pasture. Feed should be stored in a bird-proof building. Dogs, cats and other pets should be kept out of the hog house and lots. Under no condition house poultry in the hog house, or use the poultry houses in swine production unless they have been thoroughly disinfected.

Transmissible Gastroenteritis

Outbreaks of this disease have caused mortality as high as 100 percent among pigs in some localities. Infected older animals suffer a short-lived diarrhea, which is followed by loss in weight, but they recover in a few days.

Cause. Research indicates that the disease is caused by a virus that can spread from pig to pig by direct and indirect contact. The incubation period is very short, 18 to 24 hours.

Symptoms. The infected pigs vomit and have diarrhea. The feces appear to be partially curdled milk. The pigs become thin and die within three or four days. Postmortem examinations of the pigs show inflammation of the stomach and intestines.

Treatment and prevention. While no treatments have been proven, some veterinarians have prepared a solution containing electrolytes, dextrose, antibiotics and water which is injected into the pig in three to four areas. The solution replaced the water lost in dehydration, provides some energy, and while the antibiotics have no effect on TGE, it may control other diseases which may be present.

A means of prevention is to scatter the farrowing places over a wide area. Sows should be moved to clean ground away from other hogs. Any bred sows brought onto the farm should be isolated, and sanitation, as recommended for the control of other diseases, should be practiced. New stock should never be placed in the central houses with other hogs.

Sows which contract TGE during gestation and have recovered two to three weeks before farrowing will farrow litters which usually are immune or resistant to the disease. These sows should be kept for future breeding purposes. Multiple farrowing will aid in the control of TGE.

Some hog producers have obtained and frozen the intestines of infected pigs. These intestines were cut into pieces and fed to sows for several days about three and one-half weeks before the sows were due to farrow. The sows were sick for a few days but recovered and farrowed normal litters that were quite immune to TGE.

One TGE vaccine is available, TGE-VAC, and while it is not 100 percent effective, many breeders have found it effective in helping control the disease. Sows and gilts should be vaccinated twice, six weeks and two weeks before time to farrow.

Prevention is the best control. Dogs, birds, cats, vermin and humans may be carriers so care must be taken to keep them from bringing the disease to the farm.

Trichinosis

According to U.S.D.A. officials about one hog per thousand is infested with trichinae. About 105,000 infested hogs are marketed each year. Nearly $12 million dollars is spent each year by packers to comply with trichinosis regulations.

Cause. The disease is caused by a microscopic worm parasite called trichinae. The disease is not contagious. Humans can become infected by eating raw or uncooked pork containing trichinae.

Symptoms. There are no visual symptoms to be observed by the swine grower. Tests of tissue are made in the packing plants.

FIGURE 11-6. The stomach and intestinal tract of a pig infected with transmissible gastroenteritis. (*Courtesy University of Illinois.*)

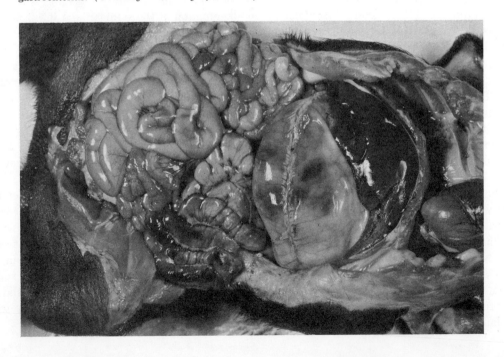

Treatment and prevention. There is no treatment for diseased hogs. Eradication of the disease will be possible only when identification of all slaughter animals is maintained until results of tests are available and laws are enacted to permit complete confiscation of infested herds. Only about 6 percent of the cases in 1968 were traced to feeding raw garbage. Most of the animals were farm raised.

A compulsory identification program will aid greatly in identifying the sources of trichinosis infected animals received at slaughter plants. A test is available which will permit identification of infected carcasses.

Extreme care should be exercised by farmers in handling animals at farrowing and castration time to avoid contact with the trichinosis organisms. The disease causes lock-jaw which is usually fatal in humans.

Ulcers

From 20 to 40 percent of the hogs slaughtered in packing plants have lesions indicating ulcers. The percentage of hogs with ulcers is increasing. Usually pigs with ulcers do not die, but do poorly and are inefficient users of feed. The ulcer condition also makes them more susceptible to disease and parasite problems.

Cause. No organism has been found to cause ulcers. The condition occurs largely because of the ration fed, the size of particles in the ration, and by environmental stress factors.

Symptoms. Pigs with ulcers usually become pale and there may be traces of blood in the feces. The feces may be black with a putrid odor. While most infected pigs will show diarrhea, some may be constipated. Pigs are usually listless and lack appetite. Massive hemorrhaging may cause death within a few days.

Treatment and prevention. It has been found that the feeding of pelleted and finely ground feeds increases the percentages of pigs with ulcers. The feeding of corn is more conducive to ulcer conditions than the feeding of oats. Pigs fed shelled corn and supplement have fewer ulcer lesions than pigs fed ground, complete rations. Feeding one-third oats in the ration aids in control of the condition.

Pigs confined in crowded, unsanitary and poorly ventilated conditions are likely to have ulcer conditions develop. While ulcers may develop while pigs are on pasture, they are less likely to do so. The best treatment for ulcers is to remove the causes. A veterinarian should be consulted and the conditioner ration recommended in Table 11-1, page 268, should be fed.

Virus Pig Pneumonia (PPLO)

Virus pig pneumonia is a wide spread disease among swine herds and has been with

FIGURE 11-7. It is a good idea to have a veterinarian post a pig when the cause of disease or death is uncertain. (*Courtesy Dave Huinker.*)

us for a long time. It has been recently found, however, that it is not caused by a virus, but by a pleuro-pneumonia-like organism of the mycoplasma family. The disease is often called PPLO. Usually few infected animals die, but losses due to slow gains, lameness, and inefficient use of feed are great.

Cause. *Mycoplasma hyopneumonia* is the specific cause. Another mycoplasmic organism may serve as a secondary invader. It is *mycoplasma hyorhinis*. These organisms are neither true bacteria or virus. They are more like the bacterium, however. The disease is passed from animal to animal through contact or inhalation of the organism. The sow can pass the disease to the pigs shortly after farrowing.

Symptoms. Pigs start coughing when two to three weeks of age. They later start thumping and breathing hard. Some die in a few days. Those that live do not do well.

Treatment. Sows should farrow in clean, disinfected, isolated areas. Litters showing signs of the disease should be isolated from other litters. All affected litters should be sold.

A test is being developed which can be used in testing breeding stock to determine the presence of antibodies in the blood for the mycoplasma causing the disease. Animals with these antibodies are carriers of the disease and can be culled from the herd. Seriously infected herds should be depopulated, and after thorough disinfection and clean-up of premises, be replaced with SPF breeding stock, or other stock free from the disease. When SPF stock is used, no non-SPF animals should be brought on to the farm.

Infected animals being fed for market will respond quite well to feeding Tetracycline in the ration at the rate of 100 to 200 grams per ton of feed, or by feeding 100 grams of tylosin * and 100 grams of sulfamethazine * per ton of feed. These materials should be fed continuously.

SWINE PARASITES

A parasite is something which lives on, and gets its food from some other plant or animal. Hogs have a number of parasites. Some live on or under the skin and are called *external parasites*. Others live within the organs of the body and are called *internal parasites*. The latter group of organisms are most injurious to swine.

Roundworms

Large intestinal roundworms, or ascarid, cause hog raisers heavy losses every year. These losses are due to stunting, development of a potbelly, general weakness, and sometimes death of the pigs.

In tests conducted by the U.S.D.A. it was found that worm-free pigs fed a well-fortified ration gained an average of 161 pounds in 169 days. Littermate worm-infested pigs fed

FIGURE 11-8. Worms in the small intestine of a growing-finishing pig. (*Courtesy American Cyanamid Co.*)

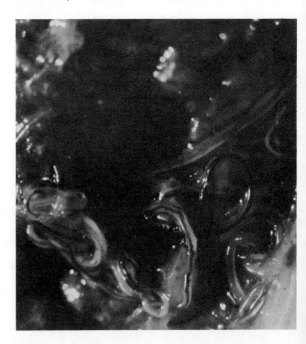

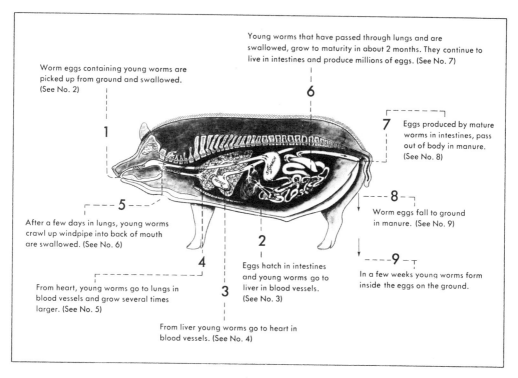

Worm eggs containing young worms are picked up from ground and swallowed. (See No. 2)

1

Young worms that have passed through lungs and are swallowed, grow to maturity in about 2 months. They continue to live in intestines and produce millions of eggs. (See No. 7)

6

7 Eggs produced by mature worms in intestines, pass out of body in manure. (See No. 8)

5

After a few days in lungs, young worms crawl up windpipe into back of mouth are swallowed. (See No. 6)

2

8

Worm eggs fall to ground in manure. (See No. 9)

4

Eggs hatch in intestines and young worms go to liver in blood vessels. (See No. 3)

3

9

In a few weeks young worms form inside the eggs on the ground.

From heart, young worms go to lungs in blood vessels and grow several times larger. (See No. 5)

From liver young worms go to heart in blood vessels. (See No. 4)

FIGURE 11-9. Life cycle of the large swine roundworm. (*Courtesy U.S.D.A. Bureau of Animal Industry.*)

FIGURE 11-10. The result of the use of a medicated wormer on a badly infested gilt. (*Courtesy Moorman Manufacturing Co.*)

the same ration gained an average of only 119 pounds.

Life history. The roundworm is a large, thick, yellow or pink worm, about the size of a lead pencil. The adult normally lives in the small intestine. The female produces thousands of eggs daily that are eliminated from the body in droppings. Pigs become infested with roundworms by swallowing the eggs with feed or water. The young worms hatch out in the pig's intestines. They penetrate the wall of the intestine and travel in the bloodstream to the liver, and on to the lungs. From the lungs they move up to the mouth where they are swallowed. They return to the intestine where they mature in about two or two and one-half months.

Damage. The roundworms in the intestine consume nutrients which the hog needs and cause digestive disturbances. Worms are more injurious to pigs on poor rations than to pigs on good rations. During the time that the young worms are in the lungs, the pig has difficulty breathing, and if exposed to dust or changes in weather, is subject to pneumonia. The worms weaken the pigs, making them susceptible to the many swine diseases.

Treatment and prevention. A number of worm remedies are on the market. Most of these remedies may be mixed and administered with feed or water. The wormer selected should be easy and economical to use. It should eliminate the worms but should not be toxic and upset the digestion of the animal.

There are at least nine worm remedies on the market. *Sodium fluoride* is efficient but toxic. *Cadmium anthranilate* ° and *cadmium oxide* ° in Nebraska tests were slightly toxic and fair in efficiency.

Hygromycin B ° was low in toxicity and efficient in Nebraska tests. It is mixed in feed and fed continuously until the pigs weigh at least 100 pounds. If fed in too large amounts or for too long a period, hygromycin B may cause loss of hearing. This wormer will control nodular and whipworms as well as roundworms. It must be fed according to directions.

Piperazine compounds in Nebraska tests were very low in toxicity and good as wormers. They are mixed with water and given over a 24-hour period after pigs have been fasted for 18 to 24 hours. Pigs should be treated twice, once shortly after weaning and again 4 to 6 weeks later, to eliminate later hatched worms. Piperazines also control nodular and whipworms.

Thiabendazole is a new drug recently approved as an aid in prevention of infestations of large roundworms. It is fed at the rate of 1 pound per 1,000 pounds of feed for a 14-day period. In Florida tests it was very effective in eliminating the eggs of *Stron-*

gyloides ransomi worms from the feces of weanling pigs.

Levamisole HC1 ° (Tramisol) is a broad spectrum wormer that will control the three major internal parasites in the Corn Belt, roundworms, lung worms, and nodular worms. It also controls thread worms, or strongyloides. It can be used with feed, or with water, and can be used with young pigs, weanlings, or with bred sows and gilts. It is a single treatment wormer.

Dichlorvos (Atgard) is another recently developed broad spectrum wormer that is very effective. It can be fed to pigs four or five weeks old, and to bred gilts and sows. It will kill both the adult parasites and the young larvae before they mature and lay eggs. It is a one treatment product, but a second treatment after four or five weeks is recommended. It will control round, whip and nodular worms.

Pyrantel tartrate (Banminth)° is the latest wormer to be approved by the Food and Drug Administration. It controls the round and nodular worms by killing both the adult and larvae. The larvae are killed before they can lay eggs.

The preventive measures for controlling worms have been described in previous chapters. Proper sanitation and rotated pastures help prevent losses due to roundworms.

Intestinal Threadworms

(Strongyloides ransomi) This is the smallest of the 10 species of roundworms that infest pigs and is considered one of the most serious parasites in North Carolina and Georgia.

Life history. The worm is very small and resembles a small piece of white thread. The parasite infests the pig through the skin and enters the blood stream. It passes through the liver, heart, lungs, and small intestine. The eggs can mature both in the body of the pig and on the ground. The worm can be trans-

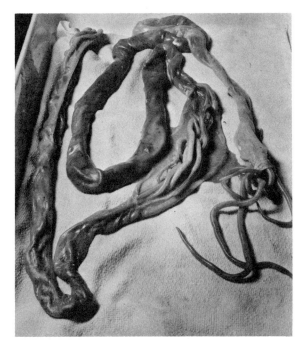

FIGURE 11-11. The roundworms almost completely fill the small intestine of a small pig. (*Courtesy Eli Lilly & Co.*)

mitted to the pig in the womb at time of birth or in colostrum milk immediately after birth.

Damage. There is evidence that the threadworm may be the cause of much pig scours. The worm has been found in the feces of four-day-old pigs. The pig passes off the worms at about 8 weeks of age. The infested pigs are unthrifty. Some may die between 10 and 14 days.

Treatment and prevention. Dichlorvos and thiabendazole have been found to be effective in removing threadworms. Regular treatment of the sows before breeding and before farrowing is recommended.

Lungworms

These worms are long, slender, and whitish in color, from one-half to two inches long, and threadlike in diameter. They are found in the windpipe, but more often in the two branches of the lower windpipe, and in the lungs.

Life history. The female lungworm produces large numbers of eggs which are coughed up, swallowed, and eliminated with the droppings. The lungworm eggs are swallowed by angleworms. They hatch and develop in the angleworms, and the pigs become infested when they eat the angleworms while rooting. In the intestines of the pig the lungworm penetrates the intestinal wall and follows the bloodstream to the heart and back to the lungs.

Damage. Lungworms weaken the pigs and cause them to cough, to breathe with difficulty, and to grow more slowly. Severe infestations may cause death.

Treatment and prevention. Sick animals should be isolated from the herd and fed highly fortified, balanced rations.

Lungworms can be partially controlled by the use of rotated pastures and by the use of sanitary practices during farrowing time and until the pigs weigh 75 pounds.

Cyanacethdrazide, a new drug, is being marketed through veterinarians. It does not kill the worms in the lungs but causes them to leave the lungs and be coughed up. The drug is injected under the skin or into a muscle. Severely infected animals should be treated on each of three successive days.

Levamisole HC1 ° or levamisole (Tramisol)° will kill both the lungworms and the thread worms. It at present is the only wormer recommended for control of lung worms. As with all wormers, the directions of the manufacturer should be closely followed.

Kidney Worms

It is estimated that kidney worms cause more than \$72 million in losses to swine producers each year, yet the parasite is easily controlled. Kidney worms require as long as a year to attain egg-laying maturity. By using only first-litter gilts for breeding and then removing them from the weaned pigs, it is possible to eradicate these worms from heavily infested pastures. The gilts normally can be sold before they are sufficiently mature to pass kidney worm eggs in urine.

Drugs have not proven satisfactory in control of kidney worms. Kidney worm eggs and larvae in a pasture are killed by sunlight, drying, or high or low temperatures. They may live from three to six months if protected under trash. They may be swallowed by earthworms which pigs may eat.

Whipworms

Studies conducted by the University of Wisconsin and by the North Carolina State University indicate that whipworms have become a serious swine internal parasite. More than 60 percent of the pigs examined in each of the six regions in the U.S. had signs of whipworm infestation.

Life history. Pigs become infected by consuming infective whipworm eggs with dirt, food or water. The eggs hatch in the small intestine of the pig. They penetrate the mucose of the small intestine and remain there for about eight days. They then migrate to the large intestine where they grow to adulthood and lay eggs. About 42 to 48 days after the eggs are swallowed, the adult whipworms are laying eggs and reinfesting the floors and lots.

Damage. Pigs infested with whipworms will begin to scour about three weeks after being moved from the farrowing house to the nursery or pasture. The pigs become gaunt and lose weight. The hair is rough. They may die after 30 to 60 days. The worms irritate the walls of the intestines and use the food that is intended for the pigs.

Treatment and prevention. Dichlorvos (Atgard) will control whipworms if properly administered. It is suggested that sows and gilts be wormed about seven days before

breeding, and again about two weeks before farrowing. The young pigs should be wormed at weaning time, when five to seven weeks of age, and again about a month later.

The feeding of rations recommended in this book and the use of slatted floors and the maintaining of dry, sanitary quarters are essential. If pigs are placed on pasture, it should be on high ground and preferably a field on which hogs had not grazed for several years.

Hygromycin * may also be used in controlling whipworms. The directions of the manufacturer should be followed.

Mange

Mange is a highly contagious skin disease, and although very few animals die as a result of mange, it is the most serious of the external parasites.

Cause. Mange is caused by a very small mite that spends its entire life on the hogs. It feeds on the tissues of the skin and blood and burrows into the skin, causing a dry, rough, and scaly hide.

Damage. Hog mange causes intense itching, which drives the hog to bite and rub itself. The infection usually starts around the eyes and ears and along the underline where the skin is tender.

Treatment and prevention. The mite can live in infested quarters for several weeks. Control measures must include a thorough cleaning of the shelters and houses.

The chemicals Ciodrin, Lindane, Malathion, and Toxaphene are excellent for use in controlling mange. One gallon of 50 to 57 percent Malathion emulsifiable concentrate plus 1 pound of a nonfoaming detergent, or 3 quarts of 60 percent Toxaphene emulsifiable concentrate plus 1 pound of nonfoaming detergent, per 100 gallons of water are recommended.

The animals should be sprayed thoroughly using 2 to 4 quarts per animal.

FIGURE 11-12. Sanitation is a must in confinement swine production. (*Courtesy Shell Chemical Co.*)

Unweaned pigs should not be sprayed. Toxaphene should not be used within 30 days of marketing. Sows should be treated twice during gestation period previous to farrowing.

Lice

Hog lice, by their bloodsucking habits, cause some loss to swine producers, and they may be responsible for the spread of infection.

The louse is about a quarter-inch long and grayish brown in color. During the winter months it may be found in the ears, in folds of skin around the neck, and around the tail.

The female lays several eggs a day during the winter. These eggs are attached to the hair, and hatch in two to three weeks. They mature in another two weeks.

Treatment and prevention. The Malathion Lindane, Ciodrin, and Toxaphene

treatments for mange will also kill the hog lice. If there is no need for mange control, the bedding should be dusted with 5 percent ronnel (Korlan) at the rate of 0.5 pound per 100 square feet.

INJURIES AND POISONS

Paralysis of Hindquarters

Paralysis of the hindquarters is common in swine. As a rule this develops gradually. At first the animal is wobbly or unsteady. It has difficulty controlling the hindquarters. It later is unable to stand on its hind feet. In severe cases the animal cannot get up.

There may be a number of causes of the paralysis. Animals in the same lots with horses or cattle may have been injured. Small pigs may be injured by larger hogs.

Mineral and vitamin deficiencies may result in paralysis.

Paralysis sometimes follows parturition in sows if they are in poor condition at farrowing time. Malnutrition is probably a common cause of paralysis.

Because of the many causes of paralysis of the hindquarters in hogs, it is difficult to prescribe a treatment or control. In treating animals, remove them from the herd and provide comfortable, dry quarters. If there is evidence of constipation, give the animal a dose of Epsom salts or linseed oil. Feed the animal a light slop made of milk and bran. Plenty of fresh water is important.

In some cases the massaging of the loin muscles with a good liniment may be helpful. Heat may also be applied. A highly fortified ration will do much to remedy the situation. Vitamins, minerals, and proteins are essential.

Gut Edema

This condition is apparently caused by toxemia, or poisoning. It may occur under all systems of management. Vaccination, castra-

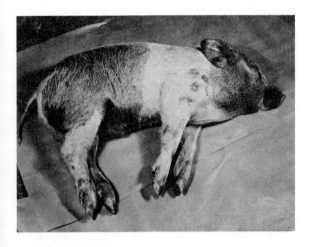

FIGURE 11-13. Swelling of the eyes and rectum is characteristic of gut edema. (*Courtesy University of Illinois.*)

tion, or other changes may cause it. Pig weed is a common cause.

Ordinarily it affects pigs eight to twenty weeks old. They usually lose their appetite suddenly, and have a swelling of the eyes and of the rectum. Some pigs become paralyzed. Some die.

A veterinarian should be called for treatment as soon as symptoms appear.

Weed Poisoning

Young cockleburs in the two-leaf stage are very dangerous to young pigs. Mature plants are not dangerous.

Weakness, unsteady gait, vomiting, and sometimes death in 24 hours are the striking features of cocklebur poisoning. A veterinarian should be called if symptoms appear. The best control is to clear the farm of cockleburs, or to keep pigs off of pasture during the early part of the growing season.

STATEWIDE SWINE HEALTH CERTIFICATION PROGRAMS

Swine breeders in several states have set up statewide swine health certification pro-

grams as a means of encouraging and recognizing breeders who maintain disease-free herds, and to provide information to prospective buyers of breeding stock in regard to the location of disease-free herds.

The programs are similar. A "Health Certified" rating is granted to breeders whose herds have passed two successive inspections by appointed veterinarians, and have passed two clean herd tests for brucellosis, or have been certified as "Brucellosis Controlled" herds. There can be no evidence of atrophic rhinitis, transmissible gastroenteritis, swine dysentery, hog cholera, anthrax, vesicular exanthema, erysipelas, brucellosis, tuberculosis, or any other contagious disease. The two inspections are made 60 to 90 days apart.

After a herd has been certified, it must be inspected every six months and maintain its "Brucellosis Controlled" rating to remain a "Health Certified Herd." These programs should be well received by producers and users of swine breeding stock.

SPECIFIC PATHOGEN FREE (SPF) SWINE

Techniques have been developed to take unborn pigs surgically from sows from two to four days before they would be farrowed. A pig at birth that has no contact with its mother is free from disease. The SPF program provides for the removal of the pigs from the uterus of the sow and the development of the pig under an environment entirely free from all disease organisms.

The original program developed by Dr. George Young of the University of Nebraska involved the following steps:

1. The dam, 112 days after breeding, is hoisted by her hind legs and lowered head first into a barrel containing a small amount of dry ice.

2. The dry ice turns to carbon dioxide which anesthetizes the sow.

3. The entire uterus is removed and passed through an antiseptic lock into an enclosed hood.

4. Within the hood the pigs are removed from the uterus. They begin breathing filtered air that contains no germs.

5. The pigs are transferred under cover to special individual brooders. They are fed modified cow's milk. After one week they are brooded in groups of 8 to 12 pigs.

6. The pigs remain in the brooders for about four weeks. They are then ready to be moved to a farm environment that has been thoroughly cleaned and disinfected for a period of at least six weeks.

7. The sow is processed for food.

8. The SPF pigs are raised on a farm where there have been no hogs for at least six weeks, and no non-SPF hogs are brought on to the farm. They are usually kept in groups of 10 to 20 pigs.

9. No one is permitted to enter the pig pens without putting on clean boots, and no trucks or other vehicles enter the lots.

10. Dogs, cats, and rats are controlled.

11. Ordinary rearing methods are used, except no new stock is introduced, and contact with other swine must be avoided by the farmer.

12. The stock reared to maturity is kept for breeding stock. Only SPF boars or gilts are brought to the farm.

FIGURE 11-14. SPF swine must be completely isolated from other swine, and from potential carriers of disease. (*Courtesy Kent Feeds.*)

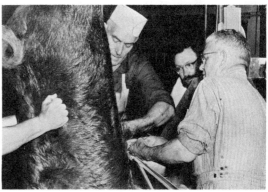

FIGURE 11-15 (*Top*). The first step in obtaining specific pathogen-free (SPF) pigs—lowering the sow into a barrel containing dry ice which anesthetizes her.   FIGURE 11-16 (*Bottom*). Removing the entire uterus. (*Figures 11-15 and 11-16 Courtesy University of Nebraska.*)

13. Normal birth is resumed.

14. The clean stock may be used to stock other farms where all non-SPF hogs have been sold, the premises have been cleaned for at least six weeks, and the herd will be completely isolated.

Laboratory techniques have been developed so that it is possible to obtain primary SPF pigs in the laboratory by either hysterectomy or by Caesarean section. The latter method does not necessitate the killing of the sow. Eight national SPF laboratories in operation in Illinois, Iowa, Minnesota, Ohio, and Texas.

A primary SPF herd is one in which all breeding animals originated in a laboratory, and all replacements also originate in an SPF laboratory. Since the cost of producing primary herd animals is quite high, most SPF herds are classified as secondary SPF herds.

In addition to a national SPF accrediting organization, there are 28 state associations.

Accreditation Standards

Advertisers of SPF breeding stock must have herds accredited with both the state and national associations. While some slight variation in standards exist among states, the basic requirements are as follows:

1. At least 10 hogs from each farrowing (90-day period) must be sent through a slaughter check. Some states have only one site. Iowa has 60 SPF herds and 7 check sites.
2. The noses and lungs of the 10 animals must show no evidence of rhinitis or mycoplasma pneumonia. Accreditation is lost if the diseases are present.
3. The producer must show proof of origin of stock. Parent stock must have originated from an accredited primary or secondary SPF herd.
4. Proof must be presented that the herd is a brucellosis-free herd.
5. The producer must submit quarterly health reports, filled out and signed by a veterinarian or by an official fieldman.
6. A uniform litter marking system must be in use.

Diseases Affected

SPF hogs are free only from the virus diseases that are passed from pig to pig by direct contact. Virus pneumonia, rhinitis, and some forms of scours are controlled.

Cost

It costs from $75 to $150 to deliver surgically a four-week old SPF pig. With high-price breeding stock, there is some loss on the sow, since she may be killed and butchered. The cost of SPF pigs farrowed normally from SPF ancestry is reasonable.

SPF Laboratories

There are eight national laboratories in operation. The laboratories hold memberships in the National SPF Swine Accrediting Agency Inc. located at Conrad, Iowa. All are under the management of licensed veterinarians.

Test Results

Records from nearly 2,400 SPF pigs in Nebraska indicated an average weight of 43.4 pounds at 56 days and 204 pounds at 5 months. The average daily gain was 1.31 pounds per day from birth to market.

There are currently nationally accredited SPF swine programs in 28 states. Nearly 500 herds are involved. Most SPF herds provide registered foundation stock to purebred breeders and commercial hog men. All breeds are involved.

During the ten year period 1958 to 1968, the percentage of SPF hogs grading U. S. No. 1 increased from 32.2 percent to 63 percent according to Nebraska data. Production testing is required for SPF accreditation in Nebraska. Nebraska standards call for an average herd weight of 155 pounds at 140 days of age. Backfat standards are 1.2 inches on boars and 1.4 inches on gilts.

In 1971, Nebraska producers performance-tested 26,646 and probed 14,560, averaging 0.89 of an inch of backfat. Individual breeders are obtaining excellent results in using SPF stock. A Nebraska Duroc breeder sent 111 barrows through a packing plant for disease check during a 12-month period. All passed the disease check. The carcasses averaged 30.4 inches long, had 1.25 inches of backfat, and an average of 5.25 square inches of loin eye.

Caesarian Section

The use of the hysterectomy operation in the SPF program results in the killing of the sow. Two problems result. Valuable breeding stock is slaughtered, and when large numbers of sows are involved, there is difficulty in marketing the carcass of the sow since the killing is not done under Federal meat inspection. Techniques have been developed permitting the pigs to be removed by Caesarian section methods. The process requires more labor and is more costly than the hysterectomy operation, but it permits the re-use of the sow in the breeding herd.

Less than one-half of the SPF laboratories in operation in January, 1974, were using only the Caesarian section methods.

Swine Identification System

It is necessary in the control of several diseases to be able to identify the farm from which hogs were produced. This is especially true of slaughter hogs. It is not possible to see in live animals the extent that they have been affected with disease. As a result the U.S.D.A. has proposed an identification system for all hogs shipped across state lines.

Ear notching systems are satisfactory for use on an individual farm, but will not permit identification of hogs when large numbers are involved. A tattoo system has been recommended which would involve a slap tattoo on each hog. A code would be used so that the producer, dealer, market agency, and packer could be identified.

An ID system is very necessary to complete the present brucellosis eradication program. It would be equally effective in other disease control programs.

The system would be very helpful in control of diseases carried by feeder pigs. Nearly a million head of these pigs are shipped into some of the Corn Belt states each year, and it is not always possible to trace pigs back to individual dealers or producers when disease outbreaks occur.

Withdrawal Times for Drugs

Pharmaceutical residues may appear in the carcasses of swine when slaughtered and as a result the carcass, or a part of the carcass is condemned. Heavy losses may result if large numbers of animals are involved. The residues may result from injections given pigs for disease prevention and control, or may result from the use of drugs in feed. They may also result from the use of wormers and external parasite controls.

Following are suggestions for making effective use of drugs in swine production and avoiding residues in the carcasses:

1. Read directions and follow them accurately.
2. Do not overdose. Most drugs are approved for use at only certain rates.

FIGURE 11-17. Removing the pigs from the uterus within the hood. The pigs begin breathing filtered air free from germs. (*Courtesy University of Nebraska.*)

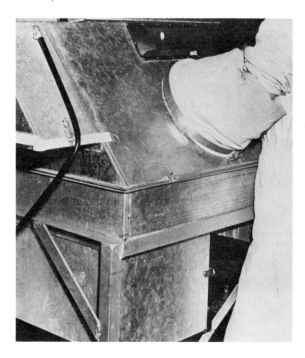

3. Do not make injections into the paritoneal cavity.
4. Irritating drugs should be injected intravenously when directions permit.
5. Do not mix or combine products in syringe.
6. Do not make more than one injection at an injection site unless the directions specifically state that it can be done.
7. When large injections are given, use more than one site.
8. Observe withdrawal times indicated on the label or carton.
9. Identify treated animals and keep a record of material, amount and date.
10. Do not market any animal unless drugs of all types have been removed from the animal for a minimum of the times recommended by the manufacturers.

Presented in Table 11-2 is a list of drugs commonly used in swine production, examples of trade names, and the withdrawal time in days for each.

SUMMARY

Swine diseases and parasites can take much of the profits out of the hog business.

Gastroenteritis, rhinitis, E. coli scours, vibrionic dysentery, MMA, and PPLO are considered most serious. Hog cholera has been eliminated as has trichinosis.

Erysipelas can best be controlled by vaccination. Vaccination of the gilts and sows for erysipelas before breeding and the erysipelas bacterin vaccination of pigs after weaning are recommended.

The feeding of arsenicals, antibiotics, highly fortified pig starters, and milk products is the best means of controlling infectious scours in pigs. Sanitation and rotated pastures are essential in controlling the diseases of the digestive tract.

Rhinitis, vibrionic dysentery, and brucellosis may be prevented by the isolation, in part, of diseased animals and by careful selection of breeding and feeding stock. All swine-breeding stock should be tested for brucellosis, and only animals which have been negative on two tests, given at least 60 days apart, should be kept.

The noses of all breeding stock should be swabbed for rhinitis and infected animals sold. New stock should be rhinitis free. The arsenicals and Mecadox (Carbodox) are the best treatments for swine dysentery. Sulfamethazine is also effective.

E. coli and smedi syndrome can best be controlled by permitting the sows to become infected through contact with animals, or bedding of infected animals, at least 30 days in advance of farrowing time. They will become immunized and the immunization will carry over to the pigs.

Breeding stock should be tested for leptospirosis and infected animals sold. Breeding animals and pigs can be vaccinated as a preventive measure.

Flu in hogs can best be controlled by the feeding of well-balanced rations and by proper housing. There is no treatment for gastroenteritis, but it can be controlled, in part, by the isolation of new animals brought onto the farm, by vaccination, and by dividing the herd and pasturing them in areas apart from other hogs. Sows which have recovered from the disease usually produce litters which are resistant to it.

When abortion or dead pigs occur, the breeding animals should be tested for leptospirosis as well as for brucellosis. The symptoms of the two diseases are similar. Infected animals should be sold or slaughtered. It is possible to vaccinate for leptospirosis.

Virus pig pneumonia is caused by a mycoplasma organism. Breeding animals should be blood tested before breeding and infected animals sold. Serious cases of infection justifies depletion of herd and purchase of new breeding animals that are free from the disease. Tetracycline and tylosin are effective in feeding diseased pigs.

T A B L E 11-2   WITHDRAWAL REGULATIONS FOR SWINE DRUGS[1]

Drug	Withdrawal Before Slaughter
ARSENICALS	
1. Arsenilic Acid	5 days
2. Carbarsone	5 days
3. Roxarsone (3-NITRO)	5 days
4. 4-Nitrophenylarsonic Acid	5 days
5. Sodium Arsanilate	5 days
SULFA DRUG COMBINATIONS	
1. Aureo Sp 250	7 days
2. Tylan plus Sulfa	5 days
3. Sulfaethoxypyridazine	10 days
4. Sulfamethazine	5 days
ANTIBIOTICS	
1. Aureomycin (chlortetracycline)	None
2. Bacitracin	None
3. Erythromycin	None
4. Neomycin	None
5. Neo-terramycin	None
6. Oleandomycin	None
7. Penicillin	None
8. Penicillin and Streptomycin	None
9. Terramycin (oxytetracycline)	None
10. Tylan (tylosin)	None
11. Virginiamycin	1 day
CHEMOBIOTICS	
1. Furazolidone	None
2. Furox (furazolidone)	None
3. Nitrofurazone	None
4. Mecadox (Carbodox)	70 days
PESTICIDES	
1. Toxaphene	28 days
2. Lindane (spray)	30 days
3. Lindane (dip)	60 days
4. Coumaphos (Co-Ral)	14 days
5. Carbaryl (Sevin)	7 days
6. Malathion	None
WORMING COMPOUNDS	
1. Atgard (dichlorvos)	None
2. Piperazine	None
3. Hygromix	2 days
4. Thiabendazole	30 days
5. Cadmium Oxide and Cadmium Anthranilate	30 days
6. Levamisole hydrochloride (Tramisol)	3 days
7. Pyrantel tartrate	1 day

[1]These recommendations are based upon the Food and Drug Administration's approval of the drugs. The FDA may approve changes in these drugs which also change withdrawal times. Read the label on your drug or feed container and follow directions. Longer withdrawal periods are necessary when drugs are injected.

Dichlorvos, levamisole HC1, piperazine compound, hydromycin B, thiabendazole, and pyrantel tartrate may be used in treating pigs for round worms. Dichlorvos will also control whipworms and nodular worms. Levamisole HC1 will also control lung worms and nodular worms. Thiabendazole is an excellent control of the thread worm.

Pigs should be treated just after weaning, and 30 days later if needed. Breeding animals should be wormed three weeks before breeding and at least three weeks before farrowing.

SPF programs appear to be an effective means of controlling virus pneumonia and atrophic rhinitis.

Mange and hog lice can be controlled by dipping or spraying with malathion or toxaphene.

Some form of identification of all pigs on individual farms must be used in control of diseases. A slap tattoo system appears to be workable. In market-hogs, it is important to abide by the recommendations that are made by swine drug companies concerning withdrawal periods.

QUESTIONS

1 Describe the symptoms of swine vibrionic dysentery and outline a program for its control.

2 What would you do if erysipelas was discovered in your home herd?

3 How does atrophic rhinitis differ from bull nose?

4 What is required to have a state-certified brucellosis-free herd in your state?

5 How does E. coli baby pig scours differ from the disease commonly called necrotic enteritis? How may E. coli be controlled?

6 What has been done to control vesicular exanthema in the United States?

7 How can transmissible gastroenteritis losses be controlled on your farm?

8 What is the extent of losses in your community due to leptospirosis and how may they be reduced?

9 How serious is the stress syndrome in your community and how can it be controlled?

10 Confinement feeding has brought about much stress in swine at all ages. What production procedures would you suggest a producer use to avoid this difficulty?

11 Can vaccination be used in controlling jowl absesses in swine?

12 How prevalent in African fever and swine pseudorabies in swine in the U.S.?

13 Outline methods of identifying and controlling lungworms, whipworms, and nodular worms in swine.

14 How does the life cycle of the roundworm differ from that of the lungworm?

15 Outline a plan for controlling roundwarms on your farm.

16 What methods are best in controlling mange?

17 Under what conditions will it pay to purchase and use SPF breeding stock?

18 How would you treat pigs with PPLO (mycoplasmosis) on your farm?

19 How serious is MMA (metritis-mastitis-agalactia syndrome) in your area? How can it be controlled?

20 Outline a disease and parasite control program for your farm.

21 Which of the drugs commonly fed to swine must be removed from the ration for a period of time before they are marketed?

REFERENCES

Becker, H. N., et al., *1972 Illinois Swine Seminar.* Urbana, Illinois: University of Illinois, 1972.

————, *1973 Illinois Swine Seminar.* Urbana, Illinois: University of Illinois, 1972.

Bradley, C. M., et al., *Missouri Feeder Pig Manual.* Columbia, Missouri: University of Missouri, 1970.

Caley, H. K., *Health Handbook for Profitable Swine Production.* Manhattan, Kansas: Kansas State University, 1970.

Dunne, H. W., *Diseases of Swine,* 3rd ed. Ames, Iowa: Iowa State University Press, 1970.

Dykstra, R.R., *Animal Sanitation and Disease Control.* Danville, Illinois: The Interstate Printers and Publishers, 1961.

Ferguson, D.L., and E. C. Howe, *Controlling Internal Parasites in Swine.* Lincoln, Nebraska: University of Nebraska, 1972.

Jones, J. R., and A. J. Clawson, *Raising Hogs in North Carolina.* Raleigh, North Carolina: North Carolina State University, 1969.

Leman, A. D., *Swine Health: Common Diseases Affecting Baby Pigs.* Urbana, Illinois: National Pork Council and the University of Illinois, 1970.

Schwartz, Benjamin, *Internal Parasites of Swine,* Farmers Bulletin No. 1787. Washington, D.C.: U.S. Department of Agriculture, 1952.

Seiden, Rudolph, *Livestock Health Encyclopedia,* 3rd ed. New York, New York: Springer Publishing Company, Inc., 1968.

Thomas, J. G., et al., *Texas Guide for Controlling External Parasites of Livestock and Poultry.* College Station, Texas: Texas A & M University, 1968.

Underdahl, N. R., et al., *Nebraska's Pathogen-Free (SPF) Swine Program Tenth Year Report,* SB499. Lincoln, Nebraska, 1968.

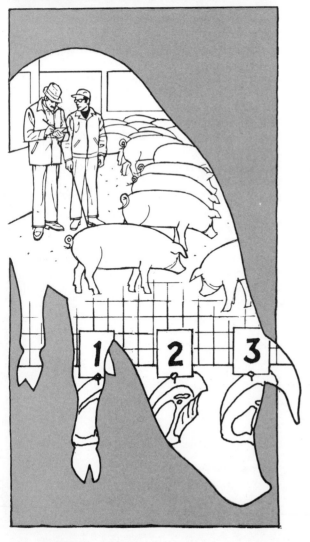

12 MARKETING ALTERNATIVES

Efficient hog raisers try to plan their breeding and feeding operations so that they will have hogs of the weight and conformation desired by the packer and consumer ready for market when the price is most favorable. By planning breeding operations and allowing a four-and-one-half to five-and-one-half-month growing-out period, it is possible to plan the time that the hogs will be ready for market.

We are not always able to accurately predict the season when the price will be best. Hogs of certain weights sell better during some months than during others. The number of hogs going to market varies seasonally, and when hogs are plentiful, the price goes down. When few hogs are being marketed, the price is higher.

Most farmers find it necessary to make changes in their breeding and feeding methods in order to have their pigs at marketable weights at the time that the price is best. Each farmer must analyze the swine enterprise and decide which production and marketing programs are most suitable.

CHANGES IN THE HOG MARKET

It was pointed out in a previous chapter that less lard and fat pork are being bought today than were being bought a few years ago. The consumers who buy meat are guiding farmers in the production of quality pork. They are demanding leaner cuts. They bypass pork chops with an inch of fat for lean chops, or they buy poultry or beef instead. As a result the demand for pork has been reduced.

With 127,540 million head of cattle on our farms on January 1, 1974, we can expect beef to be more plentiful the next few years. This supply spells cheaper beef and more competition for pork.

The fast-growing broiler industry provides another area of competition. Meat consumed in the nation in 1973 was 49.5 per-

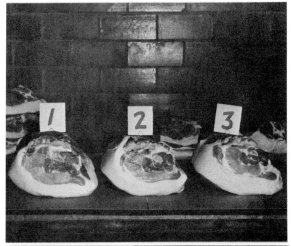

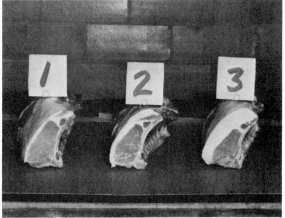

FIGURE 12-1. Hams (*top*), loins (*center*), and sides of bacon (*bottom*) from U.S. No. 1, U.S. No. 2, and U.S. No. 3 market hog carcasses. (*Courtesy Rath Packing Company.*)

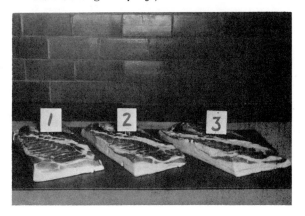

cent beef and veal, 27.7 percent pork, 17.4 percent chicken, 3.9 percent turkey, and 1.3 percent lamb or mutton.

Hog producers and pork processors must be alert to the changing meat market. The future of the enterprise is in part dependent upon the quality and quantity of pork products produced and the methods used in merchandising.

The Lard Problem

At one time lard sold wholesale for more than live hog prices. Lard had an important place in our diet, and we had a good export market. The situation has changed. Wholesale lard during the past years has been selling for considerably less than the price of live hogs. The prices of lean cuts of pork have advanced, and the prices of fats have declined.

In 1967 the retail selling price of pork chops was almost four times the retail selling price of lard. Table 12-1 shows the retail prices of various pork products for the 1964–1967 period. In 1949 the average cost of hogs in Chicago was $18.40 per 100 pounds wholesale. Lard at that time sold for $12.03 per 100 pounds.

The average price paid for hogs in Chicago in 1967 was $19.18 per hundredweight. The wholesale price of lard in bulk was about $7.80 per hundredweight.

The yield of lard from hogs slaughtered under federal inspection during the 1964 to 1967 period is shown in Table 12-2. Note that the pounds of lard per 100 pounds of live weight dropped from 10.7 to 7.8 pounds during the four-year period.

Hog raisers and packers have a joint responsibility in making available to the consumers a better quality pork product. Hogs with more lean meat must be produced, and they must be marketed at a reasonable weight. We need lean meat, but we must reduce lard production. There are two ways

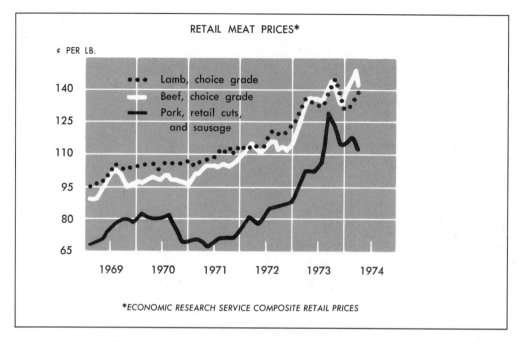

FIGURE 12-2. Retail Meat Prices (*Courtesy U.S.D.A.*)

FIGURE 12-3. Retail Prices of Selected Livestock Products. (*Courtesy U.S.D.A.*)

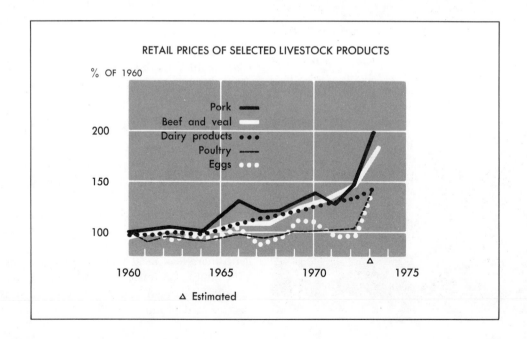

TABLE 12-1   AVERAGE RETAIL PRICES OF PORK CUTS AND LARD, 1967-1972

Year	Pork Chops Cents per Pound	Ham (Whole) Cents per Pound	Bacon (Sliced) Cents per Pound	Lard Cents per Pound
1967	87.1	69.8	76.7	22.6
1969	99.0	74.6	83.8	23.2
1970	96.9	80.6	85.6	26.2
1971	91.6	73.7	73.9	26.6
1972	107.0	83.5	89.4	27.5

Source: *Agricultural Statistics, 1973*, U.S. Department of Agriculture.

TABLE 12-2   LARD FROM COMMERCIAL HOG SLAUGHTER, 1967-1972

Year	Average Live Weight Pounds	Yield of Lard per Hog Pounds	Yield of Lard per 100 Pounds Pounds	Average Wholesale Price (Loose) Cents
1967	242.4	25.9	10.7	7.8
1969	239.5	23.2	9.7	9.7
1970	241.4	22.8	9.4	11.6
1971	239.7	21.2	8.8	10.8
1972	239.6	18.6	7.8	10.4

Source: *Livestock and Meat Statistics, 1973*, U.S. Department of Agriculture.

of doing this: (1) producing a better muscled meat-type hog and (2) marketing our hogs at lighter weights.

THE MEAT-TYPE HOG

The meat-type hog is being developed to meet the changes in demand for pork products. It is not a new breed. We have meat-type hogs in all of our standard breeds. Several new breeds, however, were developed in an effort to improve the carcass quality of our hogs. The project has not been very fruitful.

The *meat-type hog* has been described as one with more than average length and with a natural tendency to yield a carcass with a high proportion of the high-priced meaty cuts. The carcass of a 220-pound hog should measure about 30.5 inches long, and the layer of backfat should not exceed 1.1

inches. On the hook, the carcass should have a high ratio of lean meat to fat. Hams and loins should comprise 45 to 50 percent of the carcass weight.

TIME TO SELL

In the past, two pig crops were normally produced each year. One farrowed in the spring and was ready for market in the fall. The other farrowed in the late summer, and was sold in the spring. The bulk of the year's hog production was marketed during these two periods. A large share of the fall-farrowed pigs had been marketed in March, April, and May; the bulk of the spring pigs had been marketed in October, November, and December. The supply of hogs during the two marketing periods was such that packers could buy hogs at lower prices than during other seasons.

Seasonal Price Variations

The variation in monthly prices of barrows and gilts and the number purchased at eight large markets during the five-year period 1968 to 1972 are shown in Figure 12-4. The markets included were Chicago, Indianapolis, Kansas City, Omaha, Sioux City, National Stock Yards, South St. Paul, and St. Joseph.

The highest prices paid for barrows and gilts were in July, August, and June. Lowest prices paid for these slaughter hogs were in April, March, and November. The spread between the top price in July and the low in April was $3.72 per hundredweight.

The largest numbers purchased were in October (0.8 million) and April (0.79 million). Smallest numbers purchased were in July (0.6 million) and August (0.61 million). Note that the law of supply and demand was very much in evidence. As the number purchased increased, the price decreased.

Since more pigs are produced in the spring than in the fall, there was less variation in the prices received for fall pigs. The best time to market fall pigs during the 1968 to 1972 period was during the months of December and February. Pigs farrowed in July, August, and September can easily be fed out in time to get them on a high market. The spread in prices paid for hogs between February and April was about $2.12 per hundredweight.

The number of sows purchased at the eight large markets and the purchase price

FIGURE 12-4. Number of barrows and gilts purchased and price—average of 8 markets—1968–1972.

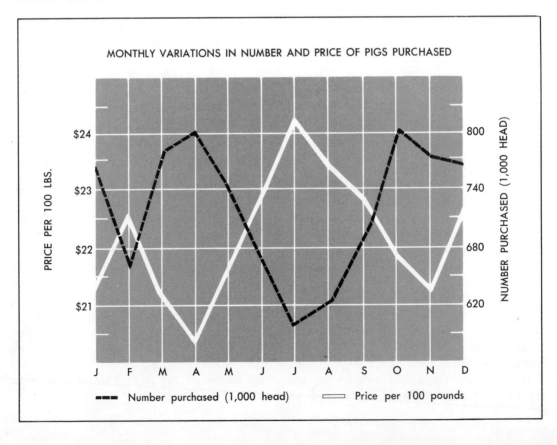

MONTHLY VARIATIONS IN NUMBER AND PRICE OF PIGS PURCHASED

PRICE PER 100 LBS.

NUMBER PURCHASED (1,000 HEAD)

J F M A M J J A S O N D

━━━ Number purchased (1,000 head)　　▭ Price per 100 pounds

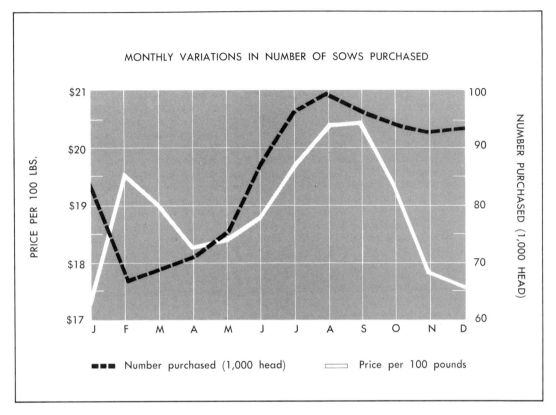

FIGURE 12-5. Number of sows purchased and price—average of 8 markets—1968–1972.

by months are presented in Figure 12-5. Sow prices were highest in August, September, and July. Low prices were paid for sows in January, December, and November. The spread between September and January was $3.03.

The law of supply and demand appeared to be functioning in December, January, February, and March. As purchases decreased, prices were increased. This was not true for the May to September period. The price paid for sows followed the trend of prices paid for barrows and gilts. The number of sows marketed each month represented only a small percentage of total hogs marketed and did not greatly affect the market price.

Commercial Hog Slaughter by Months

Shown in Table 12-3 is a summary of commercial hog slaughter by months in the United States under federal inspection. As expected, slaughter numbers are low during the months of high prices. The smallest numbers were slaughtered during that five-year period in June, February, and July; largest numbers were slaughtered in October, March, and November.

Hog producers have made considerable progress in their attempt to distribute marketings throughout the year. Marketings during June, July, August, January, February, and March have increased greatly since 1948.

TABLE 12-3 COMMERCIAL HOG SLAUGHTER UNDER FEDERAL INSPECTION BY MONTHS, 1968-1972
(in millions)

Month	1968	1969	1970	1971	1972	Average per Month
January	6.5	6.8	6.2	7.5	6.4	6.7
February	5.7	6.2	5.5	6.4	6.3	6.0
March	6.2	6.8	6.4	8.3	7.8	7.1
April	6.5	6.9	6.7	7.8	6.7	6.8
May	6.4	6.0	5.9	6.9	6.8	6.3
June	5.1	5.6	5.7	7.0	6.3	5.9
July	5.5	5.7	5.8	6.2	5.3	5.7
August	5.9	5.7	6.0	6.9	6.5	6.2
September	6.3	6.6	7.0	7.4	6.4	6.7
October	7.4	7.1	7.7	7.2	7.0	7.3
November	6.6	5.8	7.4	7.6	7.0	6.9
December	6.6	6.3	8.0	7.5	6.2	6.9
Total	74.7	75.7	78.2	86.7	78.8	78.8

Source: *Livestock and Meat Statistics*, U.S.D.A., 1973.

Marketings in December have decreased. In 1948, 6.7 million hogs were marketed in July and August as compared to 13.6 million in November and December. In 1972 11.8 million were marketed in July and August as compared to 13.2 million in November and December.

During the five-year period (1963–1967) fewest hogs were marketed in June and July. Largest numbers were marketed in March, October, and November. For the five-year period an average of 9.7 million were marketed during July and August, whereas 12.3 million were marketed during November and December.

Time of Farrowing

It usually requires a minimum of four and one-half to five months to grow out pigs to market weight. In order to market pigs when the price is high, the sows must usually be bred to farrow four and one-half to five months in advance of the anticipated market date. If "least cost" rations are fed, the length of time between farrowing and marketing is increased.

Shown in Table 12-4 is a summary of the number of sows farrowing during each of the four quarters of the year during the period from 1968 to 1972.

In 1951, 12.8 percent of the sows farrowed during the December to February period. The average for the 1968 to 1972 period was 18.8 percent. There was little variation in individual years during the period.

Forty-eight percent of the sows in 1951 farrowed during the March to May period. Only 31.6 percent farrowed during these months in 1972. For the five year average it was 33 percent.

It should be noted that 52.1 percent of the pig crop during the five-year period was farrowed during the spring and 47.9 percent in the fall.

An equal distribution of farrowings during the four quarters of the year would do

TABLE 12-4   DISTRIBUTION OF FARROWINGS BY QUARTERS 10 CORN BELT STATES, 1968-1972
Percentage of Annual Pig Crop

Year	Dec.-Feb.	Mar.-May	June-Aug.	Sept.-Nov.	Spring Crop	Fall Crop
1968	18.0	33.8	24.0	24.2	52.3	47.7
1969	19.3	33.4	23.5	23.8	52.8	47.2
1970	17.8	33.3	24.0	24.9	51.6	48.4
1971	20.2	33.1	22.9	23.8	53.0	47.0
1972	18.8	31.6	24.0	25.6	50.8	49.2
Average	18.8	33.0	23.7	24.5	52.1	47.9

Source: *Livestock and Meat Statistics*, U.S.D.A., 1973.

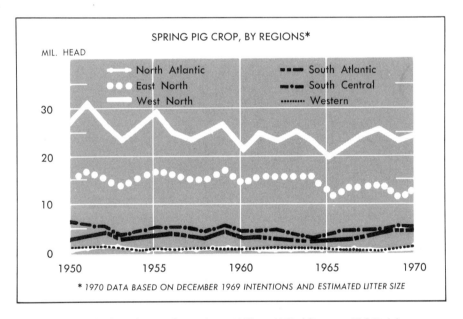

FIGURE 12-6. Spring pig crop by regions, 1950 to 1970 (*Courtesy U.S.D.A.*)

much to decrease the seasonal market price fluctuations. Multiple-farrowing programs are recommended.

WEIGHT TO SELL

The consumer wants smaller and leaner cuts of pork and less lard, and the consumer must be satisfied if pork is to meet the com-petition of other meats. Consequently, farm-ers must grow meatier hogs and market them at lighter weights.

Heavy Hogs Are Discounted

In the past there has been nearly twice as much lard in the carcass of a 300-pound hog as there was in the carcass of a 200-

pound pig. A 200-pound live hog would produce about 30 pounds of lard and backfat. The 300-pound hog would yield about 55 pounds.

A comparison of lard and hog prices over a period of years indicates the changes that have come about in the lard market. In the 1910 to 1919 period, the price of prime-steam lard in Chicago was $4.40 per hundredweight above the price of hogs. Up until the end of World War II, heavy butcher hogs sold on an average for nearly as much per pound as did 200-pound hogs.

Wisconsin and Missouri tests show that meaty hogs can be fed to heavier weights with less undesirable results than can fat-type hogs. In Missouri tests, the bacon from 300-pound hogs was almost equal to that from 200-pound pigs in leanness and in visual desirability. The pork chops of the lighter hogs had 2 percent less cooking loss and were more tender. The hams from the 300-pound hogs lost about 4 pounds of fat in trimming as compared to 2.7 pounds for the lighter hams. The heavier hogs, however, had 13.8 pounds of lean meat in the trimmed hams as compared to 9.7 pounds in the hams of the lighter hogs.

The heavier hogs had 1.7 inches of backfat and 49.4 percent of lean cuts as compared to 1.4 inches of backfat and 51.8 percent of

TABLE 12-5 PRODUCTION OF LEAN AND FAT IN HOGS OF DIFFERENT WEIGHTS

Gain in Weight (Lbs.)		Increase in Carcass (Lbs.)	
From	To	Lean	Fat
150	175	6.9	10.5
175	200	6.4	11.9
200	225	6	13.2
225	250	5.5	14.2
250	275	5	15.6
275	300	4.5	16.7

Source: Vocational Agriculture Service, University of Illinois.

lean cuts in the carcasses from the 200-pound pigs. It appears that the discount currently charged to heavy hogs may be in part unfair.

Swine specialists are indicating that today's meat-type hog can be fed to a heavier weight than the hogs of ten years a ago. The average hog marketed in 30 states in the U.S. in December 1973 weighed 7 pounds more than those marketed a year earlier. Greatest increase in weight was found in the Corn Belt states. Following are the weights of hogs marketed in December 1973 in selected states:

Iowa	258 pounds
Missouri	251 pounds
South Dakota	260 pounds
Indiana	263 pounds
Illinois	265 pounds

The average weight of all hogs marketed in the U.S. in 1972 was 239 pounds. The average weight of those marketed in December 1973 was 248 pounds, and they had only 6.4 pounds of lard per 100 pounds of live weight. Ten years ago the average hog marketed weighed 230 pounds and had 12.4 pounds of lard for each 100 pounds of live weight.

Presented in Table 12-5 is a summary of the production of lean and fat in hogs of different weights.

The average prices paid at seven midwest markets for slaughter hogs at various weights during 1971 and 1972 are shown in Table 12-6. Note that the first three U.S.D.A. grades barrows and gilts weighing 200 to 220 pounds brought only an average of 9 cents more per 100 pounds than did slaughter pigs weighing 220 to 240 pounds. There was, however, a difference of $1.03 per hundredweight between slaughter pigs weighing 200 to 220 pounds and those weighing 240 to 270 pounds.

T A B L E 12-6   AVERAGE PRICES PAID FOR SLAUGHTER HOGS AT VARIOUS WEIGHTS, 7 MARKETS, 1971-1972 AVERAGES

| | U.S. No's. 1-3 | | U.S. No's. 2-4 |
| | 200-220 Pounds | 220-240 Pounds | 240-270 Pounds |
Market			
South St. Paul	$22.96	$22.94	$21.32
National Stock Yards	23.55	23.50	22.40
Indianapolis	23.71	23.65	22.64
Interior Iowa and S. Minn.	22.65	22.51	21.82
Peoria	23.55	23.41	22.50
Omaha	23.33	23.24	22.50
Sioux City	23.18	23.10	22.54
Average	23.28	23.19	22.25

Source: *Livestock and Meat Statistics*, U.S.D.A., 1973.

Shown in Figure 12-7 are the average prices paid for barrows and gilts and for sows by months during 1971 and 1972 at interior Iowa and southern Minnesota markets. Slaughter pigs averaging 200 to 220 pounds sold for an average of $22.79 during the two-year period, whereas pigs weighing 220 to 240 pounds averaged $22.67 per hundredweight. Heavy barrows and gilts averaging 240 to 270 pounds sold for an average of $21.89. Sows weighing 330 to 400 pounds brought an average of $18.83 per hundredweight.

The spread between the U.S. 1 and 2 barrows and gilts weighing 200 to 220 pounds was only 12 cents above the price paid for barrows and gilts graded U.S. 2 and 3 weighing 220 to 240 pounds. There was a spread of 78 cents, however, between the price paid for slaughter pigs weighing 220 to 240 pounds and those weighing 240 to 270 pounds. It was more profitable to market pigs at 240 pounds and above during 1971 and 1972 than to sell them under 220 pounds.

Light Hogs Make Cheaper Gains

Swine producers are interested in economical use of feed, and it is known that young animals make cheaper gains. Tests conducted at the University of Michigan indicated that lighter hogs used less feed and made cheaper gains than did heavier hogs. A summary of these data is presented in Table 12-7.

The Michigan tests indicated that about 1.5 more pounds of feed were required to produce 10 pounds gain on a meaty 260-pound pig than on a 220-pound pig. About 5.5 pounds of additional feed were required to put 10 pound gain on a fat-type 260-pound pig. Pounds of feed and feed costs increased as the pigs became heavier. Meaty hogs made more economical gains than did average or fat-type pigs.

Swine producers should analyze carefully the feed costs and the market prices in determining the best weight to sell. During the past two years it has paid to feed meat-

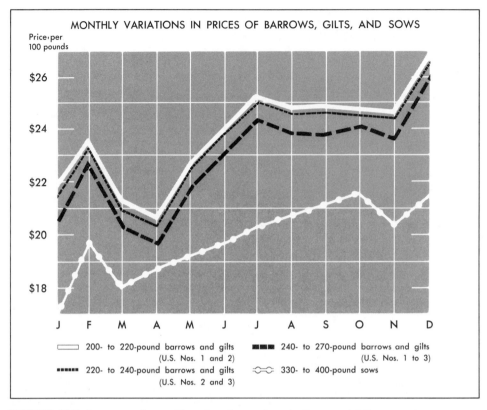

FIGURE 12-7. Average prices paid at interior markets in Iowa and Minnesota for barrows and gilts and packing sows—1971–1972 average.

T A B L E 12-7   FEED COSTS OF HOGS AT VARIOUS WEIGHTS

Weight in Pounds	Meaty Type		Average Type		Fat Type	
	Pounds Feed Per Pound Gain	Feed Cost Per Pound Gain (¢)	Pounds Feed Per Pound Gain	Feed Cost Per Pound Gain (¢)	Pounds Feed Per Pound Gain	Feed Cost Per Pound Gain (¢)
220 to 229	3.80	22.8	4.00	24.0	4.20	25.2
230 to 239	3.85	23.1	4.07	24.4	4.30	25.8
240 to 249	3.85	23.1	4.15	24.9	4.45	26.7
250 to 259	3.90	23.4	4.25	25.5	4.60	27.6
260 to 269	3.95	23.7	4.35	26.1	4.75	28.5

Source: Michigan State University, 1973.

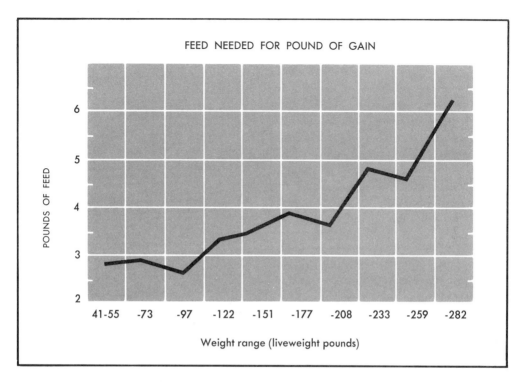

FIGURE 12-8. Feed required per pound of gain for hogs. (*Courtesy U.S.D.A.*)

type pigs to weights above 220 pounds. When feed costs are high and market prices on live hogs are low, it usually pays to sell light. When feed costs are low and slaughter prices high, it is usually more profitable to sell pigs at 240 pounds or above.

MARKET CLASSES AND GRADES OF HOGS

Market hogs are classified in terms of sex, use, weight, and value. Classes are provided for barrows and gilts, sows, stags, and boars. Animals are classified according to use, as slaughter hogs, slaughter pigs, stockers, and feeders. The weights vary with the classes according to sex and use.

Previous to 1952 most hogs were classified as choice, good, medium, and cull. On September 12, 1952, the U. S. Department of

Agriculture announced new grade standards for barrows and gilts. The five grades were Choice No. 1, Choice No. 2, Choice No. 3, Medium, and Cull. These grades were recommended for voluntary use by farmers, dealers, packers, and market specialists. Revised grades for slaughter hogs and pork carcasses were announced in July 1955.

The 1955 revision changed the names of the grades to U. S. No. 1, U. S. No. 2, U. S. No. 3, Medium, and Cull.

A further revision of live hog grades was announced by the U. S. Department of Agriculture in June, 1968. The revised standards for slaughter barrows and gilts correspond directly to recently revised grades on pork carcasses. The standards provide for four numerical grades, U. S. No. 1, U. S. No. 2, U. S. No. 3, and U. S. No. 4. The standards for the last three grades are equivalent to the requirements for the previous U. S. No. 1, U. S. No. 2, and U. S. No. 3 grades. The new U. S. No. 1 grade is used to identify only the most superior meat-type hogs. The previous U. S. Medium and U. S. Cull grades have been replaced with a U. S. Utility grade.

A systematic grading procedure makes it possible for hogs to be marketed according to the value of their carcasses or according to the value of the animals as stockers or feeders. Market grades of hogs serve the same purposes as grades of corn or grades of butter.

Shown in Table 12-8 are the market classes and grades of hogs and pigs. Variations in weight and grade classifications among markets are common.

U.S.D.A. Hog Grades

The five grades established by the U. S. Department of Agriculture apply to hogs on foot and on the hook. The degree of finish, the quantity and the quality of the lean meat, and the percentage of fat determine the grade. The four top grades are numbered from one to four according to the percentage of lean meat in the carcass. The other grade, Utility, is so called because of unacceptable lean quality and/or unacceptable belly thickness. The use of these grades is voluntary.

The thickness of backfat in relation to the length, weight, and dressing percentage of the carcass serves as the basis for determining the grade. Shown in Figure 12-9 is an illustration of the method used in determining carcass length.

The relationship between average thickness of backfat, carcass length or weight, and grade for carcasses with muscling typical of their degree of fatness is shown in Figure 12-10. The official specifications for the U. S. standards for grades follow:

U. S. No. 1. Hogs in this grade will produce carcasses with superior lean quality and belly thickness, and a high percentage

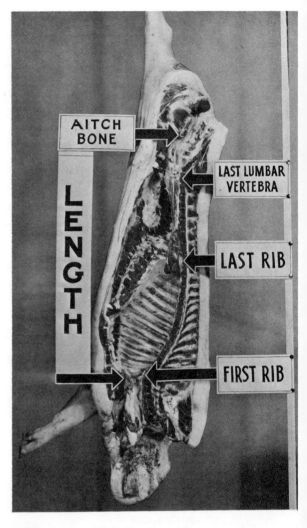

FIGURE 12-9. The method used in determining carcass length. (*Courtesy University of Wisconsin, Wisconsin State Board for Vocational Education, and Oscar Mayer and Co.*)

of lean cuts. The width through the hams is nearly equal to the width through the shoulders and both are wider than the back. The sides are long and smooth. The rear flank is slightly full and has less depth than the fore flank.

The chilled carcass should produce about 38 percent of the live weight in the four lean cuts. Backfat thickness may vary

T A B L E 12-8   MARKET CLASSES AND GRADES OF HOGS AND PIGS

Use	Sex	(Lbs.) Weights	Grades
HOGS			
Slaughter hogs	Barrows	Under 180	U.S. No. 1
	and gilts	180-240	U.S. No. 2
		240-300	U.S. No. 3
		300 and over	U.S. No. 4
			U.S. Utility
	Sows	270-300	
		300-330	U.S. No. 1
		330-360	U.S. No. 2
		360-400	U.S. No. 3
		400-450	U.S. No. 4
		450-500	U.S. Utility
		500-600	
		600 and over	
	Stags	All weights	Ungraded
	Boars	All weights	Ungraded
	Unclassified	All weights	Ungraded
Feeder and	Barrows	120-140	Choice
stocker hogs	and gilts	140-160	Good
		160-180	Medium
			Common
PIGS			
Slaughter pigs	All classes	Under 30	Ungraded
		30-60	Ungraded
		60-80	Good
		80-100	Medium
			Cull
	Barrows	100-200	Choice
	and gilts		Good
			Medium
			Cull
Feeder pigs	Barrows	Under 80	U.S. No. 1
	and gilts	80-100	U.S. No. 1
		100-120	U.S. No. 3
			U.S. No. 4
			U.S. Utility

from 1 to 1.6 inches as carcass length increases from 27 to 36 inches, or as carcass weight increases from 120 to 255 pounds. Superior muscling may compensate for slight overdevelopment of fatness.

U. S. No. 2. Carcasses from hogs of this grade have an acceptable quality of lean and a slightly high yield of lean cuts. From 36 to 37.9 percent of the live weight will be in the four lean cuts. The average backfat thickness will increase from 1.3 to 1.9 inches with an increase of carcass length from 27 to 36 inches, or increased weight of carcass from 120 to 255 pounds.

U. S. No. 3. These hogs have an acceptable quality of lean and a slightly low yield of lean cuts—34 to 35.9 percent of the live weight. The maximum average backfat thick-

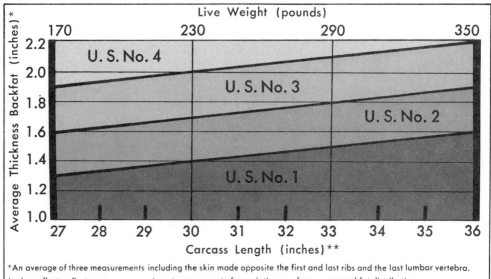

FIGURE 12-10. Relationship between average thickness of backfat, weight or carcass length, and grade for barrows and gilts with muscle typical of their degree of fatness. (*Courtesy U.S.D.A.*)

ness increases from 1.6 to 2.2 inches with increases in carcass length from 27 to 36 inches, or increases in weight from 120 to 255 pounds. A slight excess of fat may be compensated by superior muscling.

U. S. No. 4. Barrows and gilts of this grade have carcasses acceptable in quality of lean, but with a lower expected yield of lean cuts—less than 34 percent of the live weight. The average backfat thickness increases from 1.9 to about 2.5 inches as carcass length increases from 27 to 36 inches, or as carcass weight is increased from 120 to 255 pounds.

U. S. Utility

This grade includes all hogs that have characteristics indicating less development of lean quality than the minimum requirements

for the U. S. No. 4 grade. Hogs will usually have a thin covering of fat. The sides may be wrinkled and the flanks shallow and thin.

The standards presented in Figure 12-10 assume that a 170-pound live hog will yield a 120-pound carcass, a 230-pound hog will yield a 165-pound carcass, a 290-pound hog will yield a 205-pound carcass, and a 350-pound hog will yield a 255-pound carcass.

Presented in Figures 12-16 through 12-20 are pictures showing the carcasses of the top four U. S. grades.

Primal cuts. Packers are able to pay higher prices for hogs that grade U. S. No. 1 because the cuts of meat from the carcasses are worth more to the consumer. It is estimated that not more than 60 percent of the hogs delivered to the packing plants of the nation will grade U. S. No. 1. Producers of hogs and meat packers can do much to im-

FIGURE 12-11. *Above:* Side view of a U.S. No. 1 slaughter hog. *Right:* Rear view of a U.S. No. 1 slaughter hog. (*Courtesy U.S.D.A.*)

FIGURE 12-12. *Below:* Side view of a U.S. No. 2 slaughter hog. *Right:* Rear view of a U.S. No. 2 slaughter hog. (*Courtesy U.S.D.A.*)

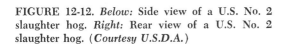

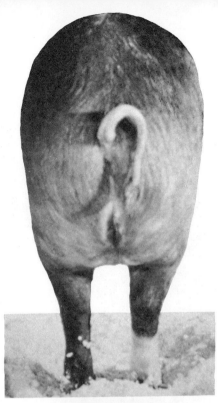

FIGURE 12-13. *Above:* Side view of a U.S. No. 3 slaughter hog. *Right:* Rear view of a U.S. No. 3 slaughter hog. (*Courtesy U.S.D.A.*)

FIGURE 12-14. *Below:* Side view of a U.S. No. 4 slaughter hog. *Right:* Rear view of a U.S. No. 4 slaughter hog. (*Courtesy U.S.D.A.*)

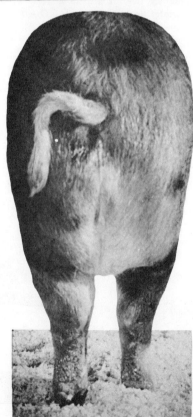

FIGURE 12-15. Roy B. Keppy, Davenport, Iowa, and his Grand Champion Barrow at the 1973 National Barrow Show. This crossbred barrow was an example of the elite of U.S. No. 1 slaughter hogs. (*Courtesy Geo. A. Hormel & Co.*)

prove the quality of pork available at the meat markets. Improvement in quality will result in greater profits.

SELLING HOGS ACCORDING TO CARCASS YIELD AND GRADE

The government grades represent an attempt to satisfy hog producers who have felt that packers were not paying for quality hogs. Market quotations have usually been based upon the weight of the hogs, rather than upon quality. It is difficult to grade hogs on foot, and most packers have preferred to buy them by the pound in droves or loads. There has been little incentive for hog breeders to improve carcass quality, for all hogs sold at the same price.

Advantages of Selling Hogs by Carcass Grade

The demand of the housewife for leaner pork cuts and the loss of our lard market make it imperative that we improve the quality and popularity of pork. Selling hogs on the basis of carcass grade and yield encourages farmers to produce quality hogs and eliminates the wasteful practice of "filling" hogs to get the maximum market weight. Selling by carcass grade provides the producer with an unbiased evaluation of the quality of his hogs, and it also permits the tracing of diseased, injured, and inferior pork carcasses to the producers who are responsible.

Hogs have been sold by carcass grade in Denmark, Sweden, Great Britain, and Canada, and the method has proved efficient and practical. Many packing plants in the Corn Belt have purchased hogs on this basis for several years. In general, breeders with high quality hogs profit by selling their hogs on a carcass-grade-and-yield basis, whereas hog producers with below-average grade hogs profit by selling them in the traditional manner.

During 1972 an Illinois producer marketed 1,244 slaughter pigs on a carcass grade and yield basis. They sold for over $3,100 more than they would have brought on the hoof.

Marketing by grade and yield provides the seller with valuable information for use in selecting breeding stock and in managing the swine enterprise.

Disadvantages of Selling Hogs by Carcass Grade

One of the chief reasons for delay in selling hogs by carcass grade is that the

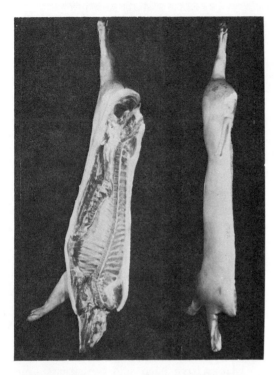 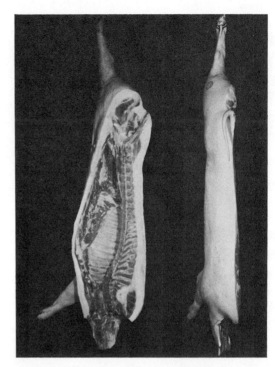

FIGURE 12-16 (*Above*). Carcass of U.S. No. 1 slaughter barrow or gilt. **FIGURE 12-17** (*Below*). Carcass of U.S. No. 2 slaughter barrow or gilt. (*Figures 12-16 and 12-17 courtesy U.S.D.A.*)

FIGURE 12-18 (*Above*). Carcass of U.S. No. 3 slaughter barrow or gilt. **FIGURE 12-19** (*Below*). Carcass of U.S. No. 4 slaughter barrow or gilt. (*Figures 12-18 and 12-19 courtesy U.S.D.A.*)

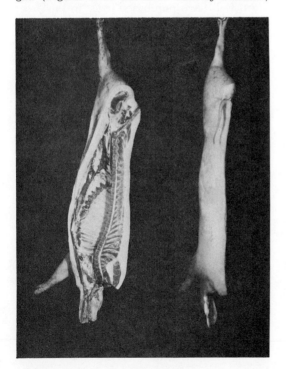 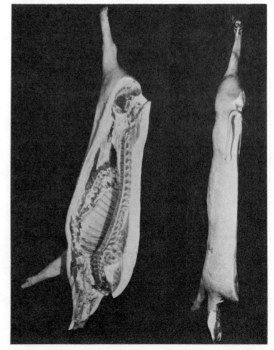

FIGURE 12-20. Degrees of muscling in pork carcasses. (*From left to right*) Very thick, thick, moderately thick, slightly thin, and thin. (*Courtesy U.S.D.A.*)

packer buyers have had neither the facilities nor the personnel to do the job. The procedure requires more time and physical facilities than does the traditional system. Buying by carcass grade is less flexible than the present system. There is less opportunity for both the buyer and seller to bargain. Usually the grading is done in the absence of the seller, who may later question the results. Many farmers are not sufficiently informed in regard to carcass quality to accept the packers' judgment. Another serious objection to the system is that the seller must wait until the hogs have been slaughtered and processed before he can receive his check.

Selling by grade and yield is more costly than other methods, and the seller has to agree with the standards set by the packer. The system is more complicated and a uni-

form tattoo system must be used. The advantages of the system, however, far outweigh the disadvantages.

The Mechanics of Selling by Yield and Grade

Shown in Figure 12-21 is a copy of a hog carcass grade and yield report being used by one packing plant when the producer chooses to sell his hogs on the basis of yield and grade. While the account of sale is fictitious, it illustrates the procedures followed.

Presented in Figure 12-22 is the average backfat guide used by this packing company with its butcher hog carcass grades. In the following paragraphs Carroll Plager of Geo. A. Hormel & Co. of Austin, Minnesota, describes the marketing of hogs by carcass grade and yield. The original material is included in a Hormel publication entitled *Marketing Hogs by Carcass Grade and Yield.* It is reprinted by special permission.

Actual yield can be defined as the amount of carcass produced for the live weight purchased. A yield of 72 per cent means the hog hangs up 72 pounds of carcass for each 100 pounds of live weight.

(Example—200 lb. live wt. hog yields a carcass of 144 lbs., or $\frac{144}{200}$ = 72 per cent).

Standard yields are based on experience gained by securing actual yields on large numbers of each class and weight of hogs.

Different Standard Yields

There are different standard yields because the heavier the animal, the more it will tend to dress or yield—each weight class, therefore, has a different standard yield. Butcher hogs tend to yield higher than packing sows of the same weight because butcher hogs, generally, carry a higher degree of finish than a comparable weight of packer. The underline of many sows requires trimming. Early stages of pregnancy are not uncommon.

For example, if it were established that the standard yield for a 200-pound hog is 72 per cent of 144 pounds of carcass; then, if a 200-pound hog has a carcass weighing 148 pounds, the actual yield would be 74 per cent, and this carcass would be more valuable than the 144-pound standard carcass. It is desirable to have more pounds of carcass weight per live cwt. than standard, thus increasing the value of the hog.

Carcass Weight Range

For each live weight range, there is an equivalent carcass weight range, and each carcass is weighed and then priced in its appropriate carcass weight range. The top live price range of butchers is ordinarily grouped together as one equivalent carcass range with an average standard yield for the range. The base prices for carcass Grade & Yield hogs are the same as the Hormel live price card in use when the hogs are sold to the buyer.

The live hog prices are converted to carcass prices by use of the standard yield, which is the live to carcass price conversion factor. For example, if the base market is $30.00 per cwt. for a 200–230 pound hog and the standard yield is 72 per cent, the value of the carcass is $41.66 per cwt. This is the carcass price for a No. 3 base grade carcass weighing 144 to 168 pounds.

(Example: Live Price ÷ Standard Yield = Meat Price OR $\frac{\$30.00}{.72}$ = $41.66.)

The value for each per cent of yield over or under the standard in this example is 41⅔ cents per cwt. on a $30.00 base market. The value of one per cent of yield will be higher or lower, according to the live market. In this example, a hog yielding two per cent over standard will bring an additional 83 cents per cwt.

Price differentials for grade differences are revealed in the carcass prices. The No. 3 regular grade is the base grade and receives the base price. The price differentials between grades reflect the quality and the cutout value differences between grades and are based on many tests which have been conducted over a period of time.

In Figure 12-22 is an example of 10 head sold on the grade-and-yield basis. The appearance of the Hog Carcass Grade-and-Yield Report may vary depending on which Hormel plant slaughtered the hogs. The hog carcass Grade & Yield report may be completely calculated and printed out by a computer or it may be calculated and the report prepared by hand. While the appearance of the report will vary, the content and results are identical.

These conclusions can be made from the hog carcass Grade & Yield report:

Conclusions

1. $30.00 No. 3 market.
2. Live weight 2,150 pounds.
3. Carcass weight 1,591 pounds.
4. Yield difference of plus 2.00 per cent.
5. Four carcasses graded No. 1; three graded No. 2; two graded No. 3; and one graded No. 4.
6. 83 cents per cwt. overage paid for yield.

AVERAGE BACKFAT GUIDE HORMEL BUTCHER HOG CARCASS GRADES

	1 Premium	2 Desirable	3 Regular (Base Grade)	4 Over-Finished	6 Medium And Cull
180-200	1.2-Under	1.3-1.5	1.6-1.9	2.0-Over	Value Based On Condition
200-240	1.3-Under	1.4-1.6	1.7-2.0	2.1-Over	
240-300	1.3-Under	1.4-1.6	1.7-2.1	2.2-Over	
300-360	1.4-Under	1.5-1.7	1.8-2.2	2.3-Over	
360-450	1.5-Under	1.6-1.8	1.9-2.3	2.4-Over	

HORMEL PACKING SOW CARCASS GRADES

	1 Premium	2 Desirable	3 Regular	4 Over-Finished	6 Medium And Cull
330-Under	1.4-Under	1.5-1.7	1.8-2.0	2.1-Over	Value Based On Condition
330-400	1.5-Under	1.6-1.8	1.9-2.2	2.3-Over	
400-Up	1.6-Under	1.7-1.9	2.0-2.4	2.5-Over	

FIGURE 12-21 (*Above*). Average backfat guide Hormel butcher hog carcass grades.

FIGURE 12-22 (*Below*). Hog Carcass Grade and Yield Report (*Figures 12-21 and 12-22 courtesy John Phillips, Geo. A. Hormel & Co.*)

Geo A Hormel & Co General Office Austin Minn

✤Hormel✤

HOG CARCASS GRADE AND YIELD REPORT

TATTOO	DATE BOUGHT	DATE KILLED	COUNT	MARKET	LIVE WEIGHT	HOW BOUGHT	CARD NO	BUYER NO
11-100	1-2-74	1-3-74	10	30.00	2150	G & Y	10	01

LIVE WEIGHT RANGE	COUNT	GRD	HOT WEIGHT	MEAT PRICE	EXTENDED AMOUNT	LIVE SORT	GRADE CODE DESCRIPTION
210-220	4	1	638	43.41	276.95	10 B 215	No. 1 · Premium
210-220	3	2	477	42.66	203.48		No. 2 · Desirable
210-220	2	3	318	41.66	132.47		No. 3 · Regular
210-220	1	4	158	39.66	62.66		No. 4 · Overfinished
							No. 6 · Medium

LIVE RANGE CODES
B · BUTCHERS
S · SOWS

AVOID LIVESTOCK BRUISES
ADD TO YOUR PROFITS
02

COUNT BY GRADES ▶	NO 1	NO 2	NO 3	NO 4	NO 6		
	4	3	2	1			

ACTUAL YIELD	STD YIELD	YIELD DIFF	TOTAL HOT WGH	TOTAL VALUE	REMARKS	COST	
74.00%	72.00%	2.00%	1591	675.56		675.56	
STATE WEIGHING	CHECK OFF	PICK UP & TRK	OTHER DED	NET VALUE		YLD P&L	CWT
	.50			675.06		17.91	.83
TOTAL VALUE CWT	NET VALUE CWT	**PAY THIS AMOUNT** ➤				GRD P&L	CWT
31.42	31.40					12.77	.59

CUT-OUTS

7. 59 cents per cwt. overage paid for grade.

8. Total value per cwt. equals $31.42. The base No. 3 market was $30.00 and, thus, this lot earned $1.42 per cwt. over the base No. 3 market.

9. Checkoff—Five cents per hog, of which four cents goes to the National Pork Producers Council and one cent to the National Live Stock and Meat Board. We actively cooperate with these organizations and support their programs.

Federal regulations covering carcass grade and weight livestock marketing became effective in April, 1968. It was estimated at that time that 10 percent of the cattle and 2.5 percent of the slaughter hogs in the nation were marketed on the basis of grade and weight.

The regulations require all meat packers who buy livestock on the basis of carcass grade, carcass weight, or a combination of the two to do the following:

1. Make known to the seller, prior to the sale, significant details of the purchase contract.
2. Maintain identity of each seller's livestock and carcasses.
3. Maintain sufficient records to substantiate settlement for each purchase transaction.
4. Buy and make payment on the basis of carcass prices.
5. Make payment on the basis of U.S.D.A. carcass grades, or make available to the seller detailed written specifications for any other grades used in determining final payment.
6. Make payment on the basis of actual (hot) carcass weight before carcasses are shrouded.
7. Carcasses must be graded by the close of the second business day following slaughter.
8. Use hooks, rollers, gambrels, and similar equipment of uniform weight in weighing carcasses from the same species of livestock in each packing plant, and include only the weight of this equipment in tare weight.

CHOOSING A MARKET

Assume that during early March you have 200 fall-farrowed pigs ready for market as U.S. No. 1 slaughter hogs. Where should you sell them? Several times each year almost every hog raiser is confronted with this or a similar problem regarding the best hog market. There is no stock answer to the question. The answer depends upon where you are located.

Markets Available

In every community there are local hog buyers, dealers, and livestock auctions. In many communities you can consign your hogs and sell through a cooperative shipping association. Some meat packers have concentration yards or buying plants in your communities. You can truck your hogs to interior packing companies located 10 to 100 miles away, or you can truck them to public stockyards at a central market, such as Joliet, Illinois, or South St. Paul, Minnesota. You can ship direct to the packer, or you can sell them through a commission firm. Which is best?

FIGURE 12-23. A group of 70 barrows and gilts at a slaughter plant in East St. Louis. Sixty-seven head graded U.S. No. 1. (*Maxwell photograph. Courtesy Harold Lucie.*)

Factors in Choosing a Market

A number of factors are involved in the selection of a market. Distance, transportation problems, shrinkage, methods of grading, handling and selling charges, dependability, and price quotations are perhaps the most important.

To market effectively you need to know the price at each available market of the grade you have to sell. You may have a half dozen or more markets to choose from. The U. S. Department of Agriculture Market News Service, the various cooperating radio stations, and the newspapers supply market information. At times a telephone call may result in several dollars more in profits.

In addition to the kind of hogs that you have to sell, you need to know the cost of transporting them and the shrinkage you may expect. If you consign to a terminal market, there will be a commission or handling charge. This may amount to 20 to 30 cents per hundredweight. The transportation cost will vary with the number of hogs to be transported and with the distance to be covered. The shrinkage and loss due to injury or death will vary with the method of transporting and with the distance. These losses may be as little as 10 cents or as large as 60 cents per hundredweight.

Major Markets

Public stockyards. We emphasized in Chapter 1 that Iowa, Illinois, and neighbor states dominate in the production of hogs. It is logical that the most and the largest public stockyards and packing plants are located in this area. Shown in Table 12-9 is a list of the major public stockyards in the Corn Belt area and the numbers of salable hogs received by each stockyard during 1969, 1970, 1971, and 1972. The public stockyards at Sioux City, Omaha, and South St. Paul, and the National Stockyards at St. Louis received the largest numbers of salable hogs in 1972. The St. Joseph and Peoria yards were next in order of salable hogs received.

Interior Packing Plants

At one time a large percentage of the hogs marketed in this country were shipped to public stockyards where they were purchased by packers for immediate slaughter or for reshipment. The growth of the interior packing industry has changed our sys-

T A B L E 12-9 SALABLE RECEIPTS OF HOGS AT LEADING PUBLIC STOCKYARDS BY MARKETS

Market	1969 Mil.	1970 Mil.	1971 Mil.	1972 Mil.
Sioux City	1.86	1.90	2.00	1.76
Omaha	1.78	1.82	1.84	1.50
National Stockyards	1.64	1.57	1.65	1.44
South St. Paul	1.55	1.50	1.60	1.36
St. Joseph	1.27	1.30	1.28	1.17
Peoria	.73	1.11	1.30	1.09
Sioux Falls	.80	.86	1.08	1.04
Kansas City	1.06	.97	.98	.78
Indianapolis	.94	.93	.95	.74
Mexico, Mo.	.35	.36	.43	.39

Source: *Agricultural Statistics 1973*, U.S.D.A., 1973.

tem of marketing. The interior packing plants in Iowa and Southern Minnesota received, during 1972, a total of 20,506,000 head of hogs. The largest receipts at any of the public stockyards for any one year during the 1969 to 1972 period was at the Sioux City yards in 1971, when 2,000,077 hogs were received.

The receipts of hogs at the Iowa and Southern Minnesota interior packing plants during the 1969 to 1972 period are shown in Table 12-10. The receipts of hogs at the

T A B L E 12-10 HOG RECEIPTS AT IOWA AND SOUTHERN MINNESOTA INTERIOR PACKING PLANTS, 1963 TO 1972

Year	Receipts (Mil.)	Year	Receipts (Mil.)
1963	17.8	1968	20.9
1964	18.0	1969	20.2
1965	17.3	1970	21.2
1966	18.4	1971	22.7
1967	20.4	1972	20.5

Source: *Livestock and Meat Statistics*, U.S.D.A., 1973.

interior plants in Iowa and Southern Minnesota in 1972 totaled 20,506,000 head. The combined receipts at the St. Paul, St. Louis, Indianapolis, Omaha, Sioux City, and St. Joseph public stockyards in 1972 amounted to 7,970,344 hogs.

A large percentage of the hogs produced in this country are sold directly to the packing plant without passing through a public market.

Nearly 70 percent of all slaughter hogs produced in the nation in 1971 were purchased by packers direct, or from country dealers. Approximately 17 percent went through terminal markets, and 13 percent were purchased in auction markets.

USING THE FUTURES MARKET IN HOG MARKETING

The futures market on pork bellies and on live hogs offers producers, packers, and others a means of taking much of the price risk out of the business and a means of estab-

FIGURE 12-24. Consumer demand for pork depends upon quality and price. These hogs are being sold in the Central Livestock Association market in South St. Paul, Minnesota. (*Courtesy Central Livestock Association, Inc.*)

lishing prices in advance. Many believe that, as pork production becomes more specialized with fewer but larger operations, increased participation in the futures market will take place.

Factors Affecting Participation in the Futures Market

Three factors will determine the extent that hog producers will use the futures market.

1. The magnitude or scope of the hog enterprise and its relationship to the total farm business.
2. The willingness and ability of the producer to bear the risk of falling prices.
3. The margin or per unit profit expected and necessary for successful operation of the enterprise.

In the past a swine enterprise has been a part of a diversified farm program. Even in the Corn Belt, the average farmer sold fewer than 250 hogs per year. It has been estimated that in the next decade nearly 50 percent of the hogs produced will be from farms marketing 1,000 to 5,000 or more hogs per year. These operators will require considerable capital, and smaller margins of profit per unit of production will be expected. These operators are vulnerable to market fluctuations. They may find the futures market a valuable tool in their swine marketing program.

Advantages of Using the Futures Market

The futures market can protect the swine grower against declines in the cash market for slaughter hogs. It can provide the feeder-pig producers with an indication of the profitability of selling pigs as feeder pigs or as slaughter hogs.

Disadvantages of Using the Futures Market

Only a small percentage of the swine producers in this country are sufficiently knowledgeable about the futures market to use it well. Serious mistakes can be made. There is a cost in using the market. Some small growers do not have the necessary volume of production to use the futures market effectively.

Hedging

Hedging is the primary reason for futures markets. It is the sale or purchase of a futures contract against a presently held commodity or against commodity needs to protect against price fluctuations. A farmer who will have 600 hogs to sell in January could hedge by selling these hogs on a futures market when the sows are bred, when the pigs are farrowed, when the pigs are weaned, or when feeder pigs are purchased. The farmer would sell January futures and buy back a like contract when the hogs are sold.

If the market goes down, the pigs are bought back for less than contract price. If the market goes up, they bring the guaranteed price but nothing more.

Contract Provisions

The live hog futures contract is an agreement to deliver a 30,000-pound unit of U.S.D.A. No. 1 and 2 barrows and gilts averaging 200 to 230 pounds. Delivery is made at Chicago with delivery at other specified markets allowable at a 75 cents per hundredweight discount. Other approved delivery points are Peoria, St. Louis, Omaha, Sioux City, Kansas City and St. Joseph.

Major delivery months are February, April, June, July, August, October, and December.

Trading is done through the Chicago Mercantile Exchange. The contract allows up to

90 hogs of U. S. No. 3 and 8 head of U. S. No. 4 grade hogs in the 30,000-pound unit. The present U. S. No. 3 grade of hogs is the same as the U. S. No. 2 grade previous to June, 1968.

Live hogs contracts are traded every month in sequence. The commission charge is $35.00 per unit of 30,000 pounds. The initial margin demanded is $400. The margin during month of delivery is $600 per unit.

Pork-belly contracts are for 36,000 pounds. The commission fee is $45.00 per unit. The initial margin demanded is $750, with the margin during month of delivery $1,000. Contracts are traded every month in sequence. Major delivery months are February, March, May, July, and August.

Extent of Use of the Futures Market

Records are available of the trading of live hog futures from the Chicago Mercantile Exchange. Trading began February 26, 1966. The actual number of contracts per month since that time have varied from a low of 430 during October of 1966 and February of 1967 to a high of 1,680 contracts in May of 1967 and 1,535 in March of 1966.

A University of Illinois study indicated that in 1966, 1967, and early 1968, 147 of 366 accounts were hedgers. The number of contracts per account averaged 3.2 or the equivalent of 320 head of hogs. There were 215 speculator customers averaging 3.1 contracts per customer.

Of the 366 accounts, 239 were from farmers or feeders. Farmers, packers, and market people accounted for about 75 percent of all accounts. Fourteen packing companies participated.

Use of the futures market has increased since 1970. In 1970, a total of 1.78 million pork-belly and .2 million live-hog contracts were traded. Contracts traded in 1971 included 1.69 million pork-belly and .26 million

FIGURE 12-25. The Tennessee Livestock Association staged a feeder-pig grading demonstration for youth in that area. (*Courtesy Tennessee Livestock Association.*)

live-hog contracts. In 1972, pork-belly contracts totaled 2.05 million and .54 million live-hog contracts were traded.

COOPERATIVE FEEDER-PIG MARKETS

The increase in demand and production of feeder pigs has resulted in the formation of a large number of cooperative feeder-pig markets. In most states assistance in organization and administration is provided by the state Department of Agriculture, and by the Cooperative Extension Services of U.S.D.A.

Missouri and Wisconsin are the leading states in feeder-pig sales. The Missouri Farmers Association sells pigs through the Tel-O-Auction program. The pigs must meet high standards in both health and conformation. The organization maintains a staff of about 13 production supervisors, each working with 75 or more feeder-pig producers. They assist in improving breeding stock and each year help purchase nearly 1,000 tested boars.

Four cooperative feeder-pig marketing associations are in operation in Texas. These associations are providing pigs for specialized pig finishers, who previously had no ready source of pigs.

The associations in east central Texas handle about 40,000 pigs annually. The sales are held monthly and about 75 consignors will sell an average of 15 to 20 pigs. The pigs are inspected, graded and ear tagged for identification. Some of these associations have extended their services to include marketing agreements of finished pigs with area packers.

The Tennessee Department of Agriculture has assisted the formation and activities of the Tennessee Livestock Association, which conducts feeder-pig sales at 18 locations in Tennessee. From one to seven sales are held at each location each month, with 300 to 3,000 pigs sold in each sale.

Consignors must be certified producers, and all pigs are farm inspected, as well as inspected and graded at time of sale. High standards for both health and meat qualities are demanded.

The U.S.D.A. is assisting low income farmers in several southeastern states develop programs of feeder-pig production and sales. In North Carolina, graded feeder-pig marketing is supervised by the Farmer Cooperative Service. An example of the marketing program is the Albemarle Cooperative Association of Edenton, N. C.

The association serves nine counties with a board of directors from each county. Two sales are held each month in a 51 by 141 foot building owned by the association. The pigs are graded by a state grader and inspected by a veterinarian. All pigs are ear tagged and consignors must have moving permits.

The pigs weigh from 40 to 130 pounds. A typical sale will include about 130 No. 1's, 320 No. 2's, and 220 No. 3's. Sales will vary from 400 to 800 head. The first sale was held in March 1973 and in its first year nearly $1 million was grossed. The association also has a leasing program for breeding stock. Most of the consignors are low-income farmers who use feeder-pig production on a small scale to improve their annual income. The Tel-O-Auction program of the N.C. Department of Agriculture is used.

A recent development in the Corn Belt is the formation of cooperative feeder-pig production associations among small groups of farmers who in turn buy the pigs that are produced. Each member agrees to buy at cost-of-production prices a specific number of pigs. Some of these cooperatives have made provision to market some pigs to other feeders. Several such organizations are in operation in Iowa and Nebraska.

CARE IN MARKETING HOGS

A study completed by Livestock Conservation, Inc. in 1973 of carcasses inspected for slaughter by the U.S. Department of

FIGURE 12-26. *Above, left:* The U.S.D.A. is assisting swine producers in North Carolina to market feeder pigs through cooperative markets. *Above, right:* A typical assembly of feeder pigs for an Albemarle sale. "Tel-O-Auction" is used here. (*Courtesy Farmers Cooperative Service.*) FIGURE 12-27 (*below*). Losses to swine carcasses caused by condemnations, bruises, and transit injuries amounted to $72,012,000 in one year. (*Courtesy Livestock Conservation, Inc.*)

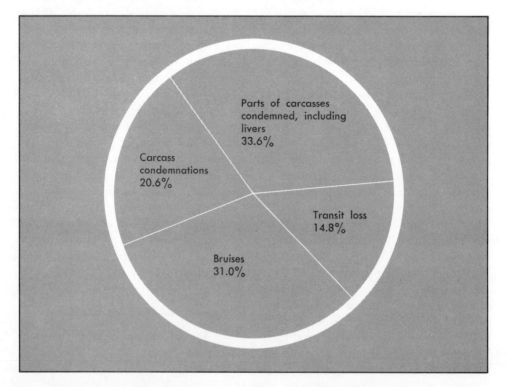

Agriculture, indicated that the pork industry loses each year about $72 million dollars in transit losses, carcass condemnations, and bruises. As shown in Figure 12-27, 33.6 percent of the losses are due to parts of carcasses being condemned, 20.6 percent due to entire carcass condemnations, 31 percent due to bruises, and 14.8 percent of losses are due to death in transit.

Causes of Bruises

The Livestock Conservation, Inc., survey showed the following to be the chief causes of bruising:

Canes, whips, and clubs 42%
Kicking, prodding 20%
Crowding, trampling 15%
Fork and nail punctures 12%
Other causes 9%
Spreaders 2%

Location of Bruises

Nearly one-half of the bruises that are found in swine carcasses are on the hams, the source of one of the highest priced cuts. More than one-quarter of the bruises are on the back and loin, the source of rib and loin chops. Shown in Figure 12-28 is a bruised ham with the bruised section removed. Hams and loins that have been trimmed because of the removal of bruised sections cannot usually be sold at the prices received for unbruised cuts. Losses due to bruises result from a loss in weight of the cut because of the removal of the bruised section, and a decrease in selling price per pound because of the irregularity of the cut.

Livestock producers and handlers can prevent most of these bruises by proper care in the handling and transportation of the hogs.

FIGURE 12-28. *Left:* Hams showing bruises.   *Right:* The removal of the bruised sections. (*Courtesy Rath Packing Co.*)

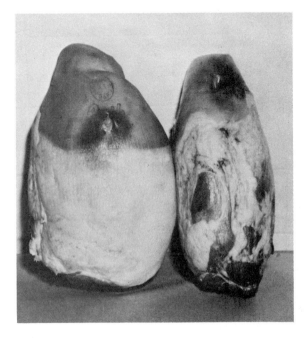

Cripples and Dead Animals

Many animals arrive at the market either dead or crippled from injuries received in loading or en route. Some animals die after reaching the market due to the condition of the animals before shipment or to injuries received en route or after reaching the market.

Dead hogs. Surveys indicate that about 14 hogs in each 1,000 were dead on arrival at the market. Based upon recent data, the loss of hogs in the nation amounted to about $10 million.

Crippled hogs. It is estimated that nearly 13 hogs in each 1,000 marketed arrive at the market as cripples. Applied to recent marketings, crippling resulted in losses to swine producers and handlers amounting to more than 4.3 million dollars.

Prevention of Losses

Care in preparing animals for shipment, in loading, and in handling hogs in shipment will reduce greatly losses due to bruised, crippled, and dead animals. Following are suggestions for the proper handling of hogs at market time:

1. Do not feed hogs heavily before shipping.
2. Clean truck or car before loading.
3. Use sand for bedding in hot weather and straw, or sand and straw, during cold weather.
4. Separate heavy from light hogs in truck or car.
5. Allow 2.2 200-pound pigs for each running foot in the livestock truck. Allow only 1.8 pigs weighing 250 pounds per foot of length of truck floor.
6. Wet or sprinkle hogs and bedding during hot weather.
7. Remove all boltheads, nailheads, and other obstructions from loading chute and truck.
8. Have adequate loading equipment.
9. Handle hogs quietly and with care.
10. Use canvass slappers in loading.
11. Do not put too few or too many hogs in a truck.
12. Separate hogs from other types of livestock when transported in the same truck.
13. Move hogs at night in hot weather.

FIGURE 12-29. Different types of livestock should be separated by partitions while being transported to avoid trampling and other losses. (*Courtesy Livestock Conservation, Inc.*)

14. Close side openings and put cover over truck during cold weather.
15. Drive carefully and avoid sudden stops.

Sprinkling Hogs in Trucks

Trucks are used to transport about 96 percent of the hogs received at slaughter plants. Shipping losses may be heavy on long-distance shipments made in hot weather. Manual and semi-automatic sprinkling systems have been tested. Tests indicate that sprinkling live hogs in transit should be considered when the temperature is 80 degrees F. or above, regardless of distance. The greater the distance, the greater the value of sprinkling. Hogs cooled by sprinkling are more quiet and contented. There is less danger of mortality from heat and less bruising from trampling and crowding. There is also less shrinkage.

Truck Capacities

Shown in Table 12-11 are the maximum numbers of hogs of various weights for a truck load.

Drug Residue Condemnations

Hog producers are responsible for any chemical residues found in the carcasses of their hogs when slaughtered. The 1970 Federal law provides for a prison sentence and/or a $10,000 fine for offenders. It is important that care be taken in drug use. The following are musts in the proper use of drugs in swine production:

1. Follow the label instructions.
2. Follow the withdrawal requirements.
3. Feed only the recommended dosages.
4. Do not mix antibiotics and drugs unless this is specifically recommended by the manufacturer.
5. Make certain that all feed bins and feeders are empty of medicated feed during the withdrawal period.
6. Follow closely recommendations in Table 11-2.

SUMMARY

Breeding and feeding operations should be planned carefully in order to have U. S.

T A B L E 12-11 MAXIMUM NUMBER OF HOGS FOR TRUCK LOAD

Floor Length (Feet)	SINGLE-DECK TRUCKS Weight of Hogs (Lbs.)			DOUBLE-DECK TRUCKS Weight of Hogs (Lbs.)		
	200	250	350	200	250	350
8	18	14	11	29	24	18
10	22	18	14	36	30	23
12	26	22	17	43	36	28
15	33	27	21	54	45	34
18	40	33	25	65	55	41
20	44	35	28	72	61	46
24	52	44	34	87	73	55
28	61	51	39	101	86	64
30	66	55	42	108	91	68
32	70	58	44	115	97	73
36	79	66	50	130	110	82
42	92	77	55	151	128	96

Source: Livestock Conservation, Inc.

No. 1 hogs ready when the market is best. A leaner, better muscled market hog should be produced and sold when it weighs from 220 to 240 pounds. Unless hogs of higher quality are marketed, the consumer will buy beef and poultry. Because the lard market is weakened, less lard should be produced. A 300-pound hog produces nearly twice as much lard as does a 200-pound hog.

To get the best price spring pigs should be sold in July, August, and June, and fall pigs in December and February.

The U.S.D.A. grades are U. S. No. 1, U. S. No. 2, U. S. No. 3, U. S. No. 4, and Utility. Hogs should be sold as they reach the U. S. No. 1 grade, not held until they are overfat and grade down to U. S. No. 2 or U. S. No. 3. Selling hogs by grade will encourage farmers to improve the quality of hogs which they produce. Packers that buy hogs according to grade and yield usually pay higher prices for choice hogs and lower prices for lardy hogs than do buyers who do not grade the hogs. Most hogs in the Corn Belt are sold directly to the packer.

Meaty slaughter hogs can be marketed at heavier weights than fat-type hogs. The cost of gain on meaty pigs weighing 240 pounds is only slightly above the cost of gain on 200-pound pigs. The market spread has been less than 10 cents per hundredweight.

Selling pigs on the basis of grade and yield may net the producer as much as $1.50 to $2.50 per head, and will provide valuable information concerning the quality of carcasses produced by the various bloodlines used in the herd. The producer will, however, have to wait some time for the check, and must accept the data provided by the packer.

A market should be chosen carefully. The net return after transportation, shrinkage, and other costs have been deducted is the determining factor.

Nearly 70 percent of all market hogs sold in 1971 were purchased by packers direct or from country dealers. Only 17 percent were sold in terminal markets, and 13 percent went to auction markets.

Pork producers lose about $72 million a year in losses due to partial or entire carcass condemnations, bruises, or in transit losses.

More than 25 percent of the hogs arriving at packing plants are bruised. There are many crippled and some arrive dead.

Care in trucking and shipping will prevent losses, injuries, and bruises. The truck should be carefully bedded down and should not be underloaded or overloaded. Each 200-pound hog should have 3½ square feet of floor space. Heavy hogs should be separated from light hogs, and hogs should be separated from other types of livestock in the truck.

Boltheads and nails should be removed from within the truck and on the loading chute. Hogs should be handled with care. Hogs should be protected from the hot sun in the summer and cold winds in the winter. It pays to drive carefully when trucking hogs.

Hedging will protect pork producers against declines in the market price of slaughter hogs. Producers of large numbers of pigs should analyze their records carefully and sell a part of their production when the futures market shows it to be profitable. About 75 percent of those who use the futures market to hedge make a profit by doing so. About the same percentage who use the futures market to speculate lose money.

Producers are responsible for any drug residues found in the carcasses of animals produced by them. Not only will the carcasses be condemned if residues are found, the producer may also be subject to a fine and prison sentence. Read all directions and labels provided by drug suppliers, and follow them to the letter.

QUESTIONS

1 Why does the packer pay more per hundredweight for a 225-pound hog than for a 280-pound hog?

2 What percentage of the carcass of a hog should be of the four primal cuts: hams, loins, picnics, and Boston butts?

3 What is the difference between the USDA grades and the old grading system?

4 What goes to make up a U. S. No. 1 hog?

5 In which months should spring pigs be sold to bring in the most money? When should fall pigs be sold?

6 How much variation occurs in the price of light hogs during the year?

7 How much less do you get per hundredweight for a 300-pound hog than for a light 200-pound hog?

8 What percentage of the hogs marketed in your state are marketed directly to the packer or packer representative in the local community?

9 How many terminal markets are there, and what is their slaughter-hog volume?

10 What part does the local auction sale play in marketing hogs in your community?

11 When is the best time to market packing sows?

12 What are the mechanics of using the futures market in hedging future profits from your home swine enterprise?

13 How will you select the market through which you will sell your hogs? What factors will you consider?

14 What precautions do you need to take in transporting market hogs?

15 Would you prefer to sell hogs by carcass grade or by weight? Why?

16 How can you improve the income from your home hog enterprise through better marketing practices?

REFERENCES

Chicago Mercantile Exchange, *Futures Trading in Live Hogs*. Chicago, Ill., 1968.

George A. Hormel & Co., *Marketing Hogs by Grade and Yield*. Austin, Minn., 1974.

Livestock Conservation, Inc., *1974 Safety Education Program*. Omaha, Nebr., 1974.

U.S. Department of Agriculture, *Agricultural Statistics, 1973*. Washington, D.C., 1973.

——, *How Do Your Hogs Grade?* Marketing Bulletin No. 16. Washington, D.C., 1961.

——, *Livestock and Meat Statistics*, Washington, D.C., 1973.

——, *Livestock Slaughter, 1973*. Washington, D.C., 1974.

——, *Livestock Slaughter*, MtAn 1-2-1. Washington, D.C., 1970.

——, *Official U.S. Standards for Grades of Slaughter Barrows and Gilts*. Washington, D.C., 1968.

——, Standards for Grades of Barrow and Gilt Carcasses. Washington, D.C., 1968.

——, *U.S.D.A. Grades for Pork Carcasses*. Marketing Bulletin No. 49. Washington, D.C., 1970.

——, *U.S.D.A. Grades for Slaughter Swine and Feeder Pigs*, Marketing Bulletin No. 51. Washington, D.C., 1970.

13 INCOME AND COST FACTORS IN PORK PRODUCTION

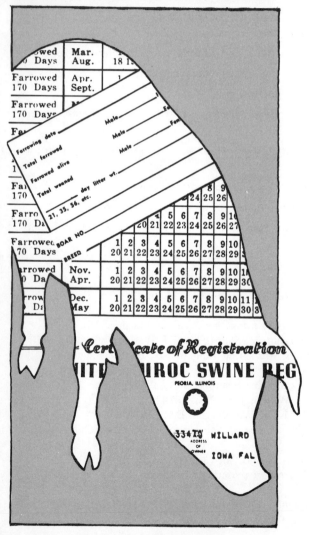

In the operation of a manufacturing or retail business, regardless of the kinds of merchandise handled, the owner keeps a careful record of the income and expenses of each part of the business.

Swine production is a major enterprise in many farm businesses, and the only enterprise on increasing numbers of highly specialized pork producing farms. If no records are kept, a producer cannot determine the extent that the enterprise is profitable, or more important in making management decisions, pinpoint the practices which contribute to the efficiency, or to the losses.

The capital necessary to finance a highly specialized hog farm has increased greatly with confinement feeding, environmentally controlled housing, the use of production tested breeding stock, and high prices demanded for interest, labor, feed and medication. Swine producers must have good records to satisfy their backers in obtaining credit.

A producer with no records has no way of comparing the efficiency of the swine enterprise with that of neighboring producers, or with other farm enterprises.

Profitable pork production is usually dependent upon (1) the production of large litters, (2) economical use of feed and labor, (3) emphasis upon meat-type hogs, (4) control of diseases and parasites, (5) economical use of buildings and equipment, and (6) the marketing of quality hogs at desired weights at the best time and place. Farmers who keep records involving these factors can analyze their enterprises. Those who do not keep records can only guess.

The kinds of records which should be kept will vary with the production program followed but usually include the following:

1. **Breeding records**
2. **Production records**
3. **Pedigree and herd records**
4. **Feed records**

5. Labor records
6. Complete enterprise record

BREEDING RECORDS

A record of the date that each sow or gilt is bred should be kept. Producers who pen-breed may tack a chart on the wall of the hog house and as each female is bred, the ear notches, ear tag number, pedigree name, or other identifying characteristics of the animal are recorded together with the date and name of the boar.

At the end of the breeding season, the information on the wall chart should be transferred to a record book. Some producers use only the record book. A number of books are available from swine breed associations. Most of the books provide space for names and pedigrees of boars, name and ear notch identification of each sow, and for a complete litter record.

FIGURE 13-1. Swine business is big business. Large amounts of capital are involved. Careful records must be kept to determine efficient volume and practices. (*Courtesy L. H. White Company, Inc.*)

A common practice in marking sows or gilts as they are bred, when ear tags are not used, is to make a mark or marks on the body of the animal by cutting the hair with a pair of scissors. The same numbering system recommended in ear notching may be used. One mark on the right shoulder for 1; one mark on the left shoulder for 3; one mark on the right hip for 10; and one mark on the left hip for 30.

Breeding date records are necessary in order to determine if each female is settled and if the boar is a breeder. Each sow or gilt should be checked about 21 days after being bred.

Farmers who practice lot breeding usually keep a record of the date that the boar is turned in with the sow herd and then leave the boar with the herd for only a two- or three-week period.

Breeding records are very helpful at farrowing time. It is possible to separate the sow from the herd and place her in farrowing quarters at an appropriate time if the probable farrowing time is known.

PRODUCTION RECORDS

Tests have shown that bloodlines within each of the various breeds of swine differ in production ability. Some lines excel in prolificness. Some lines make more rapid gains. Certain lines may make more efficient use of feed, and some lines produce more desirable carcasses. Records must be kept to discover and maintain superior lines and to weed out inferior lines.

Production Registry

The purebred swine breed associations have promoted various production improvement programs. A commonly followed program has been the Production Registry program which is based upon the size of litter farrowed, the number of pigs weaned at eight weeks of age, and the weight of the litter at weaning time.

The weight of the pigs at 21, 35, or 56 days indicates in part the productiveness of the individual sows and the effectiveness of the breeding, feeding, and management programs. The selection of gilts and sows to be retained in the herd should be based upon the litter weights. The tried sows to be kept over for second and third litters should be those which produce the best litters. Selecting gilts from litters which were heavy at 21, 35, or 56 days will remove much risk.

As pointed out in Chapter 3, the Production Registry program developed by the National Association of Swine Records set up the following requirements for production registry:

1. **A mature sow must farrow and raise eight or more pigs to a 56-day weight of at least 320 pounds.**
2. **A first litter gilt must raise the same number of pigs to a 56-day weight of at least 275 pounds.**
3. **Sows qualify for production registry after**

producing two production registry litters.
4. **To qualfiy as production registry sires, boars must sire five qualified daughters or fifteen daughters that have produced one production registry litter.**

It is necessary in keeping official production records to keep farrowing records, to ear-notch and or tattoo the pigs of each litter at birth, and to have witnesses of the weight of the pigs at or near 56 days of age. A bathroom scale, or a scale supported by a tripod, may be used in weighing.

In case the pigs are weighed a few days before or after 56 days of age, the calculated weight at 56 days may be obtained from Table 13-1.

Other forms of production testing are being used by many breeders, however nearly 270 Yorkshire breeders grew PR (production registry) litters in 1973. The Duroc and Hampshire breeders also have continued using PR testing, as have many commercial producers.

With the development of the early weaning programs, some breed associations have made additions to their testing programs. The "Herd Improvement" program of the Poland China Record Association is based upon the number of live pigs farrowed per litter, and their weight at birth. There is interest among some breeders in a testing program that would involve the weights of pigs at three, four, or five weeks of age.

The Poland China Production Registry program is based upon birth weights. However, any properly entered litter may be weighed again between the 30th and 40th day of age and reported on the 35-day weight report form. The 35-day weight equivalent will be recorded on the certificate. Eight or more pigs from a gilt must weigh 128 pounds or more at 35 days. A similar litter from a sow must weigh 152 pounds at 35 days.

The Poland China breeders supplement PR with a national boar and gilt testing pro-

FIGURE 13-2. Efficient production necessitates the farrowing of large litters of healthy pigs. (*Courtesy Elanco Products Co.*)

T A B L E 13-1   FACTORS FOR ADJUSTING WEIGHT TO A STANDARD AGE FOR PIGS[1]

Adjust to 56 Days[2]		Adjust to 35 Days[2]		Adjust to 28 Days[2]		Adjust to 21 Days[2]	
Age	Factor	Age	Factor	Age	Factor	Age	Factor
66	.82	45	.73	38	.71	31	.72
65	.83	44	.75	37	.73	30	.75
64	.85	43	.78	36	.75	29	.77
63	.86	42	.80	35	.78	28	.79
62	.88	41	.83	34	.80	27	.82
61	.90	40	.86	33	.83	26	.84
60	.92	39	.89	32	.86	25	.87
59	.94	38	.91	31	.89	24	.90
58	.96	37	.94	30	.92	23	.93
57	.98	36	.97	29	.96	22	.96
56	1.00	35	1.00	28	1.00	21	1.00
55	1.02	34	1.03	27	1.04	20	1.04
54	1.05	33	1.07	26	1.09	19	1.08
53	1.07	32	1.11	25	1.14	18	1.13
52	1.10	31	1.15	24	1.20	17	1.17
51	1.13	30	1.19	23	1.26	16	1.23
50	1.16	29	1.24	22	1.33	15	1.29
49	1.19	28	1.29	21	1.40	14	1.35
48	1.22	27	1.34	20	1.49	13	1.41
47	1.25	26	1.40	19	1.59		
46	1.29	25	1.47	18	1.70		
45	1.32	24	1.54				
44	1.37	23	1.62				
43	1.41	22	1.71				
42	1.46	21	1.81				
41	1.51						

[1]*EXAMPLE:* To adjust the weight of a pig to a standard weight at 56 days: Assume that the actual weight is 32 pounds at 52 days of age. Multiply the actual weight (32 pounds) by the factor for 52 days (1.10). The result is 35 pounds, the adjusted weight at 56 days of age.

[2] Age at weaning time as arbitrarily designated by some of the breed associations.

Source: *Selection Programs for Profitable Swine Production,* Cir. 868, University of Illinois.

gram. Currently a Wisconsin test station is used.

The Duroc breed association has a VIP (verification and identification) program for verification of date of farrow of animals. Since 1969, the Hampshire breed association has provided sonoray testing technicians and sponsors the HIT Herd Improvement Program. It supplements their Herd Improvement Program.

The Spot breed organization has continued their certification program, and pro-

motes sire performance tests. The breed also has an Elite Spotted Breeder Club for breeders participating in their age verification program.

Certified Meat Litter and Sire Program

The National Association of Swine Records and a number of participating breed associations have promoted programs for the selection of certified meat-type litters. The first certified litter of the Hampshire breed

FIGURE 13-3. The Berkshire breeders conduct group boar tests in cooperation with the University of Missouri. (*Courtesy American Berkshire Association.*)

FIGURE 13-4. This boar was tested at the Purdue station. It took him only 138 days to reach 240 pounds. He had .64 of an inch of backfat, a 5.93 square inch loin eye, and 47.21 percent of the carcass was estimated by ILIS to be ham and loin. The boar sold for $31,000 and is owned by Oral Long of Elnora, Indiana. He was bred by Harry and Bruce Lugar of Camby, Indiana. (*Courtesy United Duroc Swine Registry.*)

was selected during the summer of 1954. The first Poland China certified meat-type litter was announced early in 1955. The Duroc breed association launched a program for the selection of certified litters during the summer of 1955.

The selection program of the Ohio Swine Improvement Association was one of the first litter certification programs. In 1954 it involved 791 litters of spring pigs. Of this number 462 litters made the weight standards established by the association; 192 were Ohio certified, and they represented eight of the major breeds.

Litters were certified and certificates were awarded to each boar or gilt intended for breeding purposes, when they had met the following standards:

1. **Farrowing requirements.** At least eight live pigs must have been farrowed. The sire, dam, and pigs of the litter must have been free from swirls, ridglings, rupture, and inverted teats.

2. **Weight standards.** The weight days and standards prescribed by the breed association were applicable in some cases. The association recommended that the pigs be weighed at 35 days if the herd was not entered in a breed production registry program.

3. **Gaining ability.** Meat-type breeding stock had to be efficient in gaining ability. Two pigs of each litter were sent to the Ohio Swine Evaluation Station where they were fed out to a weight of about 200 pounds. Rate of gain and feed requirement records were kept.

4. **Slaughter test.** The two pigs tested were slaughtered when they weighed about 210 pounds. The litter certification depended on the percentage yield of the trimmed primal cuts. Certified litters also had to meet the minimums for litter size, rate of gain, and feed efficiency set by the Ohio Swine Improvement Association.

Shown in Figure 13-5 is a sow record recommended by the Texas A & M Univer-

Sow No._____	R L	Farrowing date_____
Sow's_____litter		Total farrowed ....Male_____Female_____
(1, 2, etc.)	Ear notches	Farrowed alive ....Male_____Female_____
Birth date of sow_____		Total weaned ....Male_____Female_____
180-day wt._____200 lb. probe_____Feed/100 lb. gain_____		_____ day litter wt._____ 21, 35, 56, etc.

Name_____ Reg. No._____   BOAR NO._____

Breed_____   BREED_____

| Ear Notch | | Pig No. | Sex | Birth Weight | Weaning Weight | Age weighed ____days | | Backfat Probe | | | | | Remarks |
R	L				(Optional)	Actual Wt.	Adj. 154 day wt.	Shldr. In.	Back In.	Loin In.	Av. In.	Adj. to 200 lb.	

Av. 154-day Wt. of litter_____   Feed per 100 pounds gain_____

FIGURE 13-5. A sow record recommended by Texas A & M University. (*Courtesy Texas A & M University.*)

sity. Similar record forms are used in other states.

The *certified litter* program sponsored by the National Association of Swine Records in operation in 1974 is described in Chapters 2 and 3. It involves not only the meeting of PR requirements, but standards of growth and carcass quality of pigs at 220 pounds. There has been decreased interest among breeders in the program during the past few years, with more attention given to boar and gilt testing at swine testing stations at central locations or on the home farms.

In 1972, 506 Duroc, 363 Hampshire, and 265 Yorkshire litters were certified, compared to the 592, 870, and 480 litters, respectively, which qualified in 1969. On January 1, 1973, a total of 13,444 Hampshire, 6,000 Duroc, and 5,549 Yorkshire litters had been certified.

Only 227 herd sires qualified as CMS (certified meat sires) in 1972. On January 1, 1973, 1,329 Hampshire, 806 Duroc, 459

Poland China and 385 Yorkshire boars had met CMS standards.

Some breed associations have developed "super" certified litter programs. The Hampshire *Pacesetter Certified Litter* program is an example. There must be at least nine pigs farrowed, and the litter must be recorded within 14 days of farrowing. The pigs must be free of swirls, and hernias, and the litter must not have any boars with only one testicle. The test pigs from the litter must weigh 200 pounds at 160 days or less, or weigh 220 pounds at 170 days or less. Each pig must meet the weight standards. Pigs are to be slaughtered at 190 to 220 pounds.

The slaughtered pigs must yield at least a 5-square-inch loin eye. The carcass length must be 29.5 inches or more, and the backfat thickness cannot exceed 1.4 inches.

Following is the statement published by the National Association of Swine Records concerning the requirements for certification of litters. These requirements are used by all breed associations:

ALL BREED
SWINE CERTIFICATION PROGRAM

Prepared and adopted by National Association of Swine Records and effective on all litters farrowed on or after January 1, 1970.

1. Program is based on two test pigs (Barrows or Gilts) from the same litter.

2. Litter must qualify for certification testing in accordance with testing eligibility or PR requirements of respective breed Registry. Basically these requirements are: Litter must be recorded with breed Registry within 10 to 14 days of farrowing; must be eight or more pigs farrowed in the litter; litter must be free of swirls, hernias or boars with only one testicle.

3. Two pigs from a litter so qualified for testing are then submitted to an approved and cooperating slaughter station at any weight up to 240 pounds. No pig can qualify if over 242 pounds when weighed in for slaughter. It is recommended that test pigs be slaughtered between 200 and 240 pounds. Pigs may be slaughtered under 200 pounds, however, their measurements of length, backfat and loin will be converted to a 220 pound basis as though they actually weight 200 pounds.

4. All certification standards are based on a 220 pound market weight basis. All measurements for each pig at his slaughter weight will be converted to a 220 pound standard by use of a conversion factor.

The 220 lb. standards are:	The conversion factors are:
Days to 220—180 Max.	2 lbs. per day
Length at 220—29.5" Min.	0.025 inches per lb.
Backfat at 220—1.50" Max.	0.004 inches per lb.
Loin Area at 220— 4.50 Sq." Min.	0.015 sq. inches per lb.

5. All certification reports will show the actual slaughter weight of each pig with the rate of gain, length, backfat and loin area measurements shown as an adjusted 220 pound basis.

6. Each pig will also be scored as to quality but the score will not be a requirement for certifi-

cation. The scores will enable breeders to evaluate their animals as to quality. The loin will be used for scoring marbling and color and firmness as follows:

Marbling	Color & Firmness
1. Practically devoid	1. Very pale, soft, watery
2. Traces to small amount	2. Pale, moderately soft and watery
3. Moderate	3. Grayish pink, moderately firm and dry
4. Slightly abundant	4. Somewhat dark, quite firm and dry
5. Very abundant	5. Dark, very firm and dry

7. Recognition: When both test pigs in a litter each meet the required standards, that litter will be recognized as a Certified (CL) Litter. When a boar sires five certified litters he will be recognized as a Certified (CMS) Meat Sire. The repeat mating of a boar and sow that has produced a CL litter will also be recognized as a Certified Mating if that litter meets the breed Registry's testing eligibility or PR requirements.

8. Respective breed Registries may establish more rigid standards at 220 pounds for applying additional selection pressure within the breed and to give recognition to superior performance.

Presented in Table 13-2 are data converting backfat thickness to a 220-pound equivalent. The table has been developed by the National Association of Swine Records.

Sire Evaluation and Herd Testing Program

The National Association of Swine Records has approved a sire evaluation and herd testing program that was introduced by the various breed associations beginning in the summer of 1974. This program brings together the advantages of all previous testing programs, and the testing is done on the farm of the producer. Some associations are making minor changes in the program to meet the needs of the respective breed. Following is the program available to breeders of Yorkshire swine:

T A B L E 13-2   CONVERTING BACKFAT THICKNESS TO 220-POUND EQUIVALENT

	Weight Range									
3 PROBE	240	235	230	225	220	215	210	205	200	3 PROBE
TOTAL	238 242	233 237	228 232	223 227	218 222	213 217	208 212	203 207	198 202	TOTAL
1.5	.42	.44	.46	.48	.50	.52	.54	.56	.58	1.5
1.6	.45	.47	.49	.51	.53	.55	.57	.59	.61	1.6
1.7	.49	.51	.53	.55	.57	.59	.61	.63	.65	1.7
1.8	.52	.54	.56	.58	.60	.62	.64	.66	.68	1.8
1.9	.55	.57	.59	.61	.63	.65	.67	.69	.71	1.9
2.0	.59	.61	.63	.65	.67	.69	.71	.73	.75	2.0
2.1	.62	.64	.66	.68	.70	.72	.74	.76	.78	2.1
2.2	.65	.67	.69	.71	.73	.75	.77	.79	.81	2.2
2.3	.69	.71	.73	.75	.77	.79	.81	.83	.85	2.3
2.4	.72	.74	.76	.78	.80	.82	.84	.86	.88	2.4
2.5	.75	.77	.79	.81	.83	.85	.87	.89	.91	2.5
2.6	.79	.81	.83	.85	.87	.89	.91	.93	.95	2.6
2.7	.82	.84	.86	.88	.90	.92	.94	.96	.98	2.7
2.8	.85	.87	.89	.91	.93	.95	.97	.99	1.01	2.8
2.9	.89	.91	.93	.95	.97	.99	1.01	1.03	1.05	2.9
3.0	.92	.94	.96	.98	1.00	1.02	1.04	1.06	1.08	3.0
3.1	.95	.97	.99	1.01	1.03	1.05	1.07	1.09	1.11	3.1
3.2	.99	1.01	1.03	1.05	1.07	1.09	1.11	1.13	1.15	3.2
3.3	1.02	1.04	1.06	1.08	1.10	1.12	1.14	1.16	1.18	3.3
3.4	1.05	1.07	1.09	1.11	1.13	1.15	1.17	1.19	1.21	3.4
3.5	1.09	1.11	1.13	1.15	1.17	1.19	1.21	1.23	1.25	3.5
3.6	1.12	1.14	1.16	1.18	1.20	1.22	1.24	1.26	1.28	3.6
3.7	1.15	1.17	1.19	1.21	1.23	1.25	1.27	1.29	1.31	3.7
3.8	1.19	1.21	1.23	1.25	1.27	1.29	1.31	1.33	1.35	3.8
3.9	1.22	1.24	1.26	1.28	1.30	1.32	1.34	1.36	1.38	3.9
4.0	1.25	1.27	1.29	1.31	1.33	1.35	1.37	1.39	1.41	4.0
4.1	1.29	1.31	1.33	1.35	1.37	1.39	1.41	1.43	1.45	4.1
4.2	1.32	1.34	1.36	1.38	1.40	1.42	1.44	1.46	1.48	4.2
4.3	1.35	1.37	1.39	1.41	1.43	1.45	1.47	1.49	1.51	4.3
4.4	1.39	1.41	1.43	1.45	1.47	1.49	1.51	1.53	1.55	4.4
4.5	1.42	1.44	1.46	1.48	1.50	1.52	1.54	1.56	1.58	4.5
4.6	1.45	1.47	1.49	1.51	1.53	1.55	1.57	1.59	1.61	4.6
4.7	1.49	1.51	1.53	1.55	1.57	1.59	1.61	1.63	1.65	4.7
4.8	1.52	1.54	1.56	1.58	1.60	1.62	1.64	1.66	1.68	4.8
4.9	1.55	1.57	1.59	1.61	1.63	1.65	1.67	1.69	1.71	4.9
5.0	1.59	1.61	1.63	1.65	1.67	1.69	1.71	1.73	1.75	5.0
5.1	1.62	1.64	1.66	1.68	1.70	1.72	1.74	1.76	1.78	5.1
5.2	1.65	1.67	1.69	1.71	1.73	1.75	1.77	1.79	1.81	5.2
5.3	1.69	1.71	1.73	1.75	1.77	1.79	1.81	1.83	1.85	5.3
5.4	1.72	1.74	1.76	1.78	1.80	1.82	1.84	1.86	1.88	5.4
5.5	1.75	1.77	1.79	1.81	1.83	1.85	1.87	1.89	1.91	5.5
5.6	1.79	1.81	1.83	1.85	1.87	1.89	1.91	1.93	1.95	5.6
5.7	1.82	1.84	1.86	1.88	1.90	1.92	1.94	1.96	1.98	5.7
5.8	1.85	1.87	1.89	1.91	1.93	1.95	1.97	1.99	2.01	5.8
5.9	1.89	1.91	1.93	1.95	1.97	1.99	2.01	2.03	2.05	5.9
6.0	1.92	1.94	1.96	1.98	2.00	2.02	2.04	2.06	2.08	6.0
6.1	1.95	1.97	1.99	2.01	2.03	2.05	2.07	2.09	2.11	6.1
6.2	1.99	2.01	2.03	2.05	2.07	2.09	2.11	2.13	2.15	6.2
6.3	2.02	2.04	2.06	2.08	2.10	2.12	2.14	2.16	2.18	6.3
6.4	2.05	2.07	2.09	2.11	2.13	2.15	2.16	2.19	2.21	6.4
6.5	2.09	2.11	2.13	2.15	2.17	2.19	2.21	2.23	2.25	6.5
6.6	2.12	2.14	2.16	2.18	2.20	2.22	2.24	2.26	2.28	6.6
6.7	2.15	2.17	2.19	2.21	2.23	2.25	2.27	2.29	2.31	6.7
6.8	2.19	2.21	2.23	2.25	2.27	2.29	2.31	2.33	2.35	6.8
6.9	2.22	2.24	2.26	2.28	2.30	2.32	2.34	2.36	2.38	6.9

FIGURE 13-6 (*Left*). Pigs on test at the Iowa Swine Testing Station in Ames, Iowa (*Courtesy Iowa Swine Testing Station.*)   FIGURE 13-7 (*Right*). Test pens at the Joe Tschetter farm in South Dakota. (*Courtesy American Yorkshire Club, Inc.*)

American Yorkshire Club
and
National Association of Swine Records

SIRE EVALUATION AND HERD TESTING PROGRAM

GENERAL RULES

1. Any breeder wishing to participate in the sire evaluation and herd testing program must sign up for the program with the American Yorkshire Club no later than December 1 to test his spring pig crop and June 1 to test a fall pig crop.

2. Breeding dates and registration numbers of all sows in the herd and their service sires must be reported to the breed association at least 30 days prior to farrow. Up to 20 percent of the sows in the herd may be pasture bred. Service sires and the dates sows were exposed must be reported.

3. Litter reports of the litters by all sows in the herd must be reported by the 10th of the month following the month of farrow or as specified by the breed association. (Yorkshire reports must reach office within 30 days of farrowing.)

4. Boars placed on test from a gilt litter (farrowed before dam is 15 months of age) must contain eight or more pigs farrowed.

5. Boars placed on test from a sow litter (farrowed after dam is 15 months of age) must contain nine or more pigs farrowed.

6. On-test weights and off-test weights must be sent to the breed association within 7 days after weights are taken. On-and off-test weights must be verified by a disinterested party.

7. Pedigrees or complete application and fees for pedigrees on all boars placed on test must accompany on-test weights that are submitted to the breed association. Pedigrees will be held in association office until test completion at which time they will be returned to the breeder with test information added.

8. The maximum weight for all boars placed on test is 70 pounds and each boar must have a minimum weight of at least 30 pounds.

9. A maximum of 1.1 pounds per day of age will be allowed for each boar when he is initially placed on test.

10. All boars placed on test must be accounted for at the end of the test period.

11. Type of facility is optional with the individual breeder. Either pasture or confinement conditions are allowed. It is recommended by the National Association of Swine Records that boars be fed in groups of 8 to 10 head per pen.

12. Figures on all boars placed on test will be adjusted to 240 pounds. Weight ranges from 220 to 260 pounds will be accepted for boars completing the test.

13. Backfat measurements will be accepted by either probing or sonoraying the boars on test. Sonoray and probing technicians must be approved by the breed association.

I. Requirements for a performance tested herd.

A. Each breeder must comply with all the general rules as stated above.

B. Eighty percent of all sows in the herd that farrow eligible litters (gilts that farrow 8 pigs and sows that farrow 9 pigs) must have at least 1 boar from their litter placed on test.

C. All boars in the herd testing program will be identified with the letters "HT." These letters will appear after the registration number on the pedigree.

D. Each boar placed on test in the program will have his performance ratio for average daily gain and backfat thickness as expressed to the average of his test group appear after the letters "HT" on the pedigree. A pedigree from an individual boar in the herd testing program could read: Super Stud 707609 HT (120/105). This would indicate to any person looking at the pedigree that Super Stud was a tested boar and in that test group he was 20% above average for daily gain and 5% cleaner than the average of his group. A ratio of 100 equals the average of the test group.

E. The cost for each boar placed on the program is $1.00. This fee must accompany the on-test weights of the boars placed on test.

II. Requirements to qualify a sire as a performance test sire.

A. The breeder must have his herd on the performance test herd program.

B. A sire must have at least 1 pig out of 80% of the sows that are mated to him that farrow eligible litters placed on test.

C. At least 25 boars out of 10 sows with no more than 5 of those sows being half sisters must

be placed on test. If less than 25 boars are tested in one season their performance records will be carried from season to season until at least 25 have been tested.

D. A sire's record will be based on the average daily gain and backfat thickness of 80% of the boars placed on test that are sired by him.

E. The average daily gain of the progeny determining a sire's record must be at least 1.65 pounds. The average backfat thickness must not exceed 1.0 inch.

F. A minimum of five slaughter hogs either barrows or gilts from the same age group as the boars being tested and out of five different sows must be killed. The average of the five head must meet all certification standards after adjustments for sex and weight have been made.

G. Once a boar has met the above requirements he will be designated a performance tested sire and the letters "PTS" will appear on his pedigree.

Litter or Boar Testing Programs

Litter or boar testing work has been carried out by a large number of producers on an individual basis, or as members of testing associations. The associations have been organized and conducted with the help of vocational agriculture instructors, agricultural extension directors, and specialists of state agricultural colleges and swine breed associations. Numerous programs are being conducted in Corn Belt states, and some state-wide programs have been developed.

State-wide programs which have been conducted are the *Wisconsin Swine Selection Program and the Wisconsin Pacemaker Pork Program.* They involved records of pedigree information concerning each sow and herd boar, as well as farrowing date, litter size, birth weight, litter identification, and weaning weight of each litter. All weights were adjusted to a 154-day or five-month basis.

The *Illinois Swine Herd Improvement Association* was organized in 1947. Its membership in 1959 consisted of 30 local or county herd improvement associations, 15 F.F.A.

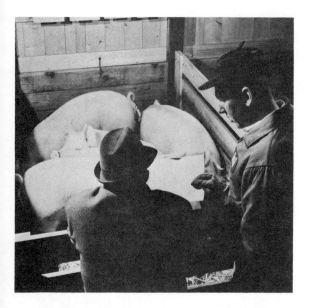

FIGURE 13-8 (*Above*). Swine producers inspecting the boars and their records in preparation for the purchase of a herd sire in a boar testing station sale. (*Courtesy American Cyanamid Co.*) FIGURE 13-9 (*Below*). The North East Iowa Swine Testing Station at New Hampton, Iowa, sponsored by Midland Cooperatives Inc. (*Courtesy Keith Olson.*)

chapter groups, and seven state breed associations. Fourteen of the associations operated testing stations in 1959.

The National Pork Producers Council has been instrumental in establishing state pork producers associations in about 30

states. These organizations have been active in initiating and supporting swine testing stations. In 1974, there were 37 testing stations in 24 states. Following is a list of states and the number of public swine testing stations:

Alabama	1
Arkansas	1
Colorado	1
Florida	1
Georgia	1
Illinois	2
Iowa	5
Kansas	2
Louisiana	1
Maryland	1
Michigan	1
Minnesota	1
Mississippi	1
Missouri	4
Montana	1
Nebraska	2
North Carolina	1
North Dakota	1
Ohio	1
Oklahoma	1
Pennsylvania	1
Tennessee	2
Virginia	1
Wisconsin	3

Many breeders conduct tests on their own farms. The summary of improvements brought about in rate of gain, loin eye area, backfat thickness, feed conversion, and percent ham and loin presented in Chapter 3 illustrates the value of tests and of records in swine improvement. Most progressive swine breeders are participating in testing station programs.

Most of the national and many of the state breed associations are conducting boar or gilt testing programs at a central location. These tests provide excellent data for use in comparing growth rate, backfat thickness, and in some cases, feed efficiency of various broodlines within the breed.

A common practice is to enter four pigs in the test stations. In some stations four boars are tested. In some tests, three boars and one barrow sired by the same sire are tested. In a few cases, gilts are tested.

Most testing stations are now using the same standards so that comparisons may be made of pigs entered in tests on other states. Following are the standards in use at the Iowa Swine Testing Station in 1974:

TEST STATION STANDARDS

Daily Gain
    1.80 lbs. or more per day
Feed Efficiency
    280 lbs. feed/100 lbs. gain in summer
    290 lbs. feed/100 lbs. gain in winter
Backfat Thickness
    If gain is 1.80–1.99, probe is under 0.9″
    If gain is 2.00–2.19, probe is under 1.0″
    If gain is 2.20–2.39, probe is under 1.1″
    If gain is 2.40 or more, probe is under 1.2″
Loin Eye
    Littermate barrow, 4.50 sq. in. or more
Index
    165 or more
Boars
    Index equals 250 plus 50 (daily gain)—50 (feed/lb. gain)—50 (backfat probe)
Health
    Boars meet requirements for interstate shipment
Vaccination Papers Provided
    All boars have been tested for brucellosis and found negative. Boars all vaccinated for leptospirosis.
Quality Score
    5–5 Dark Dry
    3–3 Best Score
    1–1 Soft Pale

A Yorkshire boar owned by Richard Bruene of Gladbrook, Iowa, set a record at the Ames, Iowa, testing station with an index of 235. The boar had a daily gain of 2.73 pounds per day while on test from 39 to 200 pounds, a backfat probe of 0.81 inch, a feed conversion of 2.34 pounds of feed per pound of gain, and a ham-loin percentage of a littermate barrow of 43.5 percent. The boar reached a 200 pound weight in 125 days. The littermate barrow had a carcass 30 inches long, with 1.27 inches of backfat, and a 6.5-square-inch loin eye.

Another Yorkshire boar owned by Richard Solberg of Armstrong, Iowa, indexed 231 to set the new record at Farmland's Ida Grove, Iowa, station. With a 2.52 pounds per day gain, the boar also had a feed efficiency of 221 pounds of feed per 100 pounds of gain, probed 0.7 inch backfat, and had a 231 index. A littermate barrow's cutout showed a 2.54 pounds per day gain, 42.3 percent ham-loin, 31.7 inches length, 1.23 inches of backfat, and a 6-square-inch loin eye.

Table 13-3 may be used in converting length of pigs of various weights to a 220-pound equivalent in meeting certification standards.

Presented in Table 13-4 is a summary of swine improvement in the Ohio Swine Evaluation Station during the 1959 to 1971 period. Similar data are available from each of the other swine testing stations.

Herd records and testing programs go hand in hand. Desirable and productive breeding stock is discovered through testing programs. Pedigree and herd records are essential if these lines are to be maintained.

Breeders of purebred hogs obtain official pedigrees for the breeding stock in their herds from the breed registry association. Shown in Figure 13-10 is a pedigree of a purebred Duroc gilt. The pedigree provides the name of the animal, the ear notches, the date of birth, and the size of the litter. It also shows the name and registration number of the sire and dam, and the names of the breeder and present owner of each.

Breeders who wish to register the progeny of registered animals may obtain pedigree blanks from the breed registry association. The applications must be filled in completely and accurately and signed. They are

TABLE 13-3   CONVERTING LENGTH OF PIGS TO A 220 - POUND EQUIVALENT

	Weight Range									
ACTUAL LENGTH	240 238 242	235 233 237	230 228 232	225 223 227	220 218 222	215 213 217	210 208 212	205 203 207	200 198 202	ACTUAL LENGTH
29.0	28.50	28.63	28.75	28.88	29.0	29.13	29.25	29.38	29.50	29.0
29.1	28.60	28.73	28.85	28.98	29.1	29.23	29.35	29.48	29.60	29.1
29.2	28.70	28.83	28.95	29.08	29.2	29.33	29.45	29.58	29.70	29.2
29.3	28.80	28.93	29.05	29.18	29.3	29.43	29.55	29.68	29.80	29.3
29.4	28.90	29.03	29.15	29.28	29.4	29.53	29.65	29.78	29.90	29.4
29.5	29.00	29.13	29.25	29.38	29.5	29.63	29.75	29.88	30.00	29.5
29.6	29.10	29.23	29.35	29.48	29.6	29.73	29.85	29.98	30.10	29.6
29.7	29.20	29.33	29.45	29.58	29.7	29.83	29.95	30.08	30.20	29.7
29.8	29.30	29.43	29.55	29.68	29.8	29.93	30.05	30.18	30.30	29.8
29.9	29.40	29.53	29.65	29.78	29.9	30.03	30.15	30.28	30.40	29.9
30.0	29.50	29.63	29.75	29.88	30.0	30.13	30.25	30.38	30.50	30.0
30.1	29.60	29.73	29.85	29.98	30.1	30.23	30.35	30.48	30.60	30.1
30.2	29.70	29.83	29.95	30.08	30.2	30.33	30.45	30.58	30.70	30.2
30.3	29.80	29.93	30.05	30.18	30.3	30.43	30.55	30.68	30.80	30.3
30.4	29.90	30.03	30.15	30.28	30.4	30.53	30.65	30.78	30.90	30.4
30.5	30.00	30.13	30.25	30.38	30.5	30.63	30.75	30.88	31.00	30.5
30.6	30.10	30.23	30.35	30.48	30.6	30.73	30.85	30.98	31.10	30.6
30.7	30.20	30.33	30.45	30.58	30.7	30.83	30.95	31.08	31.20	30.7
30.8	30.30	30.43	30.55	30.68	30.8	30.93	31.05	31.18	31.30	30.8
30.9	30.40	30.53	30.65	30.78	30.9	31.03	31.15	31.28	31.40	30.9
31.0	30.50	30.63	30.75	30.88	31.0	31.13	31.25	31.38	31.50	31.0
31.1	30.60	30.73	30.85	30.98	31.1	31.23	31.35	31.48	31.60	31.1
31.2	30.70	30.83	30.95	31.08	31.2	31.33	31.45	31.58	31.70	31.2
31.3	30.80	30.93	31.05	31.18	31.3	31.43	31.55	31.68	31.80	31.3
31.4	30.90	31.03	31.15	31.28	31.4	31.53	31.65	31.78	31.90	31.4
31.5	31.00	31.13	31.25	31.38	31.5	31.63	31.75	31.88	32.00	31.5
31.6	31.10	31.23	31.35	31.48	31.6	31.73	31.85	31.98	32.10	31.6
31.7	31.20	31.33	31.45	31.58	31.7	31.83	31.95	32.08	32.20	31.7
31.8	31.30	31.43	31.55	31.68	31.8	31.93	32.05	32.18	32.30	31.8
31.9	31.40	31.53	31.65	31.78	31.9	32.03	32.15	32.28	32.40	31.9
32.0	31.50	31.63	31.75	31.88	32.0	32.13	32.25	32.38	32.50	32.0
32.1	31.60	31.73	31.85	31.98	32.1	32.23	32.35	32.48	32.60	32.1
32.2	31.70	31.83	31.95	32.08	32.2	32.33	32.45	32.58	32.70	32.2
32.3	31.80	31.93	32.05	32.18	32.3	32.43	32.55	32.68	32.80	32.3
32.4	31.90	32.03	32.15	32.28	32.4	32.53	32.65	32.78	32.90	32.4
32.5	32.00	32.13	32.25	32.38	32.5	32.63	32.75	32.88	33.00	32.5
32.6	32.10	32.23	32.35	32.48	32.6	32.73	32.85	32.98	33.10	32.6
32.7	32.20	32.33	32.45	32.58	32.7	32.83	32.95	33.08	33.20	32.7
32.8	32.30	32.43	32.55	32.68	32.8	32.93	33.05	33.18	33.30	32.8
32.9	32.40	32.53	32.65	32.78	32.9	33.03	33.15	33.28	33.40	32.9
33.0	32.50	32.63	32.75	32.88	33.0	33.13	33.25	33.38	33.50	33.0
33.1	32.60	32.73	32.85	32.98	33.1	33.23	33.35	33.48	33.60	33.1
33.2	32.70	32.83	32.95	33.08	33.2	33.33	33.45	33.58	33.70	33.2
33.3	32.80	32.93	33.05	33.18	33.3	33.43	33.55	33.68	33.80	33.3
33.4	32.90	33.03	33.15	33.28	33.4	33.53	33.65	33.78	33.90	33.4
33.5	33.00	33.13	33.25	33.38	33.5	33.63	33.75	33.88	34.00	33.5
33.6	33.10	33.23	33.35	33.48	33.6	33.73	33.85	33.98	34.10	33.6
33.7	33.20	33.33	33.45	33.58	33.7	33.83	33.95	34.08	34.20	33.7
33.8	33.30	33.43	33.55	33.68	33.8	33.93	34.05	34.18	34.30	33.8
33.9	33.40	33.53	33.65	33.78	33.9	34.03	34.15	34.28	34.40	33.9
34.0	33.50	33.63	33.75	33.88	34.0	34.13	34.25	34.38	34.50	34.0

Prepared by National Association of Swine Records.

returned to the breed registry association with the proper registration fee.

Herd registry books are available from the respective breed associations. They may be used in maintaining pedigree and production records of commercial as well as purebred herds.

FEED RECORDS

Feed is the major expense item in the production of pork, yet few swine producers keep feed records, for they are difficult to maintain. Often grain is kept in the same bin for cattle, poultry, and hogs.

There are, however, ways of keeping feed records, and these records are very helpful in analyzing the efficiency of the enterprise.

1. Some farmers store corn and other grains in separate bins so that the amount of grain fed can be obtained by measurement of the contents of the bin.

T A B L E 13-4 TRENDS IN SWINE IMPROVEMENT AT THE OHIO SWINE EVALUATION STATION, 1959 TO 1971

Factor	1959-1965	1966-1971	1971
Percent lean cuts	53.77	58.67	60.00
Backfat thickness (inches)	1.46	1.30	1.24
Length (inches)	29.72	30.40	30.61
Loin eye area (sq. inches)	4.05	4.64	4.79
Average daily gain (lbs.)	1.83	1.80	1.74
Feed efficiency (lbs./100 lbs.)	300.00	298.00	303.00

Source: Ohio State University.

FIGURE 13-10. The certificate of registration for a gilt sired by Tex (*Figure 3-20B*). Note provision is made to indicate certified sire and litter status. (*Courtesy Klein and Hartl.*)

2. With the increased emphasis upon the feeding of a balanced mixed ration, it is quite easy to record the quantities of feed mixed at each time. These amounts may be totaled at the end of the production period.
3. Another method that is used by many producers who self-feed their hogs is to record the amounts of feed put in the feeders.
4. Some farmers who hand-feed weigh the feed fed during one day each month and calculate the total feed fed during the month. No special record form is needed in keeping feed records. The important consideration is that the date, amount, and price of each kind of feed are recorded.

The profit from an enterprise may be reflected by a saving of 40 or 50 pounds of feed in producing 100 pounds of pork. Unless feed records are maintained, the cause for profit or loss may never become known.

LABOR RECORDS

Labor today ranks second or third to feed as an expense item.

Farmers who have large, intensive swine enterprises as well as those with small herds find labor an important item.

Most labor records involve the keeping of a record of the time spent caring for the hogs during one day each week or month and then calculating the total hours spent in feeding and caring for the herd during the month. Labor involved in doing special jobs during the month, such as cleaning the hog house, mixing feed, castration, vaccination, etc., is recorded separately.

Labor records are necessary if comparisons are to be made of the net returns per hour of labor from the swine and the other enterprises on the farm.

Shown in Table 13-5 is a summary of labor requirements of swine enterprises under various systems of management.

Feed costs were extremely high in 1973, and as a result represented about 73 percent of total costs. Labor represented 10 to 12 percent. Facilities, depreciation, insurance, interest, and taxes accounted for 14 to 16 percent, while veterinary and health supplies and services, fuel, and electricity combined represented from 3 to 5 percent of the total production costs.

COMPLETE ENTERPRISE RECORD

A complete analysis of an enterprise can be made only when both production and financial records are available. A complete enterprise record includes a record of cash income and expense, beginning and closing inventories, feed and labor data, and a record of overhead costs, in addition to production records.

Only a small percentage of the swine producers of the nation maintain complete record systems. Such a system would be a wise investment for most producers. A careful analysis of the swine enterprise at the close of each year would bring to light many oportunities for improvement in the efficiency and profit of the enterprise.

While complete records may be kept by individuals, there is merit in participating in a cooperative record keeping program. Such programs are sponsored by vocational agriculture or agricultural extension personnel in most states.

The value of a complete set of records is shown in Table 13-6. The feed cost per 100 pounds of pork produced on the average farms in the Illinois Farm Bureau Farm Management Service program in 1973 was $22.31. The cost on 200-litter farms was $21.83. One hundred pounds of pork were produced on 440 pounds of feed on average farms as compared with 437 pounds of feed required on 200-litter farms. There was a difference of $5.00 in the returns per $100 worth of feed fed to the hogs on the two groups of farms.

By participating in a cooperative program, it is possible to compare your swine enterprise with the enterprises of below average, average, and above average producers in the same area of the state.

The use of records in comparing the income from various types of livestock is shown in Table 13-7, which summarizes the returns from well-managed farms in Illinois in 1972. Note the variation in income from $100 worth of feed fed to various types of livestock. An efficient producer invests feed in the enterprise that will net the most income per $100 feed invested. Hogs and dairy cows usually rated high in this respect.

Vocational agriculture students and 4-H club members have been encouraged or required to keep records of production, costs,

T A B L E 13-5   LABOR REQUIREMENTS IN HOG PRODUCTION

| | | | Annual Hours of Labor Per Unit | |
	Unit	Average	High Mechanization, Efficient Work Methods	Low Mechanization, Poor Work Methods
1 to 14 litters	1 litter	30	20	50
15 to 39 litters	1 litter	23	15	35
40 litters or more	1 litter	20	12	30
1 to 100 feeder pigs	1 pig	2.2	1.8	4.5
100 to 249 feeder pigs	1 pig	1.6	1.2	3.0
250 or more feeder pigs	1 pig	1.4	1.0	2.7

Source: *Farm Management Manual*, University of Illinois, 1968.

FIGURE 13-11. Weight for age scale for certification. (*Courtesy University of Illinois.*)

Farrowed	170 Days			
Jan. / June	1 2 3 4 5 6 7 8 9 10 11 12 13 14 15 16 17 18 19 20 21 22 23 24 25 26 27 28 29 30 31	20 21 22 23 24 25 26 27 28 29 30 1 2 3 4 5 6 7 8 9 10 11 12 13 14 15 16 17 18 19 20	Jan. / July
Feb. / July	1 2 3 4 5 6 7 8 9 10 11 12 13 14 15 16 17 18 19 20 21 22 23 24 25 26 27 28	21 22 23 24 25 26 27 28 29 30 31 1 2 3 4 5 6 7 8 9 10 11 12 13 14 15 16 17	Feb. / Aug.
Mar. / Aug.	1 2 3 4 5 6 7 8 9 10 11 12 13 14 15 16 17 18 19 20 21 22 23 24 25 26 27 28 29 30 31	18 19 20 21 22 23 24 25 26 27 28 29 30 31 1 2 3 4 5 6 7 8 9 10 11 12 13 14 15 16 17	Mar. / Sept.
Apr. / Sept.	1 2 3 4 5 6 7 8 9 10 11 12 13 14 15 16 17 18 19 20 21 22 23 24 25 26 27 28 29 30	18 19 20 21 22 23 24 25 26 27 28 29 30 1 2 3 4 5 6 7 8 9 10 11 12 13 14 15 16 17	Apr. / Oct.
May / Oct.	1 2 3 4 5 6 7 8 9 10 11 12 13 14 15 16 17 18 19 20 21 22 23 24 25 26 27 28 29 30 31	18 19 20 21 22 23 24 25 26 27 28 29 30 31 1 2 3 4 5 6 7 8 9 10 11 12 13 14 15 16 17	May / Nov.
June / Nov.	1 2 3 4 5 6 7 8 9 10 11 12 13 14 15 16 17 18 19 20 21 22 23 24 25 26 27 28 29 30	18 19 20 21 22 23 24 25 26 27 28 29 30 1 2 3 4 5 6 7 8 9 10 11 12 13 14 15 16 17	June / Dec.
July / Dec.	1 2 3 4 5 6 7 8 9 10 11 12 13 14 15 16 17 18 19 20 21 22 23 24 25 26 27 28 29 30 31	18 19 20 21 22 23 24 25 26 27 28 29 30 31 1 2 3 4 5 6 7 8 9 10 11 12 13 14 15 16 17	July / Jan.
Aug. / Jan.	1 2 3 4 5 6 7 8 9 10 11 12 13 14 15 16 17 18 19 20 21 22 23 24 25 26 27 28 29 30 31	18 19 20 21 22 23 24 25 26 27 28 29 30 31 1 2 3 4 5 6 7 8 9 10 11 12 13 14 15 16 17	Aug. / Feb.
Sept. / Feb.	1 2 3 4 5 6 7 8 9 10 11 12 13 14 15 16 17 18 19 20 21 22 23 24 25 26 27 28	18 19 20 21 22 23 24 25 26 27 28 1 2 3 4 5 6 7 8 9 10 11 12 13 14 15 16 17 18 19	Sept. / Mar.
Oct. / Mar.	1 2 3 4 5 6 7 8 9 10 11 12 13 14 15 16 17 18 19 20 21 22 23 24 25 26 27 28 29 30 31	20 21 22 23 24 25 26 27 28 29 30 31 1 2 3 4 5 6 7 8 9 10 11 12 13 14 15 16 17 18 19	Oct. / Apr.
Nov. / Apr.	1 2 3 4 5 6 7 8 9 10 11 12 13 14 15 16 17 18 19 20 21 22 23 24 25 26 27 28 29 30	20 21 22 23 24 25 26 27 28 29 30 1 2 3 4 5 6 7 8 9 10 11 12 13 14 15 16 17 18 19	Nov. / May
Dec. / May	1 2 3 4 5 6 7 8 9 10 11 12 13 14 15 16 17 18 19 20 21 22 23 24 25 26 27 28 29 30 31	20 21 22 23 24 25 26 27 28 29 30 31 1 2 3 4 5 6 7 8 9 10 11 12 13 14 15 16 17 18 19	Dec. / June

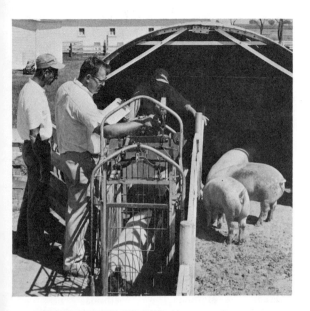

FIGURE 13-12. The scale is essential in keeping growth and feed conversion records in farm testing programs. (*Courtesy Land O Lakes, Felco Division.*)

and income in connection with their productive projects. Some of these records are very carefully kept and are used in analyzing the enterprise and in planning for next year's production. In many cases the production and financial data of several projects are compared as is done in many farm record associations.

Record Analysis

Swine enterprise records are of little value unless they are analyzed, interpreted, and used. Shown in Table 13-8 is a form used by vocational agriculture students and their instructors in comparing their swine production enterprises.

In Table 13-9 is a comparison of 168 swine enterprises managed by producers participating in the Southwest Minnesota

T A B L E   13-6   ANALYSIS OF HOG ENTERPRISES

Hog Enterprise	Average Farm	Litters Farrowed	
		10-49	200+
Number of Farms	1,145	415	94
Pounds of Pork Produced	154,428	50,425	458,452
Total Returns	66,406	21,453	196,735
Value of Feed Fed	34,447	11,209	100,064
Returns per $100 Feed Fed	192	191	196
Returns above Feed per Litter	322	341	310
Number of Litters Farrowed	99	30	311
Total Pigs Weaned	709	215	2,233
Pigs Farrowed per Litter	8.8	8.8	9.0
Pigs Weaned per Litter	7.2	7.2	7.2
Death Loss: Number	35	10	112
Pounds	2,817	903	9,280
Percent of Pounds Produced	1.8	1.7	2.0
Weight per Hog Sold	232	234	228
Price Received per 100 Pounds	39.81	39.36	40.09
Total Return per 100 Pounds Produced	43.00	42.54	42.91
Feed Cost per 100 Pounds Produced	22.31	22.23	21.83
Pounds Feed per 100 Pounds Produced:			
Farm Grains	357	358	352
Commercial Feeds	83	82	85
Total Concentrates	440	440	437
Pasture Days	.3	.4	.1
Cost per 100 lbs. of Commercial Feeds	11.36	11.47	10.83
Cost per 100 lbs. of Concentrates	5.04	5.02	4.98

Source: *Illinois Farm Business Records, 1973*, Preliminary edition. University of Illinois, 1974.

T A B L E 13-7   RETURNS PER $100 FEED FED TO DIFFERENT CLASSES OF LIVESTOCK

Year	Beef-cow Herds	Dairy-cow Herds	Feeder Cattle Bought	Native Sheep Raised	Feeder Pigs	Hogs	Poul-try	Yearly Price of Corn
1958	162	199	144	98	144	180	142	$1.10
1959	147	191	112	102	92	114	123	1.10
1960	129	200	117	108	143	164	157	1.03
1961	139	196	116	110	132	164	150	1.01
1962	149	190	148	126	129	159	144	.98
1963	117	171	88	126	108	131	141	1.11
1964	107	174	112	124	122	142	141	1.12
1965	127	174	151	143	176	210	143	1.15
1966	132	190	117	129	140	178	168	1.23
1967	138	199	119	117	123	154	128	1.17
1968	156	210	142	133	134	170	167	1.02
1969	162	205	152	146	171	212	203	1.14
1970	150	199	118	128	104	142	186	1.26
1971	180	200	156	122	122	150	135	1.27
1972	208	212	161	134	171	214	134	1.16
1958-72 aver.	147	194	130	123	134	166	151	1.12

Source: *Illinois Farm Business Records, 1972.*

Vocational Agriculture Farm Business Management Education Program.

Note that the 55 high-income producers farrowed 12 more litters of pigs averaging 0.7 more pigs per litter, and weaned 1 more pig per litter than the 55 low-income farmers. The high-income producers had a 2.3 percent lower death loss, and marketed the pigs when they weighed 213 pounds rather than when they weighed 242 pounds as did the low-income producers. As a result the high-income producers received $5.40 more per hundredweight for hogs marketed than did the low-income group.

It took 75 more pounds of feed for the low-income farmers to produce 100 pounds of gain, and the cost of gain was $2.78 higher than that of the high-income farmers. The returns per 100 pounds of pork produced were $8.16 higher for the high-income group.

Individual producers can easily see from this summary any shortcomings in their production procedures. They can also determine their strengths. Their task, then, is to determine the factors that caused the differences and correct them.

S U M M A R Y

Profitable pork production is usually dependent upon (1) the production of large litters, (2) the economical use of feed and labor, (3) the emphasis upon meat-type hogs,

	1	2	3	4	5
Number of litters					
Breed of pigs					
Month farrowed					
Number pigs farrowed per sow					
Wt. per pig farrowed					
Number pigs weaned per sow					
Wt. per pig at 35 days					
Number pigs raised/litter					
Month marketed					
Selling price/100#					
Age at marketing (days)					
Av. wt. at 150 days					
Lbs. pork/litter at 150 days					
Av. daily gain					
Lbs. feed/100# pork					
Cost of grain/100# pork					
Cost of protein/100# pork					
Cost of pasture/100# pork					
Total feed cost/100# pork					
Housing and equipment cost/100# pork					
Veterinary costs/100# pork					
Labor cost/100# pork					
Cost of marketing/100# pork					
Net income/$100 feed fed					
Net return/$100 invested					
Labor income/sow & litter					
Labor income/hour of work					

(4) control of diseases and parasites, (5) economical use of buildings and equipment, and (6) the marketing of quality hogs at desired weights at the best time and place. Farmers who keep records can analyze their enterprises and discover the strong and weak points.

For efficient production the following kinds of records should be kept: (1) breeding records, (2) production records, (3) pedigree and herd records, (4) feed records, (5) labor records, and (6) a complete enterprise record.

A record should be kept of the date each sow is bred. Sows must be properly marked. Farrowing records should be kept which include the date, litter size, weight of pigs farrowed, and sex of the pigs.

To qualify as a P.R. litter there must be at least eight pigs weaned at 56 days, and the litter must weigh 320 pounds if produced by a mature sow, or 275 pounds if it is from a first litter gilt.

Some breeders weigh the pigs at three, four, or five weeks of age. The weight of pigs at these ages is a good indication of the productiveness of the breeding stock.

Many sow test associations have been organized to aid in obtaining information concerning the productiveness of individual lines of breeding stock. Commercial producers, purebred breeders, and youngsters who are engaged in sow and litter projects should be encouraged to participate in these organizations.

Most purebred breed associations sponsor production registry and litter certification programs approved by the National Association of Swine Records. To qualify as a certified meat litter, the litter must meet the breed requirements for production registry and in addition meet requirements concerning rate of gain, freedom from defects, and carcass quality. Two pigs from the litter must be slaughtered and must produce desirable meat-type carcasses.

T A B L E 13-9   COSTS AND RETURNS FROM COMPLETE SWINE ENTERPRISES
(Southwest Minnesota Vocational Agriculture Farm Business Management Education Program, 1972)

	Average 168	High 55	Low 55
Pounds of hogs produced	74803	66597	75720
	PER CWT.	PER CWT.	PER CWT.
Total value produced	$ 28.62	$ 31.63	$ 26.23
Pounds of feed fed			
Corn	309.9	282.0	343.8
Small grain	27.9	39.7	20.7
Protein, salt and mineral	69.7	61.5	78.7
Complete ration	37.2	26.5	41.7
Total concentrates	444.7	409.7	484.9
Forages			
Feed costs			
Concentrates and forages	13.82	12.58	15.38
Pasture	.01	.02	
Total feed costs	$ 13.83	$ 12.60	$ 15.38
Return over feed cost	$ 14.79	$ 19.03	$ 10.85
Supplemental costs			
Misc. livestock expense	.30	.30	.33
Veterinary expense	.37	.43	.35
Custom work	.21	.25	.19
Total suppl. costs	$ .88	$ .98	$ .87
Return over feed and supplemental costs	$ 13.91	$ 18.05	$ 9.98
Allocated costs			
Power & mach. costs	.42	.40	.47
Livestock equipment costs	.44	.44	.43
Building and fences	.68	.70	.73
Total allocated cost	1.54	1.54	1.63
Return over all listed costs	12.37	16.51	8.35
Supplementary Management Information			
Return for $100 feed fed	206.96	251.04	170.57
Price received per cwt.	$ 26.22	$ 27.07	$ 25.61
No. of litters farrowed	41	47	35
No. of pigs born per litter	9.3	9.2	8.5
No. of pigs weaned per litter	7.4	7.5	6.5
Per cent death loss	13.4	14.2	11.9
Aver. weight of hogs sold	233.2	213.4	242.9
Price per cwt. concentrate fed	3.10	3.06	3.17
Pounds of pork purchased	3622	2002	7046

The Sire Evaluation and Herd Testing Program introduced by the National Association of Swine Records in 1974 permits a producer to conduct tests on his home farm. It embodies the features of production registry, litter certification, and central swine testing programs.

Pedigree and herd records are necessary if productive lines of hogs are to be maintained and undesirable lines are to be weeded out. Commercial producers as well as pure-bred breeders need to keep herd records.

Purebred breeders should participate in boar testing programs. Commercial growers should use boar testing station results in purchasing breeding stock.

Feed is the major item in pork production. Feed records are essential.

Labor records can be maintained by recording the labor used during one day each week or month in caring for the hogs and then calculating the total hours spent on the enterprise during the month. Labor involved in doing special jobs should be recorded separately.

A complete enterprise record of production, income, and expenses is essential if a careful analysis is to be made of the enterprise.

Regardless of the kinds of records which are kept, they are of little value unless they are carefully analyzed.

Feed costs usually represent 60 to 70 percent of total production costs. Labor accounts for 10 to 12 percent; facilities, depreciation, insurance, interest, and taxes combined account for 14 to 16 percent; and health supplies, veterinary expenses, fuel, and electricity combined represent from 3 to 5 percent of costs.

QUESTIONS

1 What use can you make of swine records on your farm?

2 What is the average number of pigs farrowed per sow on your farm, and what do they weigh at farrowing time?

3 What is the weight of the average litter on your farm at 35 days of age?

4 Explain the meaning of the term Production Registry.

5 Describe the value and place of a swine herd improvement association in your community.

6 What are the requirements for a certified meat litter?

7 Explain how you would obtain pedigrees for the pigs produced by a purebred sow which you purchased as a bred sow.

8 What kind of a swine herd record should you keep on your farm? Explain.

9 Describe the best method of keeping feed records on your farm.

10 What use could you make of a labor record on the home farm swine enterprise? Explain.

11 Justify the keeping or not keeping of a complete swine enterprise record.

12 Which of the following items do you think most important in analyzing a hog enterprise and why?
(a) Pigs farrowed per litter.
(b) 35-day litter weight.
(c) Number of pigs marketed per litter.
(d) Pounds of feed per 100 pounds of gain.
(e) Cost of producing 100 pounds of gain.
(f) Selling price per 100 pounds of pork.
(g) Net returns per $100 of feed fed.

13 Of what value are boar testing stations to hog producers in your community?

14 How many swine testing stations are located within 60 miles of your farm?

15 What does the index mean that is assigned each animal in swine testing stations, and how is it determined?

16 What do you think is the major contribution of the Sire Evaluation and Herd Testing Program sponsored by the National Association of Swine Records? Describe its features.

REFERENCES

Iowa State University, *1973 Farm Business Summaries for North Central Iowa.* Ames, Ia.: Cooperative Extension Service, 1974.

Iowa Swine Testing Station, *Summary 1973 Fall Tests.* Ames, Ia., 1973.

Maynard, Cecil, and Jim Tomlinson, *Swine Production Costs and Returns.* Stillwater, Okla.: Oklahoma State University, 1967.

Minnesota Department of Education and University of Minnesota, *1972 Annual Report Southwest Minnesota Vocational Agriculture Farm Business Management Education Program,* J.A.V.T.I., Jackson, Minn., 1973.

Moyer, A. A., W. G. Olson, and J. P. Murphy, *On-Farm Testing for Swine Herd Improvement.* Manhattan, Kans.: Kansas State University, 1973.

Pierce, E. A. *The New York Swine Improvement Program.* Ithaca, N.Y.: Cornell University, 1968.

Purdue University, *1972 Farm Business Summary.* Lafayette, Ind.: Cooperative Extension Service, 1973.

University of Illinois, *Keeping and Using Swine Records.* Urbana, Ill.: Vocational Agricultural Service, 1961.

———, *1973 Summary of Illinois Farm Business Records.* Urbana, Ill.: Cooperative Extension Service, 1974.

Woodard, J. R., and O. W. Robinson, *North Carolina Swine Evaluation Station, Eleven Years of Progeny Testing,* Cir. 511. Raleigh, N.C.: North Carolina State University, 1973.

14 EFFECTIVE SWINE EXHIBITS

The value of advertising and the use of attractive containers in the selling of foods, hardware, and clothing have been well demonstrated. The fitting and showing of hogs at fairs, at sales, or at other markets can aid materially in the sale of the hogs. Fitting and showing are important in the sale of market hogs as well as in the sale of purebred hogs. The buyer of market hogs inspects the hogs before making an offer. It is important that they are well fitted and shown. We present in this chapter suggestions which should be helpful in fitting and showing swine at fairs, sales, and markets.

BENEFITS DERIVED IN SHOWING HOGS AT FAIRS AND CONSIGNMENT SALES

There are many benefits that may come from exhibiting hogs. Just how many of them will be attained by the individual exhibitor will depend upon the interest and effort put into the project. The following are some of the things that can be gained by showing hogs:

1. Fairs, sales, and shows provide excellent opportunities to study types of hogs and factors to be considered in the selection of breeding stock, and to develop swine selection ability.
2. These events give the producer an opportunity to gain new ideas concerning efficient hog production.
3. Exhibiting is looked upon by most breeders of purebred hogs as a means of advertising whereby they may sell breeding stock.
4. Fairs and sales bring the buyer and seller together. They provide excellent opportunities to make comparisons in the purchase of new breeding stock.
5. Producers of good hogs gain much in the way of personal satisfaction in seeing their hogs compared with other breeders' hogs.
6. There may be a financial gain resulting from higher selling prices and from prize winnings.

7. Exhibitors can make comparisons of their stock with animals of other exhibitors. Such comparisons can assist them in selecting breeding stock, and in herd improvement.

SELECTION OF ANIMALS

The factors to be considered in selecting breeding animals were discussed in Chapter 3, and the qualities desired in market hogs were described in Chapter 12. Your success as a showman will depend upon (1) how good the animals were at the beginning, (2) how well they were fitted, and (3) the kind of job that you do in showing them.

Select animals that have the body conformation, quality, head characteristics, feet and legs, and productiveness needed in profitable hog production. Look for the following:

Body Conformation

A long body with good width; sides smooth and deep; good spring of rib; deep and wide heart girth; strong and moderately arched back; high tailsetting; wide and curved loin; wide plump ham with a short shank; smooth shoulder; short neck.

Quality

A smooth coat of hair; firm and uniform muscling; quality bone; trimness and bloom.

Head Characteristics

Breed character; short, wide head; trim jowl; ears set well apart; moderate size of ear (proper carriage for breed); large prominent eyes.

FIGURE 14-1. The Grand Champion Yorkshire Boar at the 1973 National Barrow Show. He was exhibited by Dick Kuecker, Algona, Iowa and purchased by Soja-No-Ya, Swine Business of Japan, for $30,000. (*Courtesy Geo. A. Hormel & Co.*)

Feet and Legs (primarily for breeding classes)

Straight legs; short, strong pasterns; legs set out on the corners; ample bone; toes close together and of equal size.

Productiveness

Boars. Masculine and rugged; well developed testicles of equal size; good underline; from large litter; good weight for age; firm muscling.

Females. Feminine; refinement and quality; from large litter; good weight for age; good mammary development; firm muscling.

Do not fit an animal that is unsound or has disqualifying defects. Only animals free from disease should be shown. The show herd should be vaccinated for leptospirosis, tested and be free of brucellosis or Bang's disease, and show no signs of rhinitis, erysipelas, or other infection.

COMMON DISQUALIFICATIONS

Purebred animals are usually disqualified in the show ring if they possess any of the following characteristics:

1. Swirl appearing on the upper part of the body.
2. Hernia in males or females.
3. One testicle.
4. Upright ears in breeds with drooping ears.
5. Drooping ears in breeds with upright ears.
6. Spots of hair or skin not characteristic of the breed color.
7. Cramped or deformed feet.
8. Blindness.
9. Color markings not in line with breed standards.

SWINE SHOW RING CLASSIFICATION

The classification of breeding classes is based upon age, while market barrows are usually classified according to weight. Swine exhibitors try to plan their breeding operations so that the pigs will be farrowed at a desirable time from a show ring standpoint. It usually is best to show pigs that approach the maximum age limit of the class.

Separate classes for breeding animals are usually provided for each breed. Following is the common classification for breeding animals:

Boar Classes

1. **Junior yearling boars.** Farrowed on or after January 1 and before July 1 of the year preceding the fair.

2. **Fall or senior boars.** Farrowed on or after July 1 of the year prior to the fair and before January 1 of the year of the show.

3. **January boars.** Farrowed during the month of January of the year of the fair.

4. **February boars.** Farrowed during the month of February of the year of the show.

5. **March boars.** Farrowed during the month of March of the year of the show.

6. **Senior champion boar.** The first prize junior yearling and fall boars are eligible to compete for senior champion boar.

7. **Junior champion boar.** The first prize January, February, and March boars are eligible to compete for junior champion boar.

8. **Grand champion boar.** The junior and senior champion boars compete for the grand champion boar award.

9. **Reserve junior, reserve senior, and reserve grand champion boars.** As the champion is selected in each of the boar championship classes, the second prize animal in the class of the champion is eligible to compete with the remaining individuals in the championship class for the reserve champion award.

Sow Classes

1. **Junior yearling sows.** Farrowed on or after January 1 and before July 1 of the year preceding the fair.

2. Fall or senior gilts. Farrowed on or after July 1 of the year prior to the fair and before January 1 of the year of the show.

3. January gilts. Farrowed during the month of January of the year of the fair.

4. February gilts. Farrowed during the month of February of the year of the show.

5. March gilts. Farrowed during the month of March of the year of the show.

6. Senior champion sow. The first prize junior yearling and fall sows are eligible to compete for senior champion sow.

7. Junior champion sow. The first prize January, February, and March gilts are eligible to compete for junior champion sow.

8. Grand champion sow. The junior and senior champion sows compete for the grand champion sow award.

9. Reserve junior, reserve senior, and reserve grand champion sows. As the champion is selected in each of the sow championship classes, the second prize animal in the class of the champion is eligible to compete with the remaining individuals in the championship class for the reserve champion award.

In some fairs classes are provided for mature boars and mature sows older than junior yearlings. A few fairs now have four boar and four gilt classes for spring farrowed pigs with differing age qualifications. Some shows have discontinued classes for junior yearling boars and sows and have added a class or classes for certified litters.

Group Classes

Classes may be provided at a few fairs for groups. Most common of these are as follows:

1. Aged herd. One boar and three sows of junior yearling age or older.

2. Young herd. One boar and three sows farrowed on or after August 1 of the year prior to the fair.

3. Get of sire. Four animals of any age sired by one boar.

FIGURE 14-2. The Champion Berkshire Barrow at the 1973 National Barrow Show. He was exhibited by Ronnie Haegele, Croton, Ohio. (*Courtesy Geo. A. Hormel & Co.*)

4. Produce of dam. Four animals of any age from the same dam.

5. Breeder-feeder litter. One boar, one gilt, and one barrow from the same litter.

6. Certified litter. One boar, one gilt, and two pigs to be slaughtered from the same litter.

Market Hogs

The increased emphasis by producers in meat-type hogs during the past few years has stimulated increased numbers of market hog shows. Hogmen have been criticized in the past for placing too much emphasis upon the showing of breeding animals. The end product in hog production is pork. It is logical that more attention be given market hogs at our livestock shows.

The weight classifications at the various fairs and shows vary. The trend is to do away with the extreme heavy weight classes and narrow the range in the weights of the lighter barrow classes.

The barrow classification at one of the Corn Belt fairs is as follows:

1. All barrows must have been farrowed after February 1 of the current year.
2. Separate individual and pen classes are provided for each of the following breeds: Poland China, Duroc, Chester White, Spot, Hampshire, Berkshire, Landrace, Tamworth, Yorkshire, and Grade or Crossbred.
3. Three weight classifications are provided: Lightweight barrows—190 to 210 pounds; welterweight barrows—210 to 225 pounds; mediumweight barrows—225 to 240 pounds.
4. Pens shall consist of three barrows: pen of lightweight (190 to 210 pounds) barrows; pen of welterweight (210 to 225 pounds) barrows; pen of mediumweight (225 to 240 pounds) barrows.
5. The champion barrow of each breed will be selected from the winners of the three individual classes.
6. The champion pen of barrows of each breed will be selected from the winners of the three pen classes.
7. The Grand Champion individual barrow will be selected from the breed, grade, or crossbred champions.
8. The Grand Champion pen of barrows will be selected from the breed, grade, or crossbred champion pens of barrows.

FIGURE 14-3. All barrows exhibited in the Production Tested Class at the National Barrow Show are tested at St. Ansgar, Iowa in groups. To be exhibited they must have gained 1.3 pounds per day and meet soundness standards. (*Courtesy Geo. A. Hormel & Co.*)

Many fairs are de-emphasizing the on-foot barrow classes, and adding classes for production-tested barrows that are judged both on foot and on carcass after slaughter. Several fairs have just two live barrow classes. Lightweight barrows weigh 190 to 210 pounds; heavyweight pigs must weigh 210 pounds, but cannot exceed 240 pounds. In most cases there are separate classes for each breed and for crossbreds. In some cases all breeds show together, but the weight classification is adjusted so that no more than 25 to 35 pigs will be exhibited in any one class. Provision is usually made for exhibitors of barrows weighing within two pounds of the maximum weight for light barrows and the minimum for heavy barrows, to show in either class.

Most carcass classes are open to all breeds. Some have weight classifications, some do not. Usually the pigs are shown on foot and placed so that comparisons can be made of the live and carcass show placings. The pigs are usually taken to a packer immediately after the show for slaughter and determination of carcass quality. The placings are then based on percent ham and loin, backfat thickness, size of loin eye, carcass length, belly grade, quality score, etc.

The most recent development in barrow shows is the production tested program initiated in 1973 by the National Barrow Show. All barrows are brought to a central location where they are fed in groups until time to take them to the show. Minimum standards in growth, soundness, and health must be met before the pigs can be exhibited. This program makes growth and carcass comparisons realistic since all pigs are given the same care and management.

A total of 1,179 barrows were entered in the 1974 National Barrow Show Production Tested Barrow Contest. They came from 33 states and Canada. They had to be farrowed after March 1, 1974 and weigh at least 30 pounds when put on official test on May 12.

The pigs were on test until September 7. Those that gained 1.3 pounds a day, were sound and in good health, were tattooed and paint numbered. They then were exhibited in the 1974 National Barrow Show.

When numbers permitted, there were three classes for individual barrows in each breed with the weight classification the same for all breeds and crossbreds.

The carcasses of these barrows were judged on percentage of adjusted live weight in skinned ham, and in the carcass. In addition the size of loin eye, length of carcass, average backfat, quality of belly, and carcass quality score were used in developing a total score.

A number of state fairs and swine breed association carcass shows are being conducted in a manner similar to that followed in the National Barrow Show.

Exhibitors of barrows in the National Barrow Show are permitted to exhibit two animals in the boar and gilt classes. Senior boars and gilts must be farrowed between February 1 and February 28. The Junior boars and gilts must be farrowed after March 1 in the year of the fair.

A truckload class of six barrows farrowed after March 1 and weighing 190 to 240 pounds is provided at the National Barrow Show. Similar classes are provided at some other shows. The entries are judged on foot, but final placings are made on carcass data.

FITTING HOGS FOR SHOWS OR SALES

The best pigs at the beginning of the fitting period are quite often not the best pigs at fair time. The growing out and fitting process can greatly affect the appearance of the pigs in the show or sale ring. Weaknesses in body conformation can often be improved by conditioning. Exercise and grooming are also important. Animals are often found in the various classes at the fairs that have excellent potential but at show time do not have the

FIGURE 14-4. These crossbred hogs owned by Allen Keppy of Wilton Junction, Iowa were Grand Champion Truckload at the 1972 National Barrow Show. (*Courtesy Geo. A. Hormel & Co.*)

finish and bloom necessary to be placed in the No. 1 pen. We present in the paragraphs that follow suggestions that may help you in getting your hogs ready for the fair or sale.

Feeding

No judge will give a hog any consideration in the show ring if it is in poor flesh or lacks proper growth. Hogs in the breeding classes should be well grown out, but not fat. They should carry a great deal of bloom and smoothness. The muscling should be firm. Market hogs should be especially trim in the jowl and middle and be firm fleshed. There should be no tendency toward lardiness. These qualities can be controlled to quite an extent by proper feeding and exercise.

The best comparison of pigs can be made when they have been with the herd and have received no pampering or special attention. Too often the special attention given the animals makes them appear better than they really are. Breeders who have bred growthiness and firm muscling into their stock can leave the animals with the herd and make good competition at the fairs. Many pigs fed in confinement do well at the fairs. It is usually advisable to remove them from the confinement area a week or two before the show to permit them to exercise, and become accustomed to being in new surroundings.

High-protein rations tend to promote growth and result in a more rangy and up-standing animal. These rations produce firm fleshing and usually smooth, wrinkle-free sides.

Low-protein rations tend to make a hog appear short and blocky. The fleshing is usually not firm, and wrinkles or flabbiness may result.

The rations suggested in previous chapters should be satisfactory in fitting hogs for the fairs. The proportion of grains to proteins may be varied as the need arises. It may be advisable to hand-feed in order to keep animals from becoming too fat.

The following ration may be used in fitting barrows and breeding animals:

	POUNDS
Ground yellow corn	50
Wheat mids	15
Ground oats	15
Meat scraps	7
Linseed meal	7
Alfalfa leaf meal	5
Minerals	1
TOTAL	100

Pigs will consume larger quantities of feed when fed in wet mash or slop form than when fed dry. It is usually best, however, to feed show animals dry rations during the month previous to the show. Dry feeding tends to produce firmer muscling and trimmer middles.

It is sometimes necessary in fitting market barrows to slow up the gains so that they will meet class weight limitations. At other times it may be necessary to speed up the gains so that the hogs will meet the minimum weight requirements. The rations which follow have been used very successfully by the vocational agriculture students at Eagle Grove, Iowa, in fitting barrows for the Iowa State Fair.

Ration 1
For Rapid Gains

	POUNDS
Wheat middlings	20
Ground hulled oats	30
Ground yellow corn	35
Dried skim milk or buttermilk	4
Tankage or meat scraps	3
Soybean oil meal	4
Complete mineral mixture	2
Standard antibiotic-vitamin premix	2
TOTAL	100

Ration 1 should be hand-fed. The pigs tend to consume more feed when they hear the rattle of the pails. More rapid gains may result from the addition of increased amounts of the B-complex vitamins. Pigs get hot and restless in hot weather, however, if excess amounts of the B-complex vitamins are fed.

Ration 2
For Slow Gains

	POUNDS
Wheat bran	60
Ground oats	25
Ground yellow corn	10
Tankage or meat scraps	2
Complete mineral mixture	2
Standard antibiotic-vitamin mix	1
TOTAL	100

Ration 2 is a holding ration and should be fed dry. It may have to be limited if pigs gain too fast. Pigs being held will lose their bloom. It is advisable to keep weight records and if necessary start holding them early enough so that they may be put on full feed at least a week before show time. It takes a

FIGURE 14-5. The Grand Champion Truckload of Barrows in the carcass division at the National Barrow Show. They were owned by LaVern Weller of Dwight, Illinois. (*Courtesy Geo. A. Hormel & Co.*)

week or ten days for them to regain the lost bloom.

Exercise

Young hogs on pasture will usually get plenty of exercise. Older animals or confined animals must be taken out of their pens and driven. Exercise promotes good feet and legs, firmness of fleshing, and trimness of middle. It is especially important that barrows be exercised if they tend to be flabby and have heavy middles.

A daily walk early each morning or late each evening is recommended. Do not exercise hogs during the heat of the day. Many breeders walk show animals a quarter mile to half a mile each day.

Training

Proper training pays off in the show ring. An untrained animal usually stands poorly and dashes about the show ring when moved. Many blue ribbons have been lost due to poor training. The judge sees each animal for only a few minutes, and his first impression of the animal is very important.

Use a short whip or cane in training your hogs. Move them slowly. Use the whip or cane to give them a command: a light tap on the back to move ahead, a tap on the nose to stop, a tap on the right side of the head to turn left, or a tap on the left side to turn right. It is a good idea to be consistent in giving commands.

While exercising your hogs, look for their defects and strong points. Experiment to discover the best methods of minimizing the weaknesses and of showing the strong points. This phase of training is very important.

The animals should be trained to stand squarely on their feet, to keep their backs up and their heads down. They should walk slowly and stop on command.

Trimming Feet

The toes and dewclaws of breeding animals usually must be trimmed. When the toes become long, they spread and hinder the animal in walking. Long toes and dewclaws also tend to make the pasterns more sloping, and the pastern will appear longer than it is.

The toes should be cut back almost even with the sole of the foot. A sharp knife, pruning shears, or a pair of nippers may be used. The cut should not be made too deeply. Avoid cutting the soft, lighter colored part of the toe. There should be no bleeding.

It is usually best to trim the toes and dewclaws at least a month before the show. The animals need some time to adjust to the shorter toes just as we have to "break in" new shoes.

A trimming crate can be used in trimming, or the animals can be held or tied. Sometimes the feet can be trimmed while the hog is resting.

Removing Tusks and Rings

Rings should be removed from the noses of pigs some time in advance of the fair or sale.

The tusks should be removed from all mature boars in advance of the show. The boar may be held by putting a noose around the upper jaw. A bolt cutter or pair of nippers may be used. A smooth cut should be made.

Clipping

Before consigning an animal to a sale or taking it to a fair, clip the hair from both the inside and outside of the ears. Care should be taken to blend in the clipped area with the unclipped area.

The tail should be clipped from the base to the switch. None of the switch should be removed. The clipped area should blend in smoothly at the base of the tail.

It may be necessary to clip the entire bodies of mature animals with long hair several months in advance of the fair so that a new hair coat may be grown by show time.

Washing

Hogs that are to be shown should be washed several times in advance of the show. Warm water, soap, and "elbow grease" will be needed. The scurf and scale should be removed from the legs, back, sides, and head. Washing will leave the skin soft and mellow. Avoid washing pigs in cool weather.

Brushing

Brushing the animals regularly will aid in getting a smooth hair coat and a healthy skin. One master hogman in southern Minnesota has a brush in the movable hog house or hanging on a post in every hog lot. He uses them regularly. His hogs do well in the show ring and show up well when buyers come to his farm.

FIGURE 14-6. Improving the appearance of show pigs by clipping the ears and tails. (*Courtesy United Duroc Swine Registry.*)

FIGURE 14-7. A Spot boar being washed with soap and water at the Iowa State Fair. (*Courtesy* Wallaces' Farmer and Iowa Homestead.)

Oiling

In the past the red and black breeds were usually oiled just before show time. The oil gave gloss and bloom to the hair and softened the skin. Oil should be used sparingly. Some breeders oiled the animals the day before the show. A light brushing the morning of the show was sufficient.

A common practice during warm weather has been to sprinkle the hogs with water and then spray light oil on the wet surface. Oil is inclined to cause hogs to overheat easily. Only a fine mist of oil was necessary. No surplus of oil was permitted on the hogs at show time.

Oil may be brushed on or rubbed on with a cloth. The spray is the easiest method and perhaps the most effective.

Light mineral oil, cottonseed oil, or linseed oil thinned with wood alcohol are satisfactory oils for use in grooming show animals, if oil is to be used.

The use of oil on show animals is being discouraged, and in some cases the pigs that have been oiled have been declared ineligible. Even in shows where oil is permitted, good results can be obtained by spraying or sprinkling the pigs with water. Powdering, oiling, or any dressing of hogs other than washing will not be permitted at the National Barrow Show.

Powdering

The white breeds have been commonly dusted with powder to give the skin a whiter appearance. Powdered soapstone or talcum powder is usually used. Marine blue is often added to the powder to counteract its yellow appearance.

Powdering of show animals is not only being discouraged, but has been eliminated at many major shows. Before using powder in preparation for showing, check carefully to make certain that it is permissible.

CARING FOR HOGS AT THE FAIR OR SALE

During hot weather move the hogs to the fair or sale at night if possible. Unload them as quickly as possible and let them rest in lightly bedded pens. They should be watered if they are hot and fed if they appear to be hungry.

If the building is cool and drafty, a covering should be placed outside the gate panels to stop drafts. If animals become overheated, apply water on the nose and legs and sprinkle the body. Electric fans and ice may be used.

It is best to drive the hogs to feeding quarters at the back of the building, if quarters are available. Hogs should be fed lightly while at the fairs. This is especially true the morning of the show.

The pens and alleys should be kept clean and neat. You should stay close to the pens to care for the hogs and to visit with people interested in them.

SHOWING HOGS

It is a good idea to have your entry ready when the class is called. Some breeders like to get their animals in the ring early so that the judge can get a good look at them before the ring is crowded. Others prefer to come in at the last minute as the judge begins to tour the show ring.

Move your animal slowly about the ring. Keep the pig about 12 to 20 feet from the judge unless the judge wishes a closer view. In that case hold the animal and give the judge an opportunity to make a close inspection of underline, muscling, or eyes.

Never get between the judge and your entry. Keep the side of the entry before the judge. If the head or ham is especially good, turn the hog so the judge can see it. Do not overshadow or tire your entry. One good look by the judge may be sufficient.

FIGURE 14-8. A good exhibitor watches both the animal and the judge at all times. (*Courtesy Dave Huinker.*)

When the pig stops, make certain that it stands squarely on its feet. Pressure on the back usually causes the animal to keep its back up. Avoid turning the pig as the judge is looking at it. This is especially important if it is inclined to have creases on the shoulder or side.

Be calm and cool when showing your hog. Move easily. Your clothes should be clean and of good quality. They should also be simple. The name of the exhibitor may be displayed on your back.

The handler should use a minimum of equipment. A whip or cane and brush are needed in showing young stock and mature sows. Two or three persons should be assigned to each yearling or mature boar. At least one of them should have a hand hurdle.

The judge's decision is final. A good exhibitor is considerate and tactful whether winning or losing. There is much more to showing than winning the blue ribbon.

Health Requirements

Swine exhibitors must understand and abide by the regulations related to health of animals exhibited and transported across state lines. Diseases and parasites can spread rapidly if these requirements are not met.

1. Exhibitors must obtain official health certificate from a licensed veterinarian indicating that the animal, or animals, are free from symptoms of infectious or communicable diseases, and that no other diseases have been present on the farm of the producer for a period of 30 days.
2. Proof of negative tests for Brucellosis and leptospirosis within 30 days of the show usually are demanded, unless the herd is certified or validated free of these two diseases.
3. Animals must be from cholera-free states.
4. Some shows demand duplicates of the health certificates and negative test reports.
5. Animals from a quarantined area cannot be exhibited.

6. The certificates must accompany the animal or animals regardless of mode of transportation.

SUMMARY

Fitting and showing hogs at fairs and sales provide opportunities to study types of hogs, to study factors in selection of animals, and to develop judging ability. Information on new methods of production can be gained. Valuable publicity, personal satisfaction, and added income may result.

Select carefully the animals to be fitted and shown. Time and feed may be wasted in fitting inferior animals. Be certain that the animal has no characteristics that will disqualify it in the show or sale ring. Check color of hair, type of ear, soundness of feet, underline, reproductive organs, and presence of swirls or hernias.

Breeding stock is usually classified according to age. January 1, February 1, March 1, and July 1 are the pivotal dates.

FIGURE 14-9. There is much interest in the sale of prize winning breeding stock at the National Barrow Show. (*Courtesy Geo. A. Hormel & Co.*)

Group classes provided at some fairs are aged herd, young herd, get of sire, produce of dam, breeder-feeder litter, and certified litter.

Market barrows are usually classified according to weight. Most fairs have two or three weight classes with a minimum weight of 190 and a maximum weight of 240 or 250 pounds. Some shows provide classes for pens of three barrows in each weight class, and for a trucklot of five or six barrows.

Carcass classes are of more value to producers than live animal shows. Many shows involve exhibiting the pigs on foot, then scoring the carcasses of each following slaughter. Production tested barrow shows are becoming popular. They involve feeding the pigs at a central location, and following up on-foot placings with carcass comparisons.

Hogs in breeding classes should be well grown but not fat. They should carry a great deal of bloom and smoothness. The muscling should be firm. High protein rations promote growth, firm fleshing, and smooth sides. Low protein rations tend to make hogs more blocky and less firm in fleshing.

Feed show animals according to individual needs. Good feeding is important.

Confined pigs and mature animals must be exercised. Train the animals to stop, start, and turn on command. Use a short whip or cane.

Exhibitors must obtain health and vaccination certificates from the veterinarian. Duplicate copies may be necessary. These papers must accompany the animals regardless of mode of transportation.

The feet should be trimmed a few weeks in advance of the fair.

Rings and tusks should be removed several weeks before the fair or sale. The hair on the inside and outside of the ears and on the tail (from the tail setting to the switch) should be clipped.

Show animals should be washed several times in advance of the show with warm water and soap.

Move hogs slowly in the show ring. Keep your entry from 12 to 20 feet from the judge. Avoid turning the animal while the judge is looking at it. Keep its back up and its head down. Make certain that it is standing squarely on its feet. Do not tire your entry. Sprinkle it with water on hot days.

Be clean and well dressed. Be considerate and tactful whether you win or lose.

QUESTIONS

1 What are the advantages and disadvantages of fitting and showing hogs?

2 Describe the characteristics in the various breeds which will disqualify them in the show ring.

3 What are the common age classifications for breeding stock at most fairs?

4 Describe the weight classification for market barrows at your county or state fair.

5 Plan a ration to be used for fitting a spring litter for the county fair.

6 How can show animals best be exercised?

7 Describe the methods that should be used in training hogs for the show ring.

8 How would you trim the feet of a litter of senior spring pigs for the fair?

9 What parts of the animal should be clipped in fitting a market barrow?

10 What are the advantages of each of the following types of market hog shows: (1) live animal shows, (2) production tested barrow shows, and (3) carcass shows?

11 Should pigs be oiled or powdered in preparation for exhibition?

12 What health regulations must be adhered to by the exhibitor?

13 Describe the methods which should be used in washing the hogs when fitting them for the show ring.

14 Outline a procedure to follow in exhibiting an animal in the show ring.

15 What care should hogs receive while at the fair?

REFERENCES

Bundy, C. E., *Fitting and Showing Hogs at Fairs,* Mimeo. Ames, Ia.: Iowa State University, 1953.

Ensminger, M. E., *Swine Science,* 4th ed. Danville, Ill.: The Interstate Printers and Publishers, 1970.

Edwards, R. L., D. O. Liptrap, M. D. Whitaker, and B. T. Dean, *4-H Swine Manual.* Lexington, Ky.: University of Kentucky, 1973.

Krider, J. L., and W. E. Carroll, *Swine Production,* 4th ed. New York, N.Y.: McGraw-Hill Book Company, 1970.

Moyer, W. A., *Profitable Pig Projects.* Manhattan, Kans.: Kansas State University, 1972.

Nordby, J. E., and H. E. Lattig, *Selecting, Fitting and Showing Swine,* Revised. Danville, Ill.: The Interstate Printers and Publishers, 1961.

U. S. Department of Agriculture, *Fitting, Showing and Judging Hogs,* F.B. 1455. Washington, D.C.: